HOT EMBOSSING:
THEORY AND TECHNOLOGY
OF MICROREPLICATION

MICRO & NANO TECHNOLOGIES

Series Editor: Jeremy Ramsden

Professor of Nanotechnology
Microsystems and Nanotechnology Centre, Department of Materials
Cranfield University, United Kingdom

The aim of this book series is to disseminate the latest developments in small scale technologies with a particular emphasis on accessible and practical content. These books will appeal to engineers from industry, academia and government sectors.

HOT EMBOSSING
Theory and Technology of Microreplication

Matthias Worgull

Institute for Microstructure Technology,
Karlsruhe, Germany

William Andrew
Applied Science Publishers

William Andrew is an imprint of Elsevier
Linacre House, Jordan Hill, Oxford OX2 8DP, UK
30 Corporate Drive, Suite 400, Burlington, MA 01803, USA

First edition 2009

Notice
No responsibility is assumed by the publisher for any injury and/or damage to persons
or property as a matter of products liability, negligence or otherwise, or from any use or
operation of any methods, products, instructions or ideas contained in the material herein.
Because of rapid advances in the medical sciences, in particular, independent verification of
diagnoses and drug dosages should be made

British Library Cataloguing in Publication Data
A catalogue record for this book is available from the British Library

Library of Congress Cataloging-in-Publication Data
A catalog record for this book is available from the Library of Congress

ISBN: 978-0-8155-1579-1

For information on all William Andrew publications
visit our website at elsevierdirect.com

Printed and bound in United States of America

09 10 11 12 11 10 9 8 7 6 5 4 3 2 1

**Working together to grow
libraries in developing countries**

www.elsevier.com | www.bookaid.org | www.sabre.org

ELSEVIER   BOOK AID International   Sabre Foundation

This book is dedicated to my parents and my friends.

Contents

Series Editor's Preface xv

Preface xvii

Acknowledgment xix

1   Introduction 1
    1.1   Hot Embossing as a Replication Technology for
         Microstructures 2
    1.2   Historic Example of (Micro) Hot Embossing 3
    1.3   Development of Micro Hot Embossing 8
    1.4   Aim of the Book 11
    References 12

2   Replication Processes 13
    2.1   Overview of Micro Replication Processes 13
    2.2   Micro Reaction Injection Molding (RIM) 15
         2.2.1   Process 15
         2.2.2   Technology of Micro Reaction Injection Molding 16
         2.2.3   Materials for Reaction Injection Molding 18
         2.2.4   Cost Effectiveness 18
         2.2.5   Characteristics 19
    2.3   Micro Injection Molding 19
         2.3.1   Injection Molding Process 19
                 2.3.1.1   Requirements on Micro Injection Molding 22
         2.3.2   Technology of Micro Injection Molding 25
         2.3.3   Materials for Micro Injection Molding 26
         2.3.4   Cost Effectiveness 28
         2.3.5   Characteristics 28
    2.4   Injection Compression Molding 30
    2.5   Micro Hot Embossing 31
         2.5.1   Technology of Hot Embossing 32
         2.5.2   Materials for Hot Embossing 33
         2.5.3   Cost Effectiveness 34
         2.5.4   Characteristics 35
    2.6   Micro Thermoforming 36
         2.6.1   Process 36
         2.6.2   Technology for Thermoforming 38

		2.6.3	Materials for Thermoforming	38
		2.6.4	Cost Effectiveness	38
		2.6.5	Characteristics	39
	2.7	Nanoimprint		39
		2.7.1	Micro Contact Printing	41
		2.7.2	Nanoimprint of UV-Curable Materials	42
		2.7.3	Nanoimprint of Thermoplastic Polymers	43
	2.8	Comparison of Processes		45
		2.8.1	Design	46
		2.8.2	Materials	47
		2.8.3	Technology	49
		2.8.4	Cost Effectiveness	50
			2.8.4.1 Fixed Costs	51
			2.8.4.2 Variable Costs	51
			2.8.4.3 Total Costs	51
	2.9	Table of Characteristic Properties		52
	References			54

3	**Molding Materials for Hot Embossing**			**57**
	3.1	Classification of Alternative Molding Materials		58
		3.1.1	Hot Embossing of Glass	58
			3.1.1.1 Materials	59
			3.1.1.2 Molding Parameters	59
			3.1.1.3 Mold Inserts	60
		3.1.2	Metals and Ceramics	60
			3.1.2.1 Hot Embossing of Metals	60
			3.1.2.2 Micro Powder Molding	62
	3.2	Polymer Materials		63
		3.2.1	Advantages and Typical Properties	63
		3.2.2	Molecular Architecture	64
	3.3	Polymer Melts		66
		3.3.1	Shear Rheologic Behavior	67
			3.3.1.1 Stationary Flow	67
			3.3.1.2 Shear Viscosity/Shear Thinning	69
			3.3.1.3 Mathematical Description of Shear Thinning Behavior	70
			3.3.1.4 Mathematical Description of Temperature Function	71
			3.3.1.5 Mathematical Description of the Time-Temperature Shift	73
		3.3.2	Flow in Capillaries	74
			3.3.2.1 Newtonian Fluid	75
			3.3.2.2 Shear Thinning Fluid	75

	3.3.3	Viscoelastic Behavior of Polymer Melts	76
		3.3.3.1 Definition of Viscoelasticity	77
		3.3.3.2 Mechanical Models to Describe Viscoelasticity	78
	3.3.4	Measurement of Viscosity	79
		3.3.4.1 Melt Index Measurement	80
		3.3.4.2 Capillary Rheometer	80
		3.3.4.3 Rotation Rheometer	81
	3.3.5	Strain-Rheologic Behavior	81
3.4	Molecular Orientation and Relaxation		82
	3.4.1	Orientation	82
	3.4.2	Relaxation and Retardation	83
	3.4.3	Mathematical Characterization of Relaxation and Retardation	84
3.5	Solidification		85
	3.5.1	Amorphous and Semicrystalline Structures	86
	3.5.2	p-v-T Diagram	87
3.6	Solid Polymers		88
	3.6.1	Linear Viscoelasticity in the Solid State	89
	3.6.2	Creeping and Relaxation	90
	3.6.3	Mechanical Properties	91
		3.6.3.1 Tensile Experiments	92
		3.6.3.2 Strength of Polymer Material during Demolding	93
		3.6.3.3 Dynamic Mechanical Analysis	94
3.7	Friction		95
	3.7.1	Friction between Mold and Polymer	96
	3.7.2	Measurement of Friction between Mold and Polymer	97
		3.7.2.1 Components of the Measurement System	98
		3.7.2.2 Functioning	99
	3.7.3	Friction Force Curves	100
3.8	Thermal Aggregate States of Polymers		101
	3.8.1	Thermal Behavior of Amorphous Polymers	102
	3.8.2	Thermal Behavior of Semicrystalline Polymers	103
	3.8.3	Thermal Molding Windows	105
	3.8.4	Commercially Available Polymers	106
3.9	Thermal Properties of Polymers		107
	3.9.1	Thermal Material Data	108
		3.9.1.1 Density	109
		3.9.1.2 Heat Capacity	111
		3.9.1.3 Heat Conduction	111
		3.9.1.4 Thermal Diffusivity	112

      3.9.2   Measurement of Calorimetric Data 112
                 3.9.2.1   Differential Thermal Analysis 113
                 3.9.2.2   Differential Scanning Calorimetry 113
  References 113

4   **Molded Parts** **113**

  4.1   Components of Hot Embossed Parts 113
  4.2   Characterization of Molded Parts 115
  4.3   Measurement Systems for Characterization 117
      4.3.1   Tactile Measurement Systems 117
                 4.3.1.1   Measurement Systems Based
                            on a Probe 117
      4.3.2   Optical Measurement Systems 119
      4.3.3   Electromagnetic Measurement Systems 119
  4.4   Quality of Molded Parts 120
      4.4.1   Classification of Failures 120
      4.4.2   Surface Quality of Molded Parts 123
      4.4.3   Shrinkage and Warpage 125
                 4.4.3.1   Shrinkage of Embossed Parts 125
                 4.4.3.2   Warpage of Embossed Parts 129
      4.4.4   Stress Inside Molded Parts 130
      4.4.5   Controlling Quality of Molded Parts 132
                 4.4.5.1   Control of Quality by Process
                            Parameters 133
                 4.4.5.2   Control of Quality by Molding
                            Technology 134
  References 136

5   **Hot Embossing Process** **137**

  5.1   Hot Embossing Principles 137
  5.2   Components for Hot Embossing 139
  5.3   Process Steps 140
  5.4   Molding Parameters 142
      5.4.1   Process Parameters 143
      5.4.2   Material Parameters 147
      5.4.3   Influencing Factors 148
                 5.4.3.1   Influence of the Hot Embossing
                            Machine 148
                 5.4.3.2   Influence of the Molding Tool 149
                 5.4.3.3   Influence of the Mold Insert 149
  5.5   Elementary Process Variations 149
      5.5.1   Position-Controlled Molding 150
      5.5.2   Double-Sided Molding 150

5.5.3 Molding of Through-Holes 152
    5.5.3.1 Post-processing Methods 152
    5.5.3.2 Molding on Selected Substrates 154
5.5.4 Multilayer Molding 157
5.5.5 Hot Punching 158
5.5.6 Thermoforming of High-Temperature Polymers
by Hot Embossing 159
5.6 Micro Embossing Processes 162
5.6.1 Roller Embossing 163
5.6.2 Ultrasonic Embossing 165
5.6.3 Gas-Assisted Embossing 167
5.6.4 UV Embossing 170
5.6.5 Soft Embossing 170
5.6.6 3D Embossing 172
5.6.7 Hot Embossing of Conducting Paths—MID
Hot Embossing 173
References 175

6 Modeling and Process Simulation 179

6.1 Analytical Model—Squeeze Flow 179
6.1.1 Squeezing Flow of a Newtonian Fluid 179
    6.1.1.1 Velocity-Controlled Molding 182
    6.1.1.2 Force-Controlled Molding 183
6.2 Process Simulation in Polymer Processing 183
6.2.1 Simulation of Macroscopic Processes 184
6.2.2 Simulation of Microscopic Polymer Processes 185
6.2.3 Commercial Simulation Programs 187
6.3 Process Simulation of Micro Hot Embossing 188
6.3.1 Modeling of Typical Microstructured Parts 189
6.3.2 Modeling of Process Steps 189
6.3.3 Modeling of Material Behavior 192
    6.3.3.1 Shear Thinning Material Model for
the Flow Behavior 192
    6.3.3.2 Material Model for the Demolding
Behavior 193
    6.3.3.3 Material Strength during Demolding 195
    6.3.3.4 Friction between Mold and Polymer 195
6.4 Process Analysis 197
6.4.1 The Process Step of Heating 197
6.4.2 The Process Step of Molding 199
    6.4.2.1 Squeeze Flow Behavior 200
    6.4.2.2 Characteristic Pressure Distribution 203
    6.4.2.3 Thickness of the Residual Layer 206

6.4.2.4   Filling of Microcavities—Pressure
Drop Estimation 207
6.4.2.5   Filling of Microcavities—Flow
Analysis 211
6.4.3   The Process Step of Cooling 215
6.4.4   The Process Step of Demolding 217
6.4.4.1   Contact Stress 217
6.4.5   Stress Distribution during Demolding
of Structures 219
6.4.5.1   Demolding of Ideal, Vertical
Structures 219
6.4.5.2   Demolding of Structures with
Undercuts 221
References 223

7   Hot Embossing Technique 227

7.1   Technical Requirements 227
7.2   Technical Implementation 230
7.2.1   Mechanical Stiffness 230
7.2.2   Drive Unit 231
7.2.2.1   Spindle Drive 231
7.2.2.2   Hydraulic Drive 233
7.2.3   Measurement of Process Parameters 236
7.2.4   Control System 238
7.3   Commercially Available Machines 240
7.3.1   Jenoptik Mikrotechnik 240
7.3.2   Wickert Press 242
7.3.3   EVGroup 242
7.3.4   Suess 244
References 245

8   Hot Embossing Tools 247

8.1   Requirements on Hot Embossing Tools 247
8.2   Elements of Hot Embossing Tools 249
8.2.1   Heating Concepts 249
8.2.2   Cooling Concept 254
8.2.3   Alignment Systems 257
8.2.4   Integration of Mold Inserts 260
8.2.5   Demolding Systems 263
8.2.5.1   Adhesion to Rough Substrate Plates 264
8.2.5.2   Ejector Pins 264
8.2.5.3   Air Pressure–Assisted Demolding 265
8.3   Simple Tool for Hot Embossing 266

8.4 Basic Tool for Hot Embossing 267
8.5 Basic Tool for Industrial Applications 270
8.6 High-Precision Tool for Double-Sided Molding 271
 8.6.1 Construction 272
 8.6.2 Operation Principle 276
  8.6.2.1 Heating 276
  8.6.2.2 Molding 278
  8.6.2.3 Demolding 280
References 281

9 Microstructured Mold Inserts for Hot Embossing 283

9.1 Direct Structuring Methods 284
 9.1.1 Mechanical Micro Machining 285
 9.1.2 Laser Structuring 286
 9.1.3 Electric Discharge Machining (EDM) 288
 9.1.4 Non-conventional Molds—Alternative Methods 289
9.2 Lithographic Structuring Methods 291
 9.2.1 Electroforming of Mold Inserts 292
 9.2.2 Mold Inserts Fabricated by UV Lithography 294
 9.2.3 Mold Inserts Fabricated by X-ray Lithography 294
 9.2.4 Mold Inserts Fabricated by E-beam
  Lithography 295
9.3 Assembled Molds 297
9.4 Mold Coatings 298
9.5 Design of Microstructured Molds 301
 9.5.1 Mold Filling 301
 9.5.2 Compensation of Shrinkage 301
 9.5.3 Reduction of Demolding Forces 302
References 304

10 Hot Embossing in Science and Industry 307

10.1 Requirements for Hot Embossing in a Scientific
 Environment and Industry 307
10.2 Micro-optical Devices 308
 10.2.1 Lenses 308
 10.2.2 Optical Gratings 310
 10.2.3 Optical Benches 312
 10.2.4 Optical Waveguides 312
 10.2.5 Photonic Structures 315
 10.2.6 Micro Spectrometer 315
  10.2.6.1 Polymer Waveguide Spectrometer 317
  10.2.6.2 Hollow Waveguide Spectrometer 318
10.3 Microfluidic Devices 319

10.3.1 Capillary Analysis Systems 319
10.3.2 Microfluidic Pumps 323
10.3.3 Microfluidic Valves 324
10.4 Further Applications 327
10.4.1 Micro Needles 327
10.4.2 3D Structures in Terms of Micro Zippers 329
10.4.3 Comb Drive, Acceleration Sensor 330
10.4.4 Lotus Structures 330
10.5 Industrial Applications 331
10.5.1 Compact Disc Replication 331
10.5.2 Film Fabrication 334
10.5.3 Fresnel Lenses 336
10.6 The Future 337
References 338

Index 341

Series Editor's Preface

The concept of replication is fundamental to the era of industrial mass production, just as printing is fundamental to the era of universal knowledge dissemination. Printing is, of course, itself a form of replication, and it could well be said that this book is about the technology of extending printing to the third dimension. The range and variety of products in our contemporary world made by hot embossing is astonishing, and their very familiarity belies the extraordinary ingenuity that has gone into making the technology behind hot embossing into the sophisticated set of processes and materials that it is today. This book is an impressively comprehensive compendium of this technology. It examines every aspect likely to be of interest to the engineer who is highly charged with introducing the technology into a production environment, or further developing it for innovative new products. Furthermore, it includes clear explanations of the underlying science wherever appropriate, which the reader will find to be very valuable in order to understand the basis for making the choice of parameter, or between the different available materials and subprocesses.

Jeremy J. Ramsden
Cranfield University, UK
April 2009

Preface

Today the cost-effective replication of components in polymers is fundamental for the mass fabrication of a wide range of products, especially in the consumer market. The replication processes—especially the process of injection molding—are highly automated and optimized for the replication of products characterized by a large bandwidth of forms and dimensions. With the development of microsystem technology and, further, of nanotechnology, replication was to be faced with a new challenge—the replication of structures in the micro or even nano range. This requirement of cost-effective replication is underlined by the fabrication costs of micro and nanostructures and fuses on the complexity of the fabrication processes. The established replication technologies optimized for structures in macroscopic dimensions cannot be used for the replication of structures in microscopic dimensions without any modifications. The technological requirements are different, because a decreasing structure size is characterized by an increasing relation between surface and volume of a structure. Further, the structure sizes require replication processes and technologies based on high-precision manufacturing and precise control, which makes it necessary to adapt established techniques or to develop new molding techniques. Today hot embossing is, besides the micro injection molding process, one of the common replication technologies for the replication of microstructures in polymers. With the upcoming of nanoimprint technologies the importance of the embossing technique increases. The process of thermal nanoimprint and the hot embossing process are especially characterized by similar process steps.

This book will give the reader a fundamental background on the different aspects of hot embossing. Beginning first with an overview of the different replication technologies, the diversity of the hot embossing process for the fabrication of micro- and nanostructures will underline the flexibility of this replication technique. The implementation of this diversity will be supported by a technology of hot embossing machines, tools, and microstructured mold inserts. The fundamental background refers also to the theoretical knowledge of polymers as molding materials and also in the fundamentals of a theoretical process analysis. The versatility of hot embossing will be illustrated by applications where the process plays an important role in the fabrication line.

Hot embossing and thermal nanoimprint undergo constant development. The limits regarding structure size, molding area, complexity of the structures, and process times are not fixed. New applications requiring mass production will shift the limits in the future. With a fundamental theoretical knowledge of the hot embossing process and its technology, combined with practical experience, the reader may be part of further developments of this technique and pave the way to new applications. If the book can make a contribution to this, its aim was achieved.

Matthias Worgull
Karlsruhe, Germany
April 2009

Acknowledgment

I would like to thank all the people who helped me to write this book: Miss Schröder, for the translation and correction of the language; Dr. Bastian Rapp, for the support in all aspects of TeX and in designing a number of pictures for this book. Especially I would like to thank Dr. Mathias Heckele for his scientific education in hot embossing and his support for my doctoral thesis, with the focus on the analysis and process simulation of the hot embossing process. A special thank-you goes to Mr. Biedermann, who taught me all the technical and practical aspects of hot embossing. Further, I would like to thank Professor Dr. Saile for his support and confidence in this project. And finally, my thanks to all the friends who motivated me to write.

1 Introduction

Today microstructures are part of our life. In many applications of everyday use, and especially in sophisticated applications, microstructures are integrated and fulfill, often invisibly, essential tasks in these applications. The size of these structures is, in relation to the name "micro" structures, in the dimensions of micrometers. The range of these structures includes the range of several hundred micrometers down to the submicron range, at least a large bandwidth. But not in each of the three dimensions do these structures have to be characterized in these ranges. Often only in two dimensions structures are characterized by a resolution in the micron range; in the third geometrical dimension the size of millimeters or even centimeters are typical, e.g., microchannels with a cross-section of $50 \times 50\ \mu m^2$ and a length up to several centimeters.

The fabrication of these structures is part of the science of microstructure technology, and the further assembling into a microsystem, the science of microsystem technology [14]. The fabrication of microstructures can be done by several processes like mechanical machining or lithographic processes, which are cost- and time-intensive processes. To obtain a further distribution of structures and the corresponding microsystem, replication from master structures is recommended and also a precondition for its use in commercial applications. In macroscopic dimensions replication processes and their technologies are already established. Here especially the process of injection molding of thermoplastic polymers is responsible for mass production of a wide range of structures. The replication of structures in microscopic dimensions requires other preconditions and has developed to independent processes with their own technologies. Nevertheless, structures in the micron range can also be replicated with the established macroscopic replication processes and technologies, but with a decrease in structure size and molding area the macroscopic-oriented processes come to their limits. Here, modified technologies and adapted processes close the gap and pave the way for mass production of microstructures and mechanical microsystems.

Besides the process of micro injection molding or thermoforming, the process of hot embossing is one of the established technologies [6,12]. Hot embossing is like the other micro replication processes named above, a process that replicates a microstructured master, a so-called mold insert, in polymer. Polymers are cheap and are available in different modifications with a wide range of properties. Therefore, microstructures with a large bandwidth of properties can be replicated, suitable for the application they for which they are needed.

1.1 Hot Embossing as a Replication Technology for Microstructures

The hot embossing process is divided into four major steps. The process starts with (1) heating of a semi-finished product, a thin polymer foil, to molding temperature, followed by (2) an isothermal molding by embossing (velocity- and force-controlled), (3) the cooling of the molded part to demolding temperature, with the force being maintained, and finally (4) demolding of the component by opening the tool. Between the tool and substrate, a semi-finished product, i.e., a polymer foil, is positioned. The thickness of the foil exceeds the structural height of the tool. The surface area of the foil covers the structured part of the tool. The tool and substrate are heated to the polymer molding temperature under vacuum. When the constant molding temperature is reached, embossing starts. At a constant embossing rate (in the order of 1 mm/min), the tool and substrate are moved towards each other until the pre-set maximum embossing force is reached. Then, relative movement between the tool and substrate is controlled by the embossing force. The force is kept constant for an additional period (packing time, holding time); the plastic material flows under constant force (packing pressure). At the same time, the tool and substrate move closer towards each other, while the thickness of the residual layer decreases with packing time. During this molding process, the temperature remains constant. This isothermal embossing under vacuum is required to completely fill the cavities of the tool. Air inclusions or cooling during mold filling may result in an incomplete molding of the microstructures, in particular at high aspect ratios. At the end of the packing time, cooling of the tool and substrate starts, while the embossing force is maintained. Cooling is continued until the temperature of the molded part drops below the glass-transition temperature or melting point of the plastic. When the demolding temperature of the polymer is reached, the molded part is demolded from the tool by the opening movement, i.e., the relative movement between tool and substrate. Demolding only works in connection with an increased adhesion of the molded part on the substrate plate. Due to this adhesion, the demolding movement is transferred homogeneously and vertically to the molded part.

Why is hot embossing a recommended replication technology for microstructures? Microcavities, especially cavities for structures with high aspect ratios, are characterized mostly by small cross-sections and this results in high pressure needed for filling these cavities with a viscous polymer melt. Here, hot embossing is characterized by short flow distances from the molten semi-finished product into the cavities. Compared to injection molding the flow distances are much shorter and the velocity is much lower, which results in a significantly lower shear stress of the polymer. This reduced shear stress during cavity filling results finally in a lower residual stress of the molded parts. Further, because of the molding principle, this process is suitable for molding microstructured areas, which are mostly impossible to mold by injection molding.

1.2 Historic Example of (Micro) Hot Embossing

Replication of microstructures by an embossing process is not a new development. In history also small features, but with low quality and low resolution, were replicated by embossing. This underlies the technique of embossing as a replication technique. The microstructures that were replicated in the past cannot be compared with the high-aspect microstructures replicated, for example, from a master produced by actual methods of microstructuring technology. Nevertheless, structures with micro features have already been replicated, and this section will show some examples of the development of this process. Besides these historical examples the focus will be set on the historical development of the hot embossing process used today.

Old civilizations used a simple embossing technique for the replication of coins. These coins also show features that are relatively small, but they do not touch the micron size because of the lack of structuring methods. A further milestone in the replication technique was the development of the letterpress, with exchangeable metal letters (Gutenberg). These metal letters had the function of a mold insert, integrated in an embossing machine. The embossing step is here more a transfer of printer's ink to paper, but also an embossing of the structure into a thin sheet of paper. These examples show the use of embossing that is characterized by an embossing step without any heating. The heating step is integrated if molding materials are used whose properties can be reversed by temperature, especially the change of the aggregate state. This allows the embossing of a completely new material class, the thermoplastic polymers, and this is also responsible for the name of the process: hot embossing.

The process of hot embossing of structures in the micron size was already commercially used at the beginning of the twientieth century for the replication of records. Here already parallels to the actual hot embossing process can be found. The groove of a record as a carrier of the acoustic information was replicated on a 12-inch disc, which can be defined as large-area hot embossing. The development of the replication of records is an illustrative historical example for the development of hot embossing; a short excursion to the development of the birth of the record may help the reader imagine the mold fabrication technique and hot embossing technique in the last century.

The development of the record and its replication is linked to Emil Berliner, a multitalented inventor from Hanover who emigrated to America in 1870. There, he started to further develop microphones and manufacture telephones. After this, he empirically studied sound recording. He focused on laterally cut disc records of metal. Berliner's enthusiasm for the lateral cut, which indeed looked like a script, certainly was enhanced by the fact that the depth cut had been patented by Edison. In the case of the lateral cut, the recording and replay needle is more or less deflected laterally at the rhythm of the sound oscillations. Groove depth, however, remains the same.

First, Berliner experimented with a soot-coated glass disc into which the record-ing needle engraved its groove. After hardening the soot, the replay needle was guided stably. In further experiments, Berliner used a metal disc of pure zinc coated by a wax layer, into which the soundtrack was engraved by a cutting pin. To prevent the formation of chips, the disc rotated in a water-alcohol bath. As a result, the chips formed were rinsed up to the surface. Following rinsing, the zinc was freely exposed in the groove. Then, the record was etched in a chro-mium solution. The sound engraved in this zinc record could be reproduced with a butt needle. The record, however, was used above all as a master in reproduction processes. Berliner applied for a patent for his invention in Washington [7–9]. His apparatus for playing records was called the gramophone. By means of an electro-plating process, a negative of the zinc disc was produced. This matrix was used for the production of records from thermoplastic material. When playing the records, disturbing side noise was audible. After detailed studies, this noise was found to be due to the etching process of the zinc disc.

In 1897, the American Jones suggested the use of wax for the recording of rolls and also for the recording of records. For this, he was awarded a patent. The dis-turbing side noise disappeared and the record started to triumph. Still, it was a problem to find a suitable material for pressing the records. The materials that had been used for rolls were far too sensitive. Berliner remembered a material used in telephone construction, the main constituent of which was shellac, which had a very smooth surface in the hardened state. Tests met with success, such that shellac records were produced in series in the same year. These one-sided records were still hand-written, and their diameter was only 5 inches, i.e., 12.5 cm. The records already produced in 1900 had a diameter of 17 cm. The soundtracks were located on one side, and runtime was about two minutes. In 1901, the record diameter changed to 25 cm, with a runtime of about three min-utes. At that time, wax records exclusively were used for recording. Odeon at Berlin provided for the next sensation. At the Leipzig spring fair in 1904, it presented double-sided records.

The replication of a record in this time was described by an article of W. Kaiser in 1911 [13]. The technique used at this time shows a lot of parallels to the modern hot embossing technique. First, a metal mold was fabricated by electroplating of a structured wax master. The thin electroplated metal molds (here also called mold inserts) were separated from the wax layer and soldered onto sinks in a 2–3 mm thick copper or brass plate. The depth of the sinks cor-responded to half of the thickness of the final record. Two plates with soldered mold inserts resulted in a mold for a double-sided record. The contour of both mold halves is made up of a centering pin (the final centered hole of a record) and a circulatory cavity absorbing the surplus of molding material.

The molding material consisted of shellac and soot from lamps of this century. The other ingredients were typically a secret of each company. The raw material was mixed and pressed between two heated roles (calendar) to a sheet with a thickness of several millimeters and was cut into rectangular pieces. Several molds consisting of two halves and a large number of the rectangular shellac pieces were put on a large steam-heated plate with a size of around 1×0.75 m^2. Opposite the heating plate a large hydraulic press was arranged. The hydraulic press consisted of four massive guiding columns, a fixed crossbar on top, and a movable crossbar below. Between the crossbars two plates with a water-based cooling system were integrated. The maximal force of these presses was in the range of 100 t = 1000 kN! To replicate the records first the open mold halves and the shellac plates were heated up by the contact to the heating table. If the molding temperature was achieved, the softened shellac sheet was put between the mold halves manually and the mold halves were closed and centered by the center pin. The closed hot mold halves were put into the press. During embossing the mold and the shellac were cooled down, and after the shellac was cured the mold could be removed from the press and finally the record was demolded manually. Typically each worker worked on a heating table, a press, and with two molds so that if one mold was in the press the second mold could be prepared. By this technique typically 200 records per day could be molded in 1911.

To keep up with the increasingly popular radio broadcasts, the record industry worked on improving its recording and replaying quality. In 1922, the Deutsche Grammophongesellschaft started electric recording tests. The cutting pin was moved through the membrane by a solenoid via a microphone and amplifier. Between the poles of this solenoid, the cutting tip moved in the rhythm of the music or language. Electric recording brought more dynamics and transparency onto the record. In addition, the sound was recorded much better in the depth and height ranges. The contrast to earlier productions was clearly audible. In August 1925, America's Columbia sold its first electric record on the market. On March 1, 1926, Homophone published its first electric record in Germany under the label of Homocord. In the same month, a new company opened its first representative store, the Elektrola Gesellschaft. This company was named after this new electric recording process.

With the start of electric recording, i.e., in 1925, the number of rotations of records was standardized in the USA and Europe. From the number of rotations of the two-pole synchronous motors, an average of 78 rotations per minute was obtained at mains frequencies ranging from 50 to 60 Hz. This value was then declared the general standard.

In the early 19302, radio wished to make its own recordings to have stored programs on stock apart from live broadcasts. The conventional method of recording—i.e., recording on a wax record, manufacture of the pressed matrix up to the pressed shellac record—however, was too long and too expensive. It was finally decided in favor of using a special wax record, scanning it directly with a very light pickup unit, and archiving it. Later, these wax records were replaced by sound foils. These new records consisted of a carrier of metal, glass, or artificial resin. Above, a layer of special gelatin or pyroxylin varnish was located. As the layer cut was much harder than wax, the cutting pin had to be replacde after each cutting process.

On the way toward the long-playing record, there were three possibilities of increasing the play time: (1) increasing the record diameter, (2) arranging the soundtracks more closely to each other, and (3) increasing time by reducing the number of rotations. To get a playing time of eleven minutes onto the record and keep the track distance of the old 78-record more or less constant, a speed of less than half was required at a record diameter of 40 cm. With a synchronous motor of 60 Hz mains frequency, a speed reduction of 54:1 was chosen, which corresponded to a number of rotations of 33⅓ per minute. Peter Goldmark, a Hungarian who studied in Berlin and emigrated to the USA via England in 1936, was the initiator of today's long-playing record. In 1945, CBS gave him a small research group that wanted to develop a long-playing record of high quality. It was to replace the old shellac record. Goldmark was mainly interested in classical music and found that 90% of the symphonies could be placed onto a record of 45 minutes' playing time. Hence, he wanted his new record to have a playing time of 22 minutes per side at excellent quality.

Electric recording technology that allowed for smaller amplitudes and, hence, narrower tracks, had already been used for years. During the Second World War, CBS cut radio programs onto 40 cm foils at 33⅓ rotations per minute. However, they did not make any reproductions. The PVC "vinylite" with an extremely smooth surface was used as the record material. Due to this smooth surface, Goldmark was able to reach a track depth of 0.1 mm only, a track width of 0.13 mm, and a rather good signal-to-noise ratio. For this fine cut, he introduced the term "microgroove." His record was given the name "LP," i.e. "long-playing record."

In 1953, the 45- and 33⅓-rpm records were further improved by an optimized cutting process. Groove distance was no longer constant, but changed with the amplitude of groove width. This means that if the amplitude, i.e., groove width, was small, groove distance was reduced as well. If the groove width was large, groove distance was increased as well. This was achieved by a control unit in record cutting. By means of this method, playing time of a long-playing record was increased from 20 minutes to a maximum of 40 minutes per side, an immense advantage.

The first real stereo cutting process of a shellac record was developed by the engineer Alan Dower Blumlein on behalf of EMI. As two channels are required for stereo, he initially used the vertical cut by Edison for the left channel and the lateral cut by Berliner for the right channel. Both information levels formed a right angle, and the cutting pin therefore described a spatial curve. This type of stereo recording, however, was not monocompatible. A mono pickup scanned the lateral cut only, not the vertical cut. Blumlein then turned the entire system by 45 degrees. Now, the left channel was cut into the left groove flank with a slope of 45 degrees. The right channel was cut into the right groove flank in the opposite direction, also at 45 degrees. This cutting process was monocompatible. In 1932, the first real stereo record in the world was cut according to Blumlein's process. In 1958, the stereo LP was introduced in America and also the Goldmark LP was published in a stereo version. Blumlein's 45 degrees to 45 degrees process was applied.

In 1982, a new cutting process was used in record production, which was referred to as direct metal mastering (DMM). It was developed by the company of Telefunken-Decca. Normally, the information of the mother band is cut into a varnish foil that is provided with a thin, electrically conductive silver layer. The metal record father is grown on top by electrode position. The DMM process requires neither coating nor electro deposition. The cutting pin directly scribes into a copper foil that may be used as a mother for the pressing matrixes. The advantage in terms of sound is that metal foils do not have to be processed by a pre-heated cutting pin, contrary to the soft varnish foils. Depth resonance, which determines the bottom frequency limit, is below 20 Hz.

This short excursion into the history of record development [15] illustrates the importance of the hot embossing process to the development of this medium. Replication of long-playing records of 30 cm in diameter and with microfeatures of less than 100 μm in size is a good example of large-area hot embossing and even dates back to the middle of the twentieth century. It clearly reveals the characteristics of hot embossing. Hot embossing is suitable for molding microstructured areas on a thin layer. Today, this layer is called the residual layer and may be considered a kind of carrier layer for microstructures. Figure 1.1 shows the typical microstructures of a stereo LP.

These microstructures are characterized by grooves with sloped sidewalls of 45 degrees, which allows relatively easy demolding. The aspect ratio—the ratio between the lateral dimension and the height of a structure—is in the range of one. Nevertheless, accurate replication is necessary to obtain high-quality records. All process features, such as molding machines, tools, microstructured mold

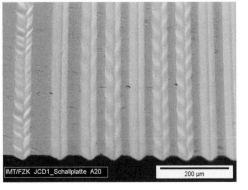
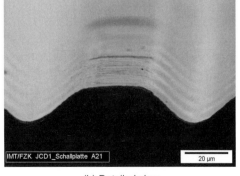

(a) Record grooves   (b) Detailed view

Figure 1.1 Micro grooves of a record. The structures are characterized by an aspect ratio in the range of one on a carrier layer of a diameter of 30 cm. To transmit the stereo information the 45-degree sloped sidewalls of the grooves are structured corresponding to the information of the left channel and the right channel. This angle also supports the demolding.

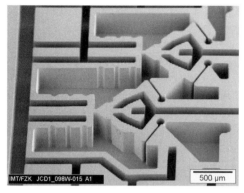 

(a) High aspect ratio structures   (b) Sub-mircon structures

Figure 1.2 In PMMA, replicated structures with high aspect ratios. The bandwidth of hot embossing is characterized by microstructures with high aspect ratios down to the sub-micron and nano range.

inserts, process parameters, and molding materials, may be transferred to today's hot embossing process for the replication of microstructures. Still, requirements made on the hot embossing of high-aspect-ratio microstructures are increasing. The structures replicated by hot embossing today are characterized by high aspect ratios, smaller feature sizes down to the nanometer range, and vertical side walls with a surface roughness below 40 nm (Fig. 1.2).

1.3 Development of Micro Hot Embossing

As described above, the replication technique of hot embossing can be traced back to the beginning of the twentieth century. Hot embossing was implemented in 1970 by a group of researchers from RCA Laboratories in Princeton, NJ, USA [4]. The objective of this work was to develop a low-cost reproduction technique of surface hologram motion pictures for television playback. A master tape was made by electroplating nickel into photo-resistant patterns. The master was run through heated rollers together with a vinyl tape and, thus, the microstructure was transferred into the vinyl. The molded structures were characterized by a depth of 0.1 μm and a lateral resolution of 1 μm. In 1978 Gale et al. [10] used the hot embossing technique for the replication of surface relief structures for color and black-and-white reproductions. These relief phase grating structures refer to a recording of light in zero-order diffraction. These structures with a height of 1–2 μm and a pacing of around 1.4 μm were replicated from a nickel mold inserted into transparent PVC by hot embossing at a molding temperature of 150°C and a pressure of 0.3 MPa. These structures were also replicated in polycarbonate (PC) and acetate.

With the birth of the microstructure and microsystem technology these replication techniques grew in importance. Especially with the development of the LIGA technique (German acronym for "Lithografie, Galvanik und Abformung" which means "lithography, electroplating, and molding") at Forschungszentrum Karlsruhe in the mid-1980s the process of hot embossing was well suited for the replication of microstructures with high aspect ratios. The replication of exemplary micron and submicron structures shown in Fig. 1.2 corresponds therefore to the development of the structuring methods to fabricate microstructures. The development of the hot embossing process therefore cannot be seen independent from the development of the LIGA process.

However, before the hot embossing process in this configuration was established previous experiments were done by reaction injection molding. This process allows the injection of mixed liquid monomers and a starter into a microstructured mold and the molding occurs by polymerization in the mold [5]. Microstructures of PMMA and PA (polyamide) were molded successfully, but at the beginning of these experiments the filling and the demolding of filigree microstructures with high aspect ratio were challenging aspects of the development. Nevertheless, because of the difficulties in controlling the polymerization and the high shrinkage of the molded part, the hot embossing technique based on polymer foils was investigated. Besides fundamental tests with micro injection molding, basic experiments in 1989 by an embossing technique of PMMA material marked the start of the use of micro hot embossing for the production of LIGA structures [11] (Fig. 1.3). With the further development of the LIGA technique, the development of the hot embossing technique also proceeded. At the beginning the technique was based on laboratory machines with a low

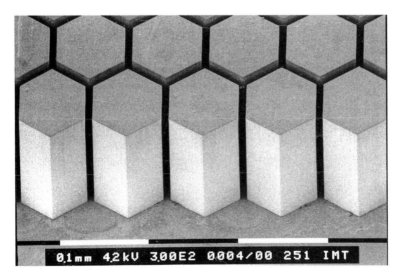

Figure 1.3 One of the first LIGA structures replicated by hot embossing in PMMA. Pillars with a height of 125 μm from a honeycomb mold insert.

grade in automation, which requires a control of every molding step by the user. Also, the precision of the molding machine regarding the press force, the molding velocity, and the temperature distribution were permanently developed and improved. A milestone in the development in the hot embossing technique was the use of computer-controlled tensile testing machines. These machines, already established in many material testing laboratories, could fulfill the requirements for micro hot embossing: a high stiffness of a machine, a precise motion, a control system for force and velocity combined with a measurement system, and finally an interface to the user. An integration of a molding tool with a heating and cooling system completed the required components for hot embossing. In the years following this concept was developed to commercially available machines. By further developments of the technique, these kinds of machines are today part of the state of the art in hot embossing technology [2]. Independent from the development based on tensile testing machines, new concepts based on hydraulic drives were also established [3]. Nevertheless, besides the commercial machines, a lot of individual machines are available in labs or industry optimized for a specific task of replication.

On the basis of the available hot embossing technology, today hot embossing is a well-established process in industry and science, with a large bandwidth of process variations. The replication of Fresnel lenses for overhead projectors or concentrating solar systems [1] are also established, like the replication of CDs by the related process of injection compression molding. The development of nanoimprint lithography with thermal and UV-curable materials opens this technology of hot embossing a way to the nanostructuring methods. A new level of different application like surface modification in the nano range or the replication of structures in wavelength dimensions of light will pave the way to a new class of applications. The development of the hot embossing technique corresponds to the development of the structuring technique for the fabrication of mold inserts. Today structures in the range below 10 nm can be fabricated and can be finally used as mold inserts for replication (Section 2.7). The hot embossing process in combination with polymers is well suited for the replication of these structures and makes a contribution to a mass production of new application based on nanostructuring.

Besides the replication under commercial aspects in industry, hot embossing is an established process for the replication of prototypes and small series. An advantage is here the flexible technique allowing one to change mold and polymer very quickly. Therefore, hot embossing is also a popular process in laboratories for microstructuring. Nevertheless, this process is characterized by a large bandwidth of process variations, like double-sided aligned molding, molding of through-holes, molding of polymer stacks, thermoforming by polymer melts, or structuring by hut punching. These variants make this process unique for developments in micro and nano replication and for the development of micro- and nanosystems.

1.4 Aim of the Book

The aim of the book is to give the reader a fundamental knowledge about hot embossing. First, an overview of micro replication process will be given to understand the meaning of the hot embossing process compared to other micro replication processes. In a later chapter, the processes and the related parameters and influence factors on the process will be discussed in detail. With this fundamental knowledge the manifold process variation can be understand. Here the basic knowledge about hot embossing will be given. The molding of thermoplastic polymers cannot be seen independent from the material behavior of thermoplastic polymers. Therefore, the basics of this material class will be discussed, with the focus on the physical behavior of the polymers during replication. Further, the mathematical description of the viscoelastic behavior and the viscosity of polymer melts will pave the way for the understanding of the process simulation later on in the book.

For the practical use of hot embossing the knowledge of the technology and its influence on the process is essential. Here, the technology is split into the description of the machine technologies, with the focus on commercially available machines and the molding tool as an intersection between a pure embossing machine and a microstructured mold insert.

Besides the machine and the tool, the microstructured mold inserts are essential. Here, the different fabrication methods and the characteristics of different mold inserts will be compared and their relevance to the hot embossing process will be discussed. The development and further improvements of process and technologies correspond to the requirements on the molded parts. To get feedback about the molding quality, the molded parts have to be characterized. The chapter on molded parts will also discuss the characteristics of embossed microstructured parts and will show the traceability of the quality of the molded parts to the process parameter. Therefore, suitable measurement methods to characterize molded parts will be described in a compact way. Also, potential failures will be analyzed and suggestions for prevention will be given.

The simulation of the process combines the theory of polymers with the modeling of material behavior and the embossing process sequence. Besides the different aspects of modeling, the simulation results allow a more detailed view of the molding process, beginning from the filling of microcavities up to the stress inside structures during demolding.

Finally, the wide range of applications in science and industry will show the manifold of this replication process and will give readers a motivation for a further development of the process by their own experiments.

References

1. Fresnel Optics. http//www.fresnel-optics.de, 2008.
2. Jenoptik Mikrotechnik. http//www.jo-mt.de, 2008.
3. Wickert Press. http//www.wickert-presstech.de, 2008.
4. R. Bartolini, W. Hannan, D. Karlsons, and M. Lurie. Embossed hologram motion pictures for television playback. *Applied Optics*, 9(10):2283–2290, 1970.
5. E. W. Becker, W. Ehrfeld, P. Hagmann, A. Maner, and D. Münchmeyer. Fabrication of microstructures with high aspect ratios and great structural heights by synchrotron radiation lithography, galvanoforming and plastic molding (ligaprocess). *Microelectron. Eng.*, 4:35–56, 1986.
6. H. Becker and U. Heim. Hot embossing as a method for the fabrication of polymer high aspect ratio structures. *Sensors and Actuators*, 83:130–135, 2000.
7. Emile Berliner. Gramophone. United States Patent No. 372,786, November 1867.
8. Emile Berliner. Process of producing records of sound. United States Patent No. 382,790, May 1888.
9. Emile Berliner. Gramophone. United States Patent No. 534,543, February 1895.
10. M. T. Gale, J. Kane, and K. Knop. Zod images: Embossable surface-relief structures for color and black-and-white reproduction. *Journal of Applied Photographic Engineering*, 4(2):41–47, 1978.
11. M. Harmening, W. Bacher, P. Bley, A. El-Kholi, H. Kalb, B. Kowanz, W. Menz, A. Michel, and J. Mohr. Molding of three dimensional microstructures by the liga process. In *Proc. MEMS '92*, Travemünde, Germany, p. 202. IEEE, 1992.
12. M. Heckele, W. Bacher, and K. D. Mueller. Hot embossing—the molding technique for plastic microstructures. *Microsystem Technologies*, 4:122–124, 1998.
13. W. Kaiser. Ueber Schallplattenfabrikation. *Kunststoffe*, 1(7), 1911.
14. W. Menz and J. Mohr. *Mikrosystemtechnik für Ingenieure*. VCH Wiley, 2nd edition, 1997.
15. David L. Morton Jr. *Sound Recording: The Life Story of a Technology*. Greenwood Press, 2004.

2 Replication Processes

The aim of this chapter is to give the reader an overview of the different kinds of processes and technologies used for the replication of microstructured molds. Each of these replication processes is specified for a special kind of molded part and uses different polymers or different molding windows of the same polymer. Therefore all these processes are not in competition with each other but are rather more complementary in replicating a wide range of microstructures with different properties for different applications. Depending on the mold design, polymer, cost effectiveness, and the requirements specified by the application, the most suitable replication process can be determined.

The chapter will first give a short listing of all processes and will specify the criteria that are used to characterize each process. Further, each process will be described briefly, including the process steps, the technology, the polymers suitable for molding, and the cost effectiveness. Finally, the advantages and disadvantages of each process will be discussed. This chapter will aid the reader in selecting a suitable replication process for a specified design.

2.1 Overview of Micro Replication Processes

Replication processes for macroscopic molded parts are well established in industrial use. Common ones are, for example, injection molding, injection compression molding, compression molding, thermoforming, blow molding, or extrusion processes. These processes are highly automated and optimized for a large number of molded parts fabricated by short cycle times. For industrial use the cost effectiveness is, besides the requirements of quality, one of the most important criteria for the choice of a replication process. In macroscopic scale the named replication processes are characterized by a large number of different molding materials and a large variation in the dimensions of the molded parts. For example, the dimensions vary from large parts for the automotive industry down to the microscopic structures of compact discs. Besides geometric dimensions, a large number of different polymers with a wide range of thermal and mechanical properties are available to cover the requirements of different applications.

The switch between macroscopic and microscopic replication processes is not strictly separated. Already with the established macroscopic replication technologies features in the sub-millimeter size can often be replicated. Nevertheless, to replicate features in the micron range and below, commercial macroscopic technologies mostly reach their limits. To overcome these limits, adaptations of process steps and modifications or new developments in technology are required. On the base of the established macroscopic replication technologies, replication processes were developed and optimized for the replication of structures in the micron and sub-micron range, down to the nano range. This development of process technology will continue, so that the state of the art regarding feature size and cost will decrease in the future.

The overview of the replication processes, especially their technology, will be limited to established processes and machines. Nevertheless, a lot of individual solutions optimized for specific tasks and designs exist in science. Here, because of the relatively small supply of commercial machines for replication on a microscopic scale, sometimes modification of commercial machines or development of individual machines is necessary to solve specific problems. Nevertheless, with the following micro replication processes, a wide range of microstructured molds can be replicated.

The established micro replication processes are [3,16]:

- Micro reaction injection molding (RIM)
- Micro injection molding
- Micro injection compression molding
- Micro hot embossing
- Micro thermoforming
- Nanoimprint lithography processes (NIL)

Each of these processes will be discussed in detail in the following sections.

To characterize these different molding principles and finally to compare the processes, criteria has to be found to evaluate their similarities and differences (Section 2.8). Which evaluation criteria can be used? The criteria can be divided into several groups—the kind of molding process with its characteristic properties, the molding material used, the process technology used for replication, and finally the cost effectiveness.

- Process
  - Typical process times
  - The length of flow path to fill microcavities
  - The kind of raw material that has to be used: monomers, granules, or semi-finished products
  - The internal stress in the molded part after replication
  - Flexibility regarding the change of polymer and mold
- Molding material
  - The kind of polymer materials that can be used in the replication process: thermoplastic polymers, thermosets, UV-curable polymers
  - The temperature range used for processing the polymer material (molding window)
  - The characteristic load on the polymer during molding (shear, strain)
- Technology
  - Commercial availability of molding machines

    — Automation of the process (handling of polymer materials and
      molded parts)
    — Set-up times (change of mold and molding material)
  • Cost effectiveness
    — Fixed costs: investments for machine and necessary infrastructure
    — Variable costs: costs of molded parts, characterized by personal
      costs, cycle times, grade of process automation

In the following section the processes named above will be described briefly
and analyzed with regard to the evaluation criteria. In the last section of this
chapter the micro replication processes will be compared with regard to these
criteria.

Besides a general review of micro molding processes [16,23], the replication processes can also be evaluated under the aspect of different applications like fluidic
applications [3] or micro- and nano-optical components [19]. In this case the replication processes shall be evaluated without any preference to an application.

2.2 Micro Reaction Injection Molding (RIM)

Reaction molding technology is one of the established replication technologies
on a macroscopic scale. For example, polyurethane reaction injection molding was
developed in the late 1960s by Bayer AG. Besides the development of this technology for a wide range of polymers and applications, micro reaction injection
molding was developed at Karlsruhe research center at the end of the 1980s
as part of the LIGA (*L*ithographie, *G*alvanik *A*bformung; German acronym for
the process steps lithography, electroplating, and replication) process [14].

2.2.1 Process

This process is characterized by an injection and mixing step of two or more
components of a polymer and finally a polymerization of these components in
the mold. The components that will be injected depend on the kind of polymer,
but in general a monomer and a starter of the polymerization will be injected in
different weight proportions (Fig. 2.1).

The liquid components are held in separate, often temperature-controlled,
tanks. From these tanks the components feed under pressure to a metering
unit and a mixer device. High precision is required to dose the single reactants
in the stoichiometric proportions. After the mixing of the reactants the valve of
the mixer opens and the injection of the liquids in the mold begins. In contrast to
the injection molding process, the mixed components will be injected under
moderate pressure into the cavities. Pressures in the range of 1 MPa are sufficient because of the time-delayed start of polymerization. The micro mold

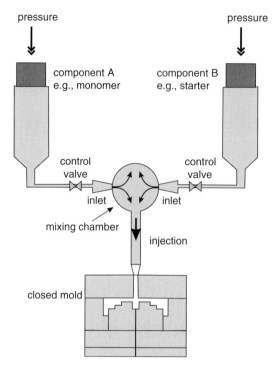

Figure 2.1 Schematic illustration of the reaction injection molding process (RIM). Two components, a liquid monomer and a starter for the polymerization reaction, are mixed and injected into the mold by an injection unit. The polymerization of the monomers to polymers occurs finally in the mold [14].

should be evacuated to be sure that all microcavities can be filled completely. Inside the mold, the mixed components finally undergo the chemical reaction to the final polymer. Characteristic for this kind of polymerization is an unavoidable shrinkage in a range up to approximately 20%, which can, without countermeasures, result in sink marks of the molded parts. To compensate for the shrinkage a dwell pressure will be effected, acting on the polymer. This pressure effects under a further delivery of liquid reactants a compensation of the volume shrinkage. To keep this delivery throughout polymerization the gate has to be kept to as low a temperature level as the mold; otherwise, the gate can be frozen by the polymerized material.

2.2.2 Technology of Micro Reaction Injection Molding

The technology of RIM processes is optimized in industry for the replication of macroscopic parts. Machines optimized for micro replication are commercially not available. The adaptation of commercial machines for micro replication may be possible in some cases. Here an important feature is the precise dosage,

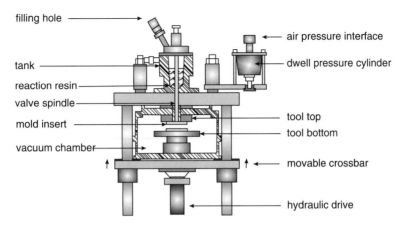

Figure 2.2 Schematic view of the technology of a reaction injection molding machine for laboratory use, adapted for replication of microstructures. Characteristic for the replication of microstructures by RIM is a tank containing the mixed resin and an injection system integrated in a press with a tool and a vacuum chamber.

mixing, and injection of volumes in the range of several cubic millimeters or even below. Especially the requirements of injection can be compared with the requirements of injection by the injection molding process. Nevertheless, the basic technology and the components of a reaction injection molding machine can be explained on the basis of a simple individual historical machine for laboratory use (Fig. 2.2).

A fundamental component is the vacuum chamber with an integrated tool that can be separated into a top half with the injection unit and a bottom half in which the microstructured mold insert is integrated. Further components are a tank for the reactants, a cylinder for the effecting of the dwell pressure, and a mechanism to close and open the tool and the vacuum chamber, in this case effected by a hydraulic drive. In a first step the vacuum chamber and the tool halves will be closed by the hydraulic drive; the typical clamp force can be set up to a range of 50 kN. A precise vertical motion of the lower crossbar will be achieved by guiding pins. In a second step the mixed components will fill the tank. Here, first the air inside the mixture can be extracted before the pressure in the tank will be increased up to a level of approximately 3 MPa. The injection into the mold occurs by opening the valve between the tank and tool and over a gate the mixture will be injected into the closed tool halves. To achieve a higher level of dwell pressure, a cylinder pin will be pressed into the gate, which allows the achievement of pressures in the range of 30 MPa. For demolding, the vacuum chamber will be vented up to a pressure of 0.1–0.3 MPa over ambient pressure. During the demolding by a precise vertical motion of one of the two tool halves against the other half the vacuum chamber is still closed. Typically the polymerization can be influenced by temperature. Therefore, the mold should be controlled precisely by an integrated heating and cooling system. In this case the temperature will be controlled by a convective oil heating and cooling unit; this allows the setting of the temperature of the mold to different levels. The temperature distribution in the mold should be very homogeneous and thermal isolation between the tank and mold is necessary; otherwise, polymerization already starts in the tank.

The principle of molding monomers was implemented by Lee et al. [20] by thermal-curing nanoimprint lithography. In this research monomer-based thermal curing lithography was used to replicate 100 nm line and space patterns on PET film. The imprint resins consists of a monomer and a thermal initiator. The polymerization can be initiated at 85°C, which is lower than the glass transition temperature.

2.2.3 Materials for Reaction Injection Molding

As mentioned above, a wide range of materials can be used for replication by reaction injection molding. Because of this molding concept, theoretically every polymer can be used that can be split into the components of a monomer fraction and a starter fraction, which will initiate the polymerization to the final polymer. This concept allows the molding of thermoplastic polymers like PMMA (monomer MMA and starter) or polyamide (PA) but also thermosets like polyurethane (PU). Besides these material classes also the UV-curable materials can be molded if the process will be completed by a UV exposure unit.

2.2.4 Cost Effectiveness

The fixed costs of this technology are mainly characterized by the investments for a molding machine. Because of the lack of commercially available machines specialized for replication of microstructures, the fixed costs of this technology cannot be determined in detail. Only individual machines developed for special requirements, mostly in laboratories, can be taken into account. But these machines are not representative for a serious determination of the costs. In general, a basic machine can be less expensive than an injection molding machine. The costs will also depend on the grade of automation. For the individual machines in science the grade of automation is normally small, reducing typical costs for laboratory RIM-machines.

The variable costs are characterized by the engineering time needed for setting up the machine, the grade in automation, the batch size of the mold that should be replicated, and finally the time for a molding cycle. The molding time is a key parameter for variable costs, especially when large series should be molded. Here, the cycle time depends on the material used and the effectiveness of the cooling system to dissipate the heat during its curing. Probably the cycle times of the laboratory machines are longer than the cycle times achieved by optimized injection molding cycles, so that this replication technology, without a high grade in automation, is under the aspect of costs only suitable for replication of prototypes and small series.

2.2.5 Characteristics

The advantage of this technology is the high flexibility regarding the different kind of polymers that can be used for replication. As mentioned above, thermoplastic polymers, thermosets, and also UV-curable polymers are suitable for molding. This wide range of polymers allows the replication of a microstructured design in different kinds of polymers. Another advantage is the low viscosity of the polymer during filling and the moderate pressure therefore required.

Nevertheless, this process is also characterized by different problems. The control of the curing process can be difficult, especially the homogeneous heat dissipation during cooling, also the high grade of shrinkage of the molded part. The mixture, dosage, and injection of resins in micro-scale dimensions require an adapted technology that will increase the cost of machines. Also the lack of commercially available machines adapted for micro replication inhibits a further use of this technology. Under commercial aspects, finally, the relatively long cycle times increase the costs of replicated parts.

2.3 Micro Injection Molding

Micro injection molding is the most established and automated process of all processes described here. This micro replication process profits from the already high grade in automation in the macroscopic injection molding technology. The development of micro injection molding therefore benefits from the multitude of commercially available machines and the experience of a large number of suppliers of this technology. Under commercial aspects this technology will be favored in industry because of the potential to mold large series cost effectively. Nevertheless, to obtain a micro replication process it is not recommended to scale down established machines; a lot of adaptations are necessary to replicate structures in microscopic dimensions [11].

2.3.1 Injection Molding Process

In general, the injection molding process is simply characterized by an injection of polymer melt into a tempered mold. The polymer in granulate form will be heated up into the plastifying unit to a temperature range above the melting range of polymers so that a viscous polymer melt will be achieved. The polymer melt will be compressed and injected via a nozzle and a runner system into the

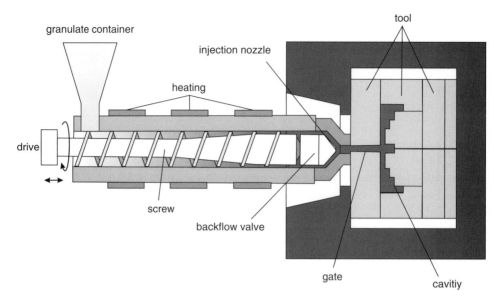

Figure 2.3 Basic components of an injection molding machine. The process is characterized by a plastifying unit, a compression and injection unit, and a mold of two halves that can be opened and closed. During the injection step both mold halves are mostly clamped by a hydraulic clamping system.

cavities of a mold. The basic component of every injection molding process is a plastifying screw, by which granulate during the rotation are dosed, melted, compressed, and finally injected into a mold. The injection occurs by a translational shift of the screw. A backflow valve inhibits the backflow of the molten polymer, enabling the polymer melt to flow under pressure via a runner system and a gate into the mold. Because of the high injection pressure in the range of several megapascals, the mold halves have to be clamped by a mechanical or hydraulic clamping unit (Fig. 2.3).

The process of injection molding can be split into the following steps (Fig. 2.4).

- *Dose of granules:* First the granulate material has to be dosed. This function is done by the plastifying unit. The main component of this unit is a screw that is responsible for the dose, transportation, and compression of the polymer. Because of the granule size (typical diameter of 2–3 mm) and the requirement of a homogeneous polymer melt, the screw has a minimum diameter of 14 mm.

- *Plastifying of granulate material:* The next step is to plastify the granulate material by a heating unit, typically implemented by circulatory-arranged electrical heating elements. The granulates are melted and heated up significantly over the softening range up to the melting state of the polymer. The temperature is set up in a range where the polymer melt is characterized by a low viscosity: A careful

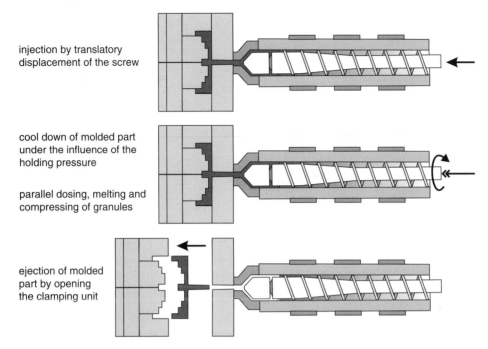

injection by translatory
displacement of the screw

cool down of molded part
under the influence of the
holding pressure

parallel dosing, melting and
compressing of granules

ejection of molded
part by opening
the clamping unit

Figure 2.4 Schematic view of the process steps of injection molding. The process is characterized by the dosage, the melting of granulate, the homogenization of melt, and the injection into the mold via nozzle, runner system, and gate. The injection of the polymer melt results in a temperated clamped mold whose temperature is below the softening temperature of the polymer. This effects a curing of the polymer melt after injection. After opening the two clamped mold halves, the demolding can be achieved by ejector pins. The injection velocity, the temperature of the polymer melt and the mold, and the dwell pressure have to be set up carefully; otherwise, the injection step fails and the (micro) cavities are not completely filled.

heating is recommended to obtain a homogeneous polymer melt, which is, in combination with a low viscosity, a precondition to fill small cavities via a runner system.

- *Sealing and compression of polymer melt:* After the melting of the granules the material must be compressed before the injection step will fill the cavities. The compression of the polymer melt is also done by the screw in the plastifying unit.

- *Injection:* The injection of the melt into the mold is done by a translational shift of the screw. To obtain the pressure necessary to fill the cavity, a seal in terms of a backflow valve is implemented. This valve inhibits the backflow of the melt during injection and guarantees the flow of the melt through a nozzle and a runner system into the cavities. The temperature of the cavities is typically in a range below the glass transition temperature of the polymer.

- *Cooling under dwell pressure:* The polymer melt is injected into a mold whose temperature is normally held to a constant value in a range near or under the glass transition or melting temperature of the polymer. Because of the temperature difference between mold and polymer melt, the melt must be injected in a short time (typically in less than 1 second) to fill all cavities completely. Otherwise, the temperature of the melt will decrease with the effect of an increase in the viscosity of the melt, resulting in the worst case in incomplete mold filling. After the cavities are filled, dwell pressure is necessary to avoid uncontrolled shrinkage of the polymer during cooling. The dwell pressure is obtained by a further compression of the melt by the screw. At this process step a controlled cooling is recommended to avoid the freezing of the polymer at the nozzle or at the gates, which effects that no material can be injected to obtain a dwell pressure profile during the whole cooling state. With ideal cooling the polymer at the gate freezes last.

- *Demolding:* Besides the filling of cavities, in particular microcavities, the demolding is one of the critical parts during the replication of microstructures. To obtain an automated process during the whole cycle an automated demolding system, typically implemented by ejector pins, is part of a complex molding tool. After the polymer melt is cooled down under dwell pressure to demolding temperature, the demolding process starts with the opening of the tool. The molded part remains in one of the mold halves. From this mold half the molded part is ejected by ejector pins. In parallel to the demolding, the next shot will already be prepared by the rotation of the screw in the plastifying unit, so that after the closing of the two mold halves the next injection step is initiated.

2.3.1.1 Requirements on Micro Injection Molding

The process steps described above characterize the process of injection molding in a general way. If the cavity size of the mold will decrease, the process has to be adapted to the requirements to fill those cavities. Especially if the size of the cavities decrease in a range of several microns, the process and the technology of the conventional injection molding process is limited. To understand the modifications in process and technology, first the requirements for micro injection molding should be defined.

- Regarding the size of microstructures and the batch sizes of microstructures on a mold insert, the typical volume of a microstructured molded part can decrease to the range of several cubic millimeters or, in extreme cases, in the range of hundreds of cubic meters. These small volumes are linked with shot volumes in the range of several milligrams (approx. 10–1,000 mg). This requires a precise and defined dosage of the polymer melt injected into the microcavities.

- Typical polymer granules are characterized by a weight of 20 up to 30 mg. Because of the small cavity sizes a homogeneous polymer melt with low viscosity is required for successful filling.

- If any air remains in the cavity during filling the gas will be compressed and heated up to high temperature. Especially if high injection velocities are required the flow fronts of the polymer are heated up over the decomposition temperature of the polymer. These locally hot areas will destroy the polymer at the flow front. Because of the similarity to the compression within a diesel engine this effect is called the diesel effect. To avoid this effect and to guarantee a complete filling in a short injection time, an evacuation of the microcavities is required.

- Further, the filling of filigree microcavities with high aspect ratios requires a modified process. The conventional process is characterized by an injection in a mold that is heated up to the range of the softening temperature of the polymer. During injection the polymer melt is cooled and the viscosity decreases. This can result, especially with fine cavities and long flow paths, in an incomplete filling of the mold. Therefore, the mold is heated up to the melt temperature and after injection the mold is cooled down to demolding temperature under the act of dwell pressure. This modified cycle is the variotherm process (Fig. 2.5).

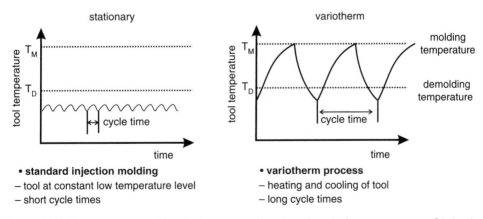

Figure 2.5 Temperature profiles during conventional and variotherm processes of injection molding. During conventional injection molding, the temperature of the mold or mold insert is kept in the range of the glass-transition temperature of the polymer, which is below the injection temperature of the melt. The variotherm injection molding process is, in contrast, characterized by a heating and cooling step between molding temperature and curing temperature of the polymer.

- Because of the small lateral sizes of the structures, the mold halves should be clamped and guided very precisely to avoid any displacement during molding.
- The fabrication of a mold with macroscopic structures typically integrates the cavities into a mold with a cooling system and runner system. The fabrication of such macroscopic molds is characterized by macroscopic mechanical machining. However, microscopic structured cavities are fabricated by completely different processes. The integration between tool and microstructured cavities result typically in the integration of the mold insert. By this concept, in special cases a tool can be used with different microstructured mold inserts.
- If the structure size decreases and the function of the structure requires an ultra-clean environment, the injection molding machine should work under clean room conditions. This requirement can be fulfilled by separation or encapsulation of the mold and further handling under clean room conditions.
- The handling of molded microstructures should be assisted by a handling system for cost-effective replication of microstructures in serial production.

A typical pressure profile of an injection cycle is shown in Fig. 2.6. In the dynamic state the pressure profile is characterized by the injection state, wherein the polymer is injected by a selected injection speed or injection profile into the cavities, and a compression state, wherein the injected melt is

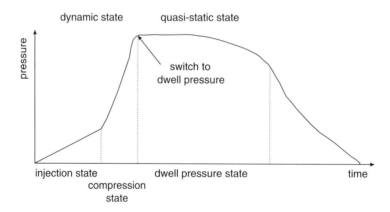

Figure 2.6 Pressure profile during an injection molding cycle. The profile is characterized by a dynamic state in which the cavities are filled and the polymer is compressed, and a quasi-static state results where the molded part was cooled under dwell pressure. The switch between both states can be achieved by a pressure control or volume control.

compressed. After a measured certain pressure of injected volume the control unit switches to the quasi-static state. In this state the polymer melt is cooled down under dwell pressure up to a temperature where the gate is frozen and no further dwell pressure can be applied. The set-up of the process parameters and the precise control of pressure, injection speed, and temperature is essential for the successful molding of microstructures. Also, metering size and holding pressure show a significant influence on the part quality [35].

2.3.2 Technology of Micro Injection Molding

The requirements named above result in adapted technologies, regarding especially the injection unit and the molding tools. The technical implementations are manifold and determined by a number of commercial suppliers for micro injection molding machines. Representative for a large number of machines the technology of micro injection molding can be explained on the basis of the injection molding machine Battenfeld Microsystem 50 [1], a machine developed especially for the replication of microstructures (Fig. 2.7). Because of the small injection volumes and the typical size of the molded part, micro injection machines can be built in a comparable compact way. Nevertheless, this machine can be equipped with a rotation unit for two molds, a handling system, and a clean room unit.

The requirement of the dosage and injection of small shot volumes is implemented by a separation of the tasks into several components. The screw with a standard diameter of 14 mm is responsible for the melting and homogenization of the granules. The dosage of small volumes is implemented by a special piston-based dosage unit, and finally the injection of the defined dose of polymer melt is processed also by a piston-based injection unit (Fig. 2.8).

The tools used in micro injection molding machines show a large bandwidth and are characterized by individual solutions for every design and material class (e.g., for powder molding of ceramics and metals) [10]. If microstructured mold inserts should be replicated, tools are developed allowing the fixation of microstructured mold inserts with defined geometries. This concept has the advantage that the mold insert can be changed quickly. Nevertheless, the ideal combination of tool, mold insert, and cooling and heating channels has to be developed individually. The characteristic molding tool consists typically of two mold halves with four guiding pins, guaranteeing the precise motion of both halves. The demolding systems also have to be adapted to the individual designs and can proceed by gas-assisted systems or ejector pins. Nevertheless, the molding of microstructures requires a tool with low tolerances in the guiding of both mold halves and the guiding of the ejector pins. During fabrication of tools, tolerances down to 1 μm should be observed [14].

Figure 2.7 Commercial micro injection molding machine Battenfeld Microsystem 50. The technology is adapted to the requirements of the replication of microstructures. The injection unit and the molding tool are encapsulated for use under clean-room conditions. Further, the machine can be equipped with an automated handling system.

2.3.3 Materials for Micro Injection Molding

Injection molding is characterized by the use of a large bandwidth of materials [22]. The proceeding requires granulate materials. Besides the conventional polymers, also the concept of powder injection molding (PIM) allows the replicatation of a completely new material class of metals and ceramics (Section 3.1.2). In detail, the following material classes are suitable:

- Thermoplastic amorphous polymers, e.g., PMMA, PC, COC, PSU
- Thermoplastic semicrystalline polymers, like PP, PE, POM, LCP, PEEK
- Compounds of different polymers

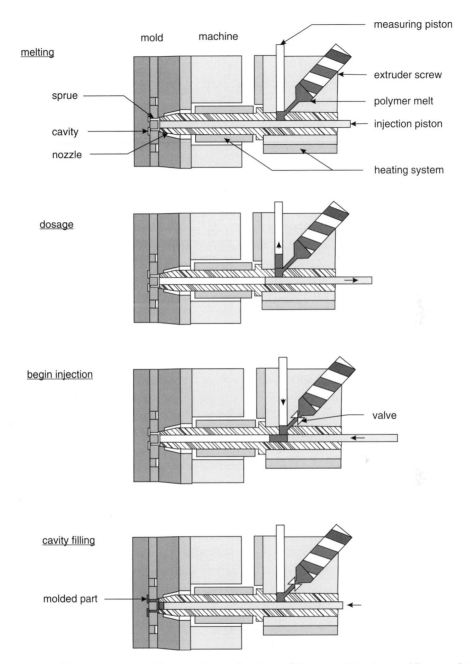

Figure 2.8 Schematic view of the injection technology of the micro injection molding machine Battenfeld Microsystem 50. The standard plastifying unit of conventional injection molding machines is separated into a screw-based plastifying unit, a piston-based dosage unit, and an injection unit. This separation allows the injection of the smallest volumes in the range of milligrams.

- Filled amorphous or semicrystalline polymers (e.g., ceramic fillers, glass fibers)
- Powder injection molding of metals (MIM), e.g., carbonyl iron powder, stainless steel (316L), or alloyed hardenable steel (17-4PH)
- Powder injection molding of ceramics (CIM), e.g., zirconium oxide powder or aluminum oxide powder [21,24,25].

Compared to injection reaction molding whereby the liquid with low viscosity is characterized by a mixture of monomers and starter, here a polymer melt or a feedstock with a higher viscosity has to be injected. Because of the molecular structure, polymer melts cannot achieve the low viscosity of monomers, which results finally in a significantly higher injection pressure. Nevertheless, the concept of injection and the flow of the polymer melt melt via runner systems into filigree microcavities requires as low a viscosity as possible, resulting in higher temperatures of the melt during molding.

2.3.4 Cost Effectiveness

As mentioned above, injection molding is well suited for the replication of large series. For small series the fixed costs like investments for the injection molding machine and the mold fabrication will increase the cost of every molded part. The investments will only amortize with an increase in molded parts. This is a result of the comparatively high fixed costs for injection molding machines and the mold. Besides the investment for a micro molding system, the costs for a mold are relatively high because of the complexity of the mold, regarding cooling channels and demolding units. The cost for a microstructured molded part can be reduced by the use of a modified mold system, which allows reversible integrated microstructured mold inserts. This concept allows the use of one mold for the replication of different mold inserts. Nevertheless, most designs require an individual solution or make it necessary to adapt existing molds. The variable costs are mainly characterized by the cycle times and the grade in automation. Commercial machines are optimized for short cycle times and show also a comfortable high grade in automation, up to integrated handling systems that result in moderate variable costs. Compared to the high investments for the machine and mold, the total costs will decrease with an increase in the number of replicated parts (Fig. 2.9).

2.3.5 Characteristics

To summarize the advantages of replication by micro injection molding:

- High grade in automation, including handling systems
- Short cycle times, down to a range of a few seconds

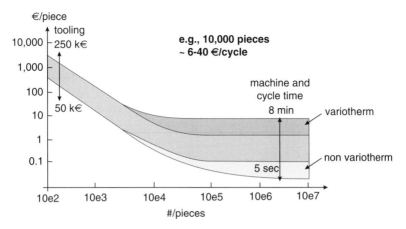

Figure 2.9 Schematic view of the cost effectiveness of injection molding. Because of the relatively high investments for an injection molding machine and the design and fabrication of molding tools, the cost will be amortized by larger series. Compared to the fixed costs, the variable costs are typically low because of short cycle times and a high grade in automation.

- Well suited for large series
- Wide range of molding materials, like polymers and powder
- Large bandwidth of commercial suppliers
- Established replication technology in industry
- Availability of commercial process simulation software (e.g., Moldflow)

Nevertheless, this molding concept shows also characteristic problems. Due to the high injection speed, high shear stress, and in combination with short cooling times, high residual stress inside molded parts can occur. During cooling a constant dwell pressure over the whole molded part is advantageous to reduce the anisotropy in shrinkage of the molded part. Due to an injection system with a sprue and a runner system, the dwell pressure during cooling vary over the molded part, which results in anisotropy of shrinkage and finally in a warpage of the molded part, especially in plane surfaces. The process of injection molding comes to a limit if, for example, the smallest cavities have to be filled at the end of a long flow path where the injection pressure is reduced by a pressure drop over the flow path. The pressure will decrease significantly by the length of the flow path and the related cross section, which makes it difficult to fill, for example, large, thin, flat geometries.

The problem resulting from large flow distances can be eliminated when the process is split into an injection step and an additional compression step for the filling of microcavities. This process is already known as injection compression molding, a process well established for replication of compact discs.

2.4 Injection Compression Molding

The process of injection compression molding combines the advantages of an injection step and a following compression step, which eliminates the disadvantages obtained by long flow paths (Fig. 2.10). First, a volume of polymer melt corresponding to the volume of the molded part is injected into a mold that is not completely closed. Because of the gap between the mold halves the injection can be done by lower pressure, resulting in a reduction of shear velocity and shear stress of the polymer melt. After the injection step the mold halves are closed by a compression step. This compression step is split into a velocity-controlled motion and finally, if the desired clamp force is achieved, in a force-controlled holding of the final press force over a defined cooling time. Regarding the filling of microcavities, both process variations of injection molding are possible: the conventional injection compression molding—injection in a mold with demolding temperature—or, recommended for filigree structures, the variotherm cycle. An advantage of this combination is the processing of polymer melt in the plastifying unit in parallel to the compression molding step, which results in a reduction of cycle times. A further advantage of this compression step is the separation of the injection step and the filling of cavities. The filling of cavities is effected only by the short flow distance initiated by the motion of the mold halves. This process variation reduces the length of flow paths of the

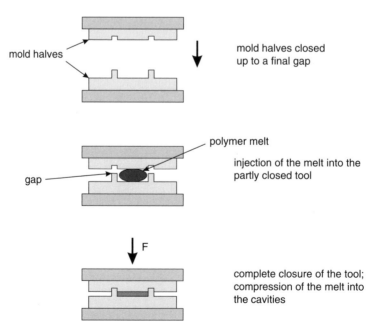

Figure 2.10 Schematic view of the injection compression molding process. This process is characterized by an injection step into a gap between two mold halves and an additional compression cycle to fill the microcavities.

polymers and allows, therefore, the molding of large microstructured areas on a layer of low thickness. Compared to the injection of the polymer melt via a runner system, the pressure distribution during compression molding is more homogeneous, which reduces the anisotropy of shrinkage, resulting finally in molded parts with lower warpage. A common example for this process is the fabrication of compact discs, a flat area with small microstructures on a carrier layer.

The technology is similar to that of micro injection molding, with the difference that the clamp mechanism is used for the compression cycle. Also, the molding materials and the costs are similar to the injection molding process.

2.5 Micro Hot Embossing

The process and technology of micro hot embossing will be briefly described in the following chapters, but for a comparison with the other replication technologies a compact description of this process will be reasonable.

Like the injection compression molding process described above, the hot embossing process also refers to a two-step compression molding cycle, with the difference that the polymer melt is not injected before. Instead of the injection step, the polymer in the form of a thin polymer film is used. This film is heated up via heat conduction up to the melting range before the compression step fills the microcavities. The schematic view of the molding of a single microstructured mold insert is shown in Fig. 2.11.

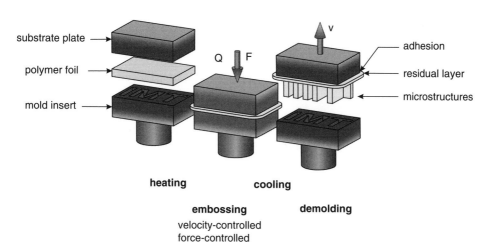

Figure 2.11 Schematic view of the hot embossing process. This process is characterized by the use of a thin polymer foil between a microstructured mold and a plate with rough surface, the substrate plate. After heating up to the melting state of the polymer, the melt is pressed into the microcavities by a two-step compression cycle. After cooling down to demolding temperature, the microstructures can be demolded vertically by the adhesion of the residual layer on the substrate plate.

Between a microstructured mold insert and a rough surface plate, called a substrate plate, a thin film of polymer is heated up to the melting range. A two-step molding cycle in the form of a velocity- and force-controlled compression step initiates the flow of the polymer melt into the microcavities. Characteristic for hot embossing is the residual layer, a carrier layer of the microstructures. This layer is generated by an excessive volume of the polymer film compared to the volume of the microcavities. This residual layer is also necessary to obtain the pressure required for filling the microcavities. The cooling of the molded part occurs under constant force, here the dwell pressure. After cooling down the mold and substrate plate to the demolding temperature of the polymer, the molded part is demolded by a relative motion between substrate plate and mold. At this step the adhesion of the residual layer on the substrate plate plays an important role. A higher adhesion of the residual layer on the substrate plate ensures that the microstructures can be demolded in an accurate vertical direction, which reduces the risk of damage.

2.5.1 Technology of Hot Embossing

The technology of micro hot embossing can be split into four main groups (Fig. 2.12).

- The embossing machine, which is responsible for the press force and the accurate molding velocity. The challenge of the technology is to obtain a high stiffness at high force and also a precise relative motion between mold and substrate plate. This requirement results in a stiff frame consisting of two crossbars and usually four massive guiding pillars. To obtain a press force one of the crossbars is fixed. The other crossbar is moved by a precise motion system, like a spindle drive or a hydraulic drive.
- The tool mounted between the two crossbars is responsible for the heating and cooling of mold, substrate plate, and polymer sheet. A typical tool consists of two halves, the top half fixed on the top crossbar and the bottom half fixed on the lower crossbar, each with a single heating and cooling system. With approximately both halves an integrated vacuum chamber is closed and isolates the microstructured mold insert and the substrate plate against the ambient pressure. Optionally, an alignment system is integrated that allows the adjustment of, for example, two mold inserts against each other or the molding of prestructured substrates.
- The microstructured mold insert is reversibly fixed in the tool. Opposite the mold insert, another mold insert or a substrate plate is positioned. A wide range of microstructured mold inserts can be

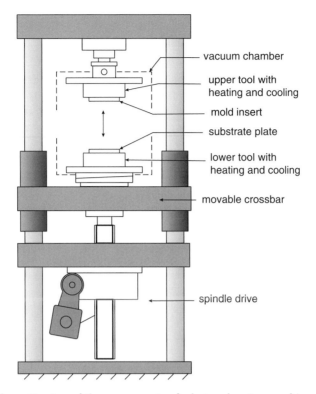

vacuum chamber

upper tool with
heating and cooling

mold insert

substrate plate

lower tool with
heating and cooling

movable crossbar

spindle drive

Figure 2.12 Schematic view of the components of a hot embossing machine. The components can be split into four main groups: a frame with a high stiffness, a tool with integrated heating and cooling unit, a microstructured mold insert, and a control unit as an interface to the user.

used, for example, inserts produced by mechanical machining or mold inserts fabricated by lithographic technologies.

- A precise controlling system is also as fundamental as the mechanical equipment. The precise control of press force, motion of cross-bars, and temperature is one of the technological challenges. Besides the control of process parameters, the measurement of press force, temperatures inside of mold inserts, and the distances between mold insert and substrate plate is the task of the control unit. As an interface to the user it is also its task to prepare and visualize all data, allowing the user to set up and control the process in an effective way.

2.5.2 Materials for Hot Embossing

Similar to the injection compression molding process, thermoplastic polymers are suitable for hot embossing. Both classes, amorphous and semicrystalline thermoplastics, are appropriate, including common low-temperature polymers like PMMA or PS, up to high-temperature semicrystalline technical polymers like PEEK or LCP. Besides the requirements on materials defined by the further application of the molded part, the use of a specific polymer is determined by the availability of foils of this polymer. Especially for technical polymers, the availability of polymer foils with defined thicknesses can be limited. To bypass this limitation polymer foils can also be molded by hot embossing, using polymer granules instead of foils.

Nevertheless, the use of polymer foils opens new perspectives in micro molding. For example, the molding of different stacked sheets of polymer foils with different glass transition or melting temperatures can be used to form a thin polymer film by a kind of thermoforming (Section 5.5.6). Further, if two foils of polymers are used showing no adhesion between them, the polymers can be separated after molding, which, for example, allows the embossing of thin films with through-holes (Section 5.5.3).

2.5.3 Cost Effectiveness

Similar to the cost effectiveness for injection molding, the fixed costs of hot embossing are generated before the production of a series starts. These costs are typically independent from the number of molded pieces. The fixed costs can be related to the investments of a hot embossing machine, the fabrication of a microstructured mold insert, and the set-up of the process. Compared to a complex injection molding tool, here only a mold insert has to be fabricated, which can easily be integrated into a hot embossing tool. The set-up of the process is mostly less time consuming, like the set-up of the process for injection molding. Here the key parameter is the engineering time responsible for the mold design, up to the set-up of the process. Compared to injection molding the fixed costs for hot embossing are much lower, because of the moderate engineering time.

Regarding the variable costs, the situation changes. The variable costs are related to every molded part of a series and increase linearly with the number of molded parts. Here the key parameter is the cycle time of the process. Typical cycle times of hot embossing are longer than injection molding, or nearly similar if the variotherm process is needed. Nevertheless, the development in hot embossing shows that an increase of the molding size and therefore the batch size of structures will reduce the variable costs. Supporting these reductions are new developments in machine technology, with more effective heating and cooling systems and automatic handling systems, which allow a significant reduction of the cycle times (Fig. 2.13).

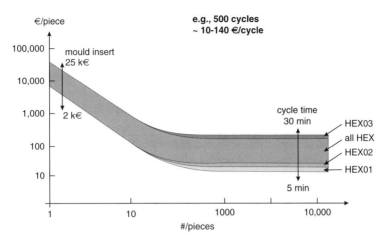

Figure 2.13 Estimation of total costs for hot embossing (one part per cycle). For small series, the fixed costs like machine investment and the cost for mold inserts are determining. For larger series, the cycle time determines the cost for a molded part.

2.5.4 Characteristics

To evaluate the replication process and compare hot embossing to the other replication processes, the main characteristics are described here only in a compact way. A detailed discussion can be found in Chapter 5.

- The molding process is characterized by short flow paths, only from the molten polymer film into the microcavities.
- Because of the moderate molding velocities, in the range of 1 mm/min, only moderate shear stress in the polymer is generated. This results in comparative low residual stress in the molded parts.
- If the molding temperature is set to the range where the relaxation times of a polymer correspond to the cycle times of molding, the stress induced by molding can be decreased by relaxation processes. This option requires the knowledge of the temperature-dependent relaxation behavior of the polymer.
- The use of standardized mold inserts allows the quick change of a microstructured mold, which underlies the flexible technology of hot embossing.
- Besides the quick change of mold inserts, the polymer can also be changed quickly. Here only a new polymer foil has to be placed between mold insert and substrate plate. This allows the replication of a mold insert into several polymers in a short time.
- The technology allows a multitude of process variations, for example, double-sided molding, molding of through-holes, multilayer

molding, and also thermoforming of a high-temperature polymer foil by a low-temperature polymer melt.

- Compared to injection molding, the process is characterized by longer cycle times. Although the development shows that with an effective technology the cycle times can be reduced, the process is economically well suited for small and medium series. Nevertheless, because of its high flexibility this process can be adapted to several requirements and shows, therefore, large potential for the development of prototypes in laboratories.

2.6 Micro Thermoforming

Thermoforming is a well-known replication process for macroscopic dimensions [26,27]. The process is characterized by the forming of thermoplastic polymer foils at elevated temperatures into a three-dimensional shape. Commercial examples are packaging for the food industry, thin, one-way coffee cups made of plastic, up to large car bumpers with a wall thickness of several millimeters. This process shows a high potential for microstructure technology, because of the capability to fabricate three-dimensional shapes with thin sidewalls in the micro range. Dreuth and Heiden [9] systematically investigate the structuring of thin polymer films of PET between 10 µm and 1.5 µm by different methods: molding between two mold inserts, molding on elastic substrates, and role-to-role forming. In 2001 Truckenmüller et al. [30] transferred the macroscopic thermoforming process into the micro range and generated a new microstructuring method—micro thermoforming [29,31]. One of the first applications is microfluidic components and cell containers for 3D tissue engineering [12,13].

2.6.1 Process

Compared to all other processes described above, thermoforming is characterized by a different forming principle. The definition of molding should be replaced by forming, because the semi-finished product, a thin polymer film, is not heated up to a polymer melt. Here the film is heated only over the transition range, in the thermoelastic, rubber-like state of the polymer. In this way the film can be three-dimensional, formed by gas pressure. The forming of this thin polymer film results in different wall thicknesses of the final microformed part. The principle of thermoforming refers to a two-dimensional stretching of a polymer foil, which is completely different from the other molding principles.

In Fig. 2.14 a schematic illustration of the basic micro thermoforming process is shown. A thermoplastic film is placed into a molding tool for micro thermoforming. The three-part molding tool consists of the mold itself, with the microcavities, a counter plate with holes for evacuation and gas pressurization, and a

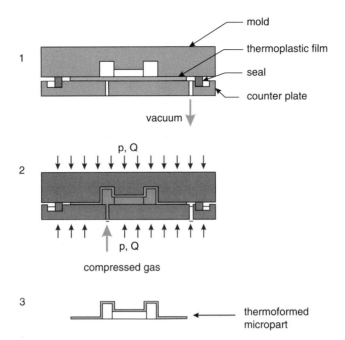

Figure 2.14 Schematic view of thermoforming process. The process is characterized by the three-dimensional forming of a thin polymer foil under the act of gas pressure.

seal between mold and counter plate. The tool halves are closed to a distance by which vacuum sealing is achieved, but the plastic film is not clamped. The tool is evacuated and completely closed, so that the film is clamped and heated up. Above the glass-transition temperature of the thermoplastic polymer the film is formed into the evacuated cavities of the mold by a compressed gas. After forming, the tool is cooled down. Approximately 20 K below the glass temperature of the polymer, the gas pressure is decreased, the tool is opened, and the thermoformed microstructure can be demolded.

Because of the different molding principles and the different temperature ranges, the combination of hot embossing and thermoforming allows the three-dimensional forming of prestructured polymer films. In a first step a thin polymer film can be structured by hot embossing, e.g., with nanostructures. In a second step these polymer foils can be formed by thermoforming without destroying the preliminary embossed structures. By this combination nanostructures on vertical sidewalls of microstructures can be achieved, structures that cannot be demolded by conventional hot embossing.

The principle of thermoforming can also be achieved by hot embossing of a stack of different polymer films. Instead of air pressure, a thin high-temperature polymer foil can be formed by a polymer melt of a low-temperature polymer. The result of this combination is the forming of the thin high-temperature foil (Section 5.5.6).

2.6.2 Technology for Thermoforming

The technology for micro thermoforming is still in development but has already been demonstrated in the laboratory. The components required for thermoforming consist of a press system to clamp the two mold halves against the gas pressure, a gas delivery system with a gas pressure of around 5 MPa, a mold with a heating and cooling system, and a vacuum unit. The task of the press system is characterized by the clamping of the tool. A precise motion, necessary for hot embossing, is not required. Therefore, the press for thermoforming can be much simpler than press systems for hot embossing. The technological challenges here are in the development of thermoforming tools with short heating and cooling times and a homogeneous temperature distribution over the molded area. Because of reduced requirements regarding the press, a thermoforming machine can be less expensive than a hot embossing system.

2.6.3 Materials for Thermoforming

For thermoforming, thin polymer films are required. The required thickness depends on the size of the cavities (cross-section) and the available gas pressure. Thin polymer film can be formed easily into small cavities. The achievable depth is limited by the maximal strength of the polymer film during forming. The forming of this film under pressure results in a decrease of the thickness of sidewalls, which finally results in a break of the film. Therefore, the depth of the thermoformed part is limited and depends also on the thickness of the polymer film. Depending on the gas pressure, films in a range between approximately 20 μm and 100 μm can be structured by thermoforming. In general, amorphous thermoplastic polymer films are also suitable for forming semicrystalline-like polymer film. The limitation here is the availability of polymer film in the desired thickness. Microfluidic structures have already been formed in 50 μm polystyrene (PS) or polycarbonate (PC) film [13]. With further development the bandwidth of material will increase.

2.6.4 Cost Effectiveness

Because of the new developments in this field a commercial micro thermoforming machine is not available. A detailed estimate of the cost effectiveness cannot be given at this time. Nevertheless, there exists a parallel between the requirements for a thermoforming machine and a hot embossing machine. This parallel is underlined by the first experiments done by a modified hot embossing machine [30]. The potential of these forming processes can also be seen by a structuring of a thermoplastic film from a coil, which also paves the way to a commercial use of this process. In the future, commercial available machines should be available, because many applications—especially in the biology and medical sector—are conceivable.

2.6.5 Characteristics

The characteristics of the micro thermoforming process may be summarized by the following items:

- Only this process shows the potential of replication of three-dimensional structures in the micron range.
- The shape of the three-dimensional formed structures is characterized by different thickness of the sidewalls. Depending on the design, the deformation of the polymer film during the forming process is not homogeneous. Areas of the polymer foil will be stretched more than other areas, which results finally in different thicknesses of the sidewalls. In extreme cases the polymer film will be overstretched and will break. This overexpansion of the thin film limits the achievable aspect ratios of the structures. The first experiments show good results with aspect ratios in the range of one [30].
- Because of the forming principle, sharp corners cannot be replicated with the accuracy known in other processes. Conditional on the process, sharp edges are transferred to edges with a certain radius.
- The concept of thermoforming allows the use of prestructured polymer films, e.g., by hot embossing. This aspect shows a high potential for new applications. For example, in a first step the surface of the polymer films can be modified by a structuring in the nano range and can be formed in a second process step into a three-dimensional shape. Because of the different temperature ranges of hot embossing and thermoforming, the structures generated by hot embossing survive during thermoforming.
- The process of micro thermoforming is still in early stages and is performed under laboratory conditions. Further developments can improve the process and the technology in commercially available machines.

2.7 Nanoimprint

Nanoimprint technologies are part of next generation lithography (NGL) [15,33], with the aim to develop new structuring methods for the patterning of nanostructures. The structuring methods refer to a structure size below the pattern size limited by current UV-lithography structuring methods. The motivation for the development of new structuring methods in the nano range was first the current developments in microelectronics, requiring new structuring methods for a further decrease in components for electronic integrated circuits. Furthermore, the structuring methods can also be used to fabricate structures in the nano range for applications in photonics and biotechnology [15].

The process of nanoimprint lithography (NIL) was first developed by Chou et al. [5–7]. In relation to lithography, a mask for further structuring steps is required. This mask can be fabricated by an imprint or molding process on a substrate, typically a silicon wafer. First a nanostructured mold insert is required, which can be fabricated, e.g., by E-beam lithography and electroplating or alternatively by casting in PDMS (polydimethylsiloxane). By a molding process the structures are transferred onto a silicon wafer which acts as a mask for further structuring methods like etching. The focus here is on the molding or imprinting methods. Three different approaches have been established (Fig. 2.15).

- Nanoimprinting by micro contact printing (MCP or μCP) with self-assembling monolayers (SAM)
- Molding of UV-curable materials
- Molding of thermoplastic polymers

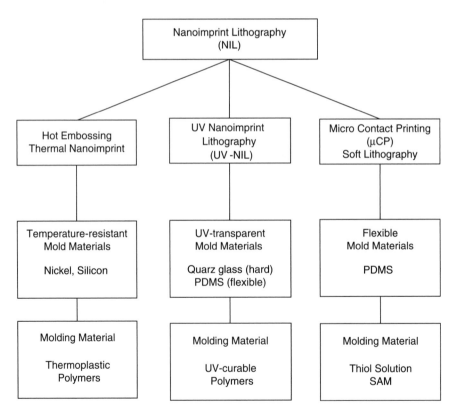

Figure 2.15 Overview of different nanoimprint techniques. (1) Micro contact printing with a flexible PDMS stamp; (2) molding of UV-curing materials; and (3) molding of thermoplastic materials—similar to hot embossing.

After structuring a polymer on a substrate by the named imprinting processes, an additional process step is needed to transfer the structures into the silicon substrate. Typically this is an additional etching step. Especially the nanoimprint approach, referring to the molding of thermoplastic polymers, is a similar process to hot embossing of nanostructures. Here only the aim is different. Hot embossing is a technique for replication of micro- and nanostructures; however, thermal nanoimprint uses this technology to replicate masks for additional structuring processes. Compared to the high aspect ratio of structures replicated by hot embossing, the structures replicated by thermal nanoimprint are typical in the range of one, which is sufficient for the use as mask. The approaches named above shall only be described in a compact way, with the aim of the replication technique. The reader will find detailed background information on each step in the literature at the end of this chapter [15,28,34].

Today the definition of nanoimprint is not limited to the classical aim of NGL. Nanoimprint defines further the replication of structures in the nano range by thermal and UV-based molding processes, in which the definition of nanostructures is not clearly defined.

2.7.1 Micro Contact Printing

The process of micro contact printing differs from the technologies described above. The molding process is, in this case, reduced to a structuring of a substrate by an ink process [18,32] (Fig. 2.16). Similar to all replication processes, a mold is also required. The difference in the replication processes described above is the mold material used. To achieve a good contact a soft material, in this case PDMS, is used as a mold. To obtain this mold, in a first step a micro or nanostructured master is required. The fabrication of a master relates to all established structuring methods, like E-beam-lithography or etching processes of silicon. From this master in a second step a liquid elastomer, conventionally polydimethylsiloxane (PDMS), is poured over the surface and cured. The solid elastomer mold is then peeled away from the master. The elastomer surface then contains the inverse features corresponding to the features in the master. This mold is then coated with a film of hexadecanethiol, a liquid organic that creates self-assembled monolayers (SAMs) on the surface of the PDMS mold. In a third step this elastomeric mold is brought into contact with the substrate. The substrate consists of a thin metal film, typically Ag or Au with a thickness of several hundred nanometers. The SAMs are then transferred from the raised features of the PDMS mold to the substrate surface, leaving a patterned organic film corresponding to the original topographic pattern on the master. This inked film acts as an effective wet-etching barrier. By etching methods the inked structures are transferred into the Ag or Au surface and may be transferred to the underlying silicon substrate by further wet- or dry-etching methods.

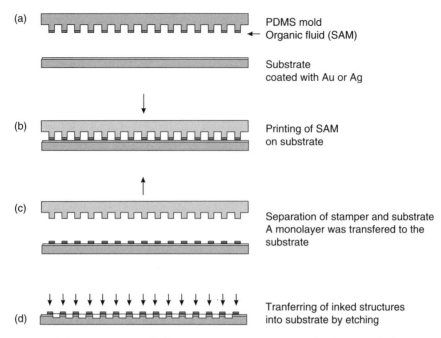

Figure 2.16 Schematic view of the micro contact printing (MCP or µCP) process. An elastomeric stamp (made of polydimethylsiloxane (PDMS)) is coated with molecules of thiols or silanes (a) and pressed against a substrate of gold or a silicon wafer (b). The structures are transferred by an ink process to the Ag or Au layer (c). By further etching processes, the structures are transferred into the silicon substrate (d).

2.7.2 Nanoimprint of UV-Curable Materials

UV-curable polymers are well known in UV-lithography, where a UV-curable polymer will by structured by UV-radiation through a micro- or nanostructured chromium mask. But this kind of material itself can also be used as a mask for further etching processes. To fabricate such a mask, spin-coated UV-curable material is structured by a micro- or nanostructured mold insert (stamp) via an embossing process (Fig. 2.17). By this process structures down to 5 nm have already been replicated [2]. The structured mold insert can be fabricated, for example, by the well-known structuring methods like E-beam lithography and etching processes. To guarantee the curing of the molded polymer, this stamp has to be transparent for UV radiation. Besides stamps fabricated by UV-transparent glass, also the elastic polymer PDMS is a well-suited material. PDMS also shows a good transparency for UV light and is also a flexible polymer that guarantees a good contact between mold and substrate, independent of the evenness of the substrate. In general, the process of nanoimprint requires a homogeneous thickness of the molded mask; otherwise, the further structuring processes by etching will cause inhomogeneous results. Compared to hot embossing of polymers, the structuring of a thin layer of this UV-curable resist is quite

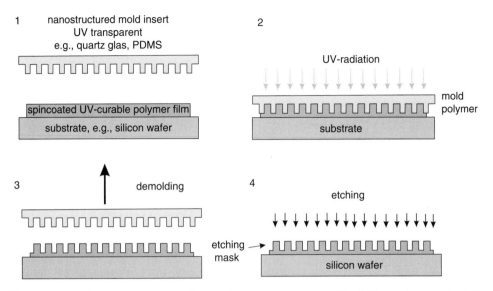

Figure 2.17 Schematic view of nanoimprint process with UV-curable materials. A precondition for molding is a UV-transparent mold or substrate plate. The mold is pressed into a UV-curable polymer and is cured by UV radiation. The mold or substrate material should be transparent for UV light. Typically, molds are fabricated from glass or PDMS.

different. A heating and cooling unit is not required; also, moderate forces are sufficient. Depending on the area and structure, design-molding pressures in the range of 0.08 MPa can be sufficient [4]. In contrast to the moderate molding pressures this, a UV-exposure unit, and, if large areas are required, a high precision alignment system are required. This system allows the creation of large structured areas by the step and repeat method, where after curing of one area the substrate is aligned to the next uncured area before the next exposure will start. Because of the different requirements, specific commercial machines for UV-nanoimprint are available. For example, mask aligners show a UV-exposure unit and an alignment system. They also can be used with some modifications for nanoimprint.

2.7.3 Nanoimprint of Thermoplastic Polymers

Besides the UV-curable materials, thermoplastic materials can also be used for the replication of structuring masks on substrates. The structuring of thermoplastic materials by embossing is well known by the process of hot embossing. Therefore, the processes of thermal nanoimprinting and hot embossing are nearly similar (Fig. 2.18). Both processes use thermoplastic material, and therefore heating and cooling systems are required. The differences can be found in the size and aspect ratio of the replicated structures. Hot embossing is characterized by the replication of a wide range of different structure sizes in the

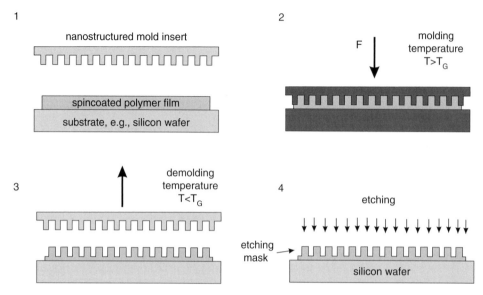

Figure 2.18 Schematic view of the thermal nanoimprint process. This process is nearly similar to the single-sided hot embossing process on a substrate. Heating, embossing, cooling, and demolding are also the fundamental process steps. But the aim of the process is different. The molded parts have a function as etching mask for further process steps.

range of several hundred micrometers down to structure sizes in the range of several hundred nanometers, with mostly high aspect ratios. Conversely, the process of nanoimprint is used to overcome the limitation of UV lithography, which results in this case in the molding of structures typically in the nano range. Molding of structures in the range of several nanometers is state of the art. Because of their function as etching masks it is sufficient to achieve thin layers, which results in typical aspect ratios of the molded structures in the range of one. Here the difference between hot embossing and thermal nanoimprint can be seen in the kind of structures. Nevertheless, a clear separation between hot embossing and thermal nanoimprint cannot be given. Here the boundaries between both processes are not strictly separated like the definitions of nanoimprint and hot embossing of micro- or sub-microstructures found in literature. Today especially the term *nanoimprint* will be used independent of the further etching steps in NIL.

The difference in structure size and aspect ratio of the structures defines the technology of nanoimprint machines developed especially therefore. Thermal nanoimprint machines require moderate molding forces and thus can be fabricated with a lower stiffness of the machine. Nevertheless, independent of the machine, hot embossing, or nanoimprinting, structures in the nano range can be replicated with both types of machines. Cui and Veres [8] demonstrate the ability of nanoimprint lithography to replicate patterns having feature sizes ranging from the nanoscale of 100 nm up to the millimeter scale of 2 mm. The replicated area for 100 nm structures can increase up to the 6-inch level [17].

2.8 Comparison of Processes

The last section of this chapter will summarize and compare the characteristics of the replication processes named above. Because of the manifold of each process regarding process variations and techniques, a comparison can only be limited to characteristic properties. Nevertheless, a comparison should give the reader an overview of the advantages of each technology and will be helpful for the choice of a suited replication process for a selected microstructured design.

To replicate a desired design, the following aspects, besides the items named at the beginning of this chapter, should be taken into account.

- First, the design to be replicated should be analyzed. Linked to the design are the microstructured area, the kind of structures (for example, freestanding structures, grooves, through-holes) and the aspect ratio of the structures, the relation between the lateral dimensions, and the height of structures. The design determines the cavities and the arrangement of the cavities on the mold and finally the size and the form.
- Once the design is determined, the material in which the mold insert should be replicated has to be determined. If only thermoplastic materials are desired, nearly all processes are theoretically suitable. If, for example, thermosets are desired, the manifold will be limited to the reaction injection molding process or UV-curing processes like UV-nanoimprint.
- Once the design and the material are determined, the process should be selected. Here the aspects of cavity filling, residual stress, shrinkage, and warpage of the molded parts can be determining factors.
- Finally, besides the selection of the process, the aspect of cost effectiveness has to be taken into account. If only prototypes or small series are desired, some replication processes are more suitable than others, specified for the replication of large series. Here the grade of automation and the cycle times are linked to the cost effectiveness.

As described above, nanoimprinting is a selected process optimized for mask fabrication with further process steps. Therefore, the comparison will be oriented on the processes of injection molding, injection compression molding, reaction injection molding, thermoforming, and hot embossing. Finally, a table will summarize all characteristic aspects named at the beginning of this chapter.

2.8.1 Design

At the beginning the application will define the design that has to be replicated. Besides the kind of structures that will result from the design, the area
on which the structures will be arranged will be defined. Regarding the structuring methods for mold inserts, the structuring areas are limited (typical sizes can
be in a range from 28 mm × 66 mm for an LIGA mold insert, up to an 8-inch
mold insert fabricated by mechanical machining). The size undergoes a continuous development that will increase the size of the mold inserts. Nevertheless, if
the design is determined and the requirements for the mold fabrication are
defined, the limitation of the structuring area of the fabrication process will
finally determine the maximum batch size of the structures. Depending on the
design, it is not recommended to always use the maximum size possible. The
influence of shrinkage will increase with an increase of structuring area, which
can result in problems during demolding (Sections 4.4.3 and 6.4.4).

The manifold of sizes of mold inserts results mostly in individual fixation systems in the different molding machines. Nevertheless, the limitation of sizes is a
combination between the structuring method of the mold insert and the individual molding machine. Typical standards for commercially available machines
are here the 4-inch and 6-inch wafer sizes; in individual solutions, larger areas
can be molded. The integration of a mold insert does not guarantee the successful
molding of the design. Especially extremely deep cavities with small cross-sections or large microstructured areas on thin carrier layers may cause problems.
The filling of microcavities or the flow on large thin areas can be limiting factors
for the replication by a desired process. For example, injection molding is characterized by long flow distances and high flow velocities. The polymer melt has to
inject in a relatively short time through a nozzle and flow via runner systems to
the microcavities. This technology effects stress inside the molded parts, which
are frozen by the short cooling times. Long flow distances complicate the filling
of filigree microcavities with high aspect ratios and the filling of large thin areas.
To avoid these problems the filling of those structures can be done by a compression cycle. Here, injection compression molding or hot embossing show advantages because the filling of cavities can be realized by short flow distances with
reduced flow velocities, which results in reduced residual stress inside the molded
parts. A compression cycle is therefore well suited for molding thin parts (like
lenses or CDs) or microcavities with high aspect ratios. Injection reaction molding is also characterized by an injection system, but normally the viscosity of the
mixture between monomer and starter is lower than a polymer melt, which
results in an easier filling of complex cavities compared to injection molding.

The structuring method of thermoforming is quite different from the other
processes. Thermoforming is a process specialized for forming three-dimensional
structures in a temperature range significantly below the temperatures of hot
embossing or injection molding. Because of this, for example, prestructured

polymer film can be formed into three-dimensional shapes, which allows the achievement of a new class of microstructured parts. Nevertheless, because of the forming principle the aspect ratios that can be achieved are limited, but a wide range of applications like fluidic applications or cell containers are well suited by the replication by thermoforming. Here, this forming principle can theoretically be fast, because of the moderate temperature cycle and the fast structuring method by gas pressure, which acts simultaneously and homogeneously over the complete structured area.

Independent of the selected process the critical part of every micro molding process is the demolding of filigree structures. Because of the difference in shrinkage between the mold (mostly metals) and polymer, the molded structures undergo high demolding forces during demolding, which often results in damage of the structure. Here, with the use of suitable process parameters the shrinkage and the demolding forces can be reduced to an uncritical limit, but not completely eliminated.

2.8.2 Materials

Considering all the processes named above, in total a wide range of material classes can theoretically be used for replication: thermoplastic polymers, thermosets, elastomers, and UV-curable polymers. But in practice not every material class can be used by every process.

Injection molding, injection compression molding, hot embossing, and thermoforming are well suited for thermoplastic polymers. These processes use granules or polymer sheets or films as semi-finished products that have to be transferred into the individual molding range of each process. Here, the processes use the same material class, but every process uses a specific molding window of the same material (Fig. 2.19). Because of the injection process and the long flow distances, micro injection molding requires a low viscosity to fill microcavities. In contrast to this, hot embossing requires for mold filling a higher viscosity of the same polymer melt, which results in a lower molding temperature. Conversely, thermoforming uses a molding window of a thermoplastic polymer in the range of the glass-transition temperature to form three-dimensional structures. In contrast to the thermoplastic polymers, the material classes of thermosets or UV-curing materials are suited for molding by reaction injection molding or UV-nanoimprint processes. Here, the material can be injected or embossed in a liquid state and the curing starts, depending on the material, by UV light or heat.

Nevertheless, regarding the multitude of different thermoplastic materials with different specific properties, it is obvious that thermoplastic polymers play a dominant role in micro replication. This role will be underlined by the wide range of applications that can be fabricated by this material class. Important properties are, for example, high transparency for optical applications (PMMA,

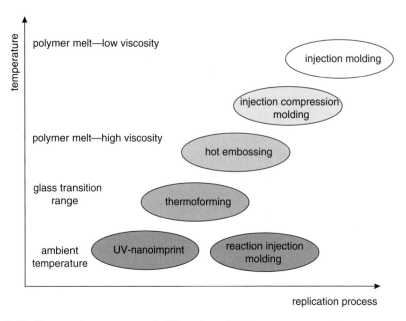

Figure 2.19 Temperature ranges of different replication processes. Every process uses a specific temperature range of a polymer material. UV-nanoimprint and reaction injection molding are characterized by a molding window at ambient temperature. Thermoforming requires a temperature window in the glass-transition range. Hot embossing is characterized by a large temperature window beginning at the softening range up to a polymer melt with high and medium viscosity. Injection compression molding and injection molding require a polymer melt with low viscosity, achieved typically at high molding temperatures.

PC) or biocompatibility for medical applications (COC, PEEK), temperature resistance (PSU, LCP, PEEK), or chemical resistance (PEEK). Especially the semicrystalline high-temperature polymers LCP and PEEK allow the fabrication of microstructures for new areas of application like lab on a chip systems or other microfluidic and medical devices. The molding of these high-temperature polymers reaches the limit of different molding systems. Injection molding machines typically can process temperatures over 340 degrees (melting range of PEEK). In contrast to this, commercial hot embossing machines can reach their limits, here individual solutions with additional heating systems that set the temperature in the molding range of the high-temperature polymers.

Finally, polymers can be modified by fillers, which allows the adaptation of the polymers to individual needs. When fillers are used it has to be taken into account that the size of the fillers—like glass fibers—should be significantly smaller than the size of the microstructures. Here, maybe the integration of fillers in the nano range allows the processing of these kind of polymers even in the sub-micron range.

2.8.3 Technology

The technology of each process is developed to different levels. Most automation is the process of injection molding and injection compression molding because of the derivation from the macroscopic dimensions. The production of microstructured compact discs or DVDs is an impressive example of automation in the replication of microstructures. Nevertheless, the structures show an aspect ratio in the range of one, which allows high-speed replication. An increase of the aspect ratio of the structures will increase the demand on molding and demolding and can also increase the cycle times. The automation of hot embossing is not developed to this high grade, but the development of automation is still in progress. The first semiautomatic handling system is linked to a hot embossing machine, which allows the handling of the polymer foils and the molded parts. The technology of reaction injection molding and thermoforming are well developed for replication in macroscopic dimensions. For the replication of parts in microscopic dimensions, only individual machines in laboratories are available. These machines typically show no high grade in automation. Because of the difficulties controlling the polymerization process the development of reaction injection molding machines for microscopic dimensions is not further in the foreground of interest. However, the development of the technology for thermoforming in microscopic dimensions is still in progress and a commercial machine for thermoforming will possibly be available soon.

Besides the automation, the complexity of the technology of the processes also determines the manifold of the attainable process variations and the possibility of the adaptation for individual needs. Here the concept of injection, especially injection molding and injection compression molding, shows a high grade of complexity. Because of the injection principle, a separate plastifying unit, a compression unit, and an injection unit with a defined dosage of polymer melt is necessary. Especially the dosage and the control of masses in the range of milligrams is a challenge. Also, the tool where the mold insert will be fixed shows a complex design. Here, runner systems, fluid channels for coolants, have also been integrated like an ejection system for demolding of the molded parts. Nevertheless, the complexity of this systems allows, after a long set-up time, the achievement of a fast process with cycle times in the range of seconds. These short cycle times are achieved by the parallel process steps. After injection and cooling down the molded part in the mold, already new material is heated up to injection temperature and can immediately be injected if the mold will close after the ejection of the molded part. Nevertheless, if a variotherm process is needed to fill filigree microcavities completely, the advantage of the parallel process steps fails and the tool has to be heated up to molding temperature before injection of the melt and has to be cooled down to demolding temperature after injection. If this process variation is used, the cycle times are in the range of minutes and are nearly similar to cycle times by hot embossing. Finally, the complexity of this concept prevents a large variety of process variations but allows, if the process has to be adjusted, an effective molding of large series.

In contrast to the high complexity of molding principles oriented by injection, hot embossing and the related concept of thermoforming refers to a wider range of complexity, beginning by simple concepts up to highly sophisticated tools. Both these processes already allow the replication of structures with basic technology concepts of heating, embossing, and cooling. Here, the use of semi-finished products like polymer sheets or polymer films instead of granules allows the neglect of any plastifying units that decrease the grade of complexity. By this delivery of the raw material an easy change of the molding material can be achieved by the change of the polymer sheet. This can be done in extreme cases after every cycle, which allows the fabrication of prototypes in different materials in a short time. Here, processes that relate to an injection step have to be clean or change the plastifying unit to use different polymer materials. Another aspect is the complexity of the tool. Because of the missing injection step, the tools for hot embossing and thermoforming can be less complex. Nevertheless, a heating and cooling unit has to be integrated, but the concept of hot embossing tools allows one to fix and change different mold inserts very easily by a simple fixation system. Here, a quick change of mold inserts is guaranteed. Nevertheless, a parallel between heating and molding in analogy to the injection molding process cannot, or only in reduced approaches, be realized. Therefore, the polymer film has to be heated up by the tool via heat conduction, which results in comparatively long heating and also long cooling times, which will increase the cycle times in the range of several minutes. Nevertheless, this molding concept shows the advantages of a large flexibility regarding different process variations. Because of the use of polymer sheets, multilayer techniques can be performed, allowing concepts like thermoforming by polymer melts or the fabrication of through-holes (Section 5.5). Here, more flexibility for individual solutions can be achieved.

2.8.4 Cost Effectiveness

To compare the cost effectiveness of the processes, the fixed costs and the variable costs have to be taken into account to achieve the total costs of the replication of a desired series of molded parts. A comparison can only be expressive if all costs can be determined successfully, especially the investments for molding machines. Here, only commercially available machines are suitable for a determination of costs. This limits the comparison to two processes, injection molding and hot embossing. Because of the different machines commercially available and the individual requirements for a serial replication, only a generalized comparison can be given. Nevertheless, the characteristic costs can be estimated and can help the reader to decide individually which process is the most cost effective for his needs.

2.8.4.1 Fixed Costs

As mentioned above, the fixed costs relate to the investments for a machine, the necessary infrastructure, the engineering time to design the mold insert and integrate it into the molding tool, and finally the engineering time to set up the process. Here, the fixed costs of injection molding vary by the configuration of the individual machines, but in general an injection molding machine requires higher investments compared to a basic hot embossing machine. Also, the engineering time to configure the mold design and the engineering time to set up the process is much higher than in hot embossing. Regarding the fixed costs before the serial production starts, the process of hot embossing can be more cost effective.

2.8.4.2 Variable Costs

In contrast to the fixed costs, the variable costs will be generated during the replication of the mold and increase linearly with the number of replicated parts. Here, the personal costs and substantially the cycle times of the processes are determined factors. The process of injection molding relates to short cycle times, which will result in lower variable costs for each replicated part. As mentioned above, the cost structure can be equalized if, for injection molding, a variotherm cycle is needed and hot embossing uses an optimized heating and cooling concept in combination with a large batch size of molds.

For replication in the microscale, the material costs for standard polymers can, compared to the other costs, be neglected. However, if technical plastics or especially high-performance plastics like LCP or PEEK are needed, the cost of these materials can be a relevant part of the variable costs. Here, injection molding is characterized by additional material used, for example, for the sprue, which can be recycled by further process steps. In contrast to this, the polymer sheets for hot embossing can be cut into the right size, which will minimize the loss of material. However, the cutting is an additional step, to be done before the serial process starts. Finally, the costs of polymer sheets will be higher than the costs for simple granules.

2.8.4.3 Total Costs

The cost effectiveness for both replication processes are shown in Table 2.1. This table can only give a coarse estimation; for each design an individual calculation is recommended.

Table 2.1 Cost Effectiveness of Micro Injection Molding and Micro Hot Embossing

Costs	Micro Injection Molding	Micro Hot Embossing
Machine cost	150,000–200,000 €	160,000–350,000 €
Engineering	Several weeks	< 8 h
Tool/mold insert	50,000–250,000 €	2000–25,000 € (LIGA)
Mold fabrication time	3–12 months	10 days to 3 months
Process setup	1–10 days	1–5 days
Cycle time	5 s to 8 min	6–30 min

Finally, the total cost will determine the resulting cost structure for a design that has to be replicated. Here, the desired number of molded parts is the determining factor. The total costs can calculated by

$$\text{Total costs } = \text{ Fixed costs } + \text{ Variable costs} + \text{Variable costs} \qquad (2.1)$$

and finally the cost for a molded part

$$\text{Cost per unit} = \left(\frac{\text{Fixed costs}}{\text{Total number of molded parts}}\right) + \text{Variable costs} \qquad (2.2)$$

Referring to the cost estimation for both processes, it is obvious that hot embossing is well suited for small and medium series because of the moderate investments for machine and tool. For large series the variable cost becomes more of an influence and the total costs for the molded parts increases to a level over the costs of molded parts fabricated by injection molding. This process is therefore characterized by high investments, but because of the shorter cycle times, the variable costs are lower, which results in reduced total costs for larger series (Fig. 2.20). Regarding the curves of costs over the molded parts, this cost structure results in an intersection of both curves at a defined number of molded parts. The number has to calculated individually regarding the investments and the variable costs. But finally, the cost effectiveness is not always a determining factor. Previously the design, the physics of the replication process, and the desired material determined the technology that is required for successful replication.

2.9 Table of Characteristic Properties

Finally, Table 2.2 summarizes selected properties of the relevant micro molding processes.

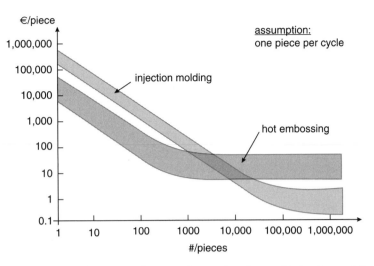

Figure 2.20 Comparison of the costs between hot embossing and injection molding. Because of the moderate fixed costs, hot embossing is well suited for small and medium series. Because of the short cycle times and moderate variable costs, injection molding is suited for large-series production. The intersection between both areas has to be defined individually and can be in a range between approximately 1,000 pieces and 20,000 pieces.

Table 2.2 Summarized Properties of Micro Replication Processes

Criteria	Injection Molding	Hot Embossing	Thermoforming	Injection Reaction Molding
Process times (typical)	Few seconds to 10 min (variotherm)	6–30 min	20 min (laboratory)	Several minutes (laboratory)
Flow path	Length of runner system	Height of microcavities	No flow/height of cavity	Runner system
Raw material	Granules	Foils approx. 50–1500 µm	Films approx. 10–100 µm	Liquid monomers
Polymers	Thermoplastic	Thermoplastic	Thermoplastic	Thermoplastic, thermosets
Residual stress	High	Low	High	High
Flexibility	Low	High	Depending on machine	Low
Set-up time	High	Low	Low	Low
Molding window	Low viscosity	Moderate viscosity	Softening range	Monomer + initiator
Automation	High grade	Laboratory and automated	Only laboratory	Only laboratory
Serial production	Large series	Small and medium	Small	Small

References

1. Battenfeld. http//www.battenfeld-imt.com, 2008.

2. M. D. Austin, H. Ge, W. Wu, M. Li, Z. Yu, D. Wasserman, S. A. Lyon, and S. Y. Chou. Fabrication of 5 nm linewidth and 14 nm pitch features by nanoimprint lithography. *Applied Physics Letters*, 84(26):5299–5302, 2004.

3. H. Becker and C. Gärtner. Polymer microfabrication technologies for microfluidic systems. *Anal. Bioanal. Chem.*, 390:89–111, 2008.

4. M. Bender, M. Otto, B. Hadam, B. Vratzov, B. Spangenberg, and H. Kurz. Fabrication of nanostructures using an UV-based imprint technique. *Microelectronic Engineering*, 53:233–236, 2000.

5. S. Y. Chou, P. R. Krauss, and P. J. Renstrom. Imprint of sub-25 nm vias and trenches in polymers. *Appl. Phys. Lett.*, 67:3114–3116, 1995.

6. S. Y. Chou, P. R. Krauss, and P. J. Renstrom. Nanoimprint lithography. *J. Vac. Sci. Technol. B*, 14(6):4129–4133, 1996.

7. S. Y. Chou, P. R. Krauss, W. Zhang, L. Guo, and L. Zhuang. Sub-10 nm nanoimprint lithography and applications. *J. Vac. Sci. Technol. B*,15(6):2897–2904, 1997.

8. B. Cui and T. Veres. Pattern replication of 100 nm to millimeter-scale features by thermal nanoimprint lithography. *Microelectronic Engineering*, 83:902–905, 2006.

9. H. Dreuth and C. Heiden. Thermoplastic structuring of thin polymer films. *Sensors and Actuators*, 78:198–204, 1999.

10. G. Fu, N. H. Loh, and S. B. Tor. A variotherm mold for micro metal injection molding. *Microsystem Technologies*, 11:1267–1271, 2005.

11. J. Giboz, T. Copponnex, and P. Mélé. Microinjection molding of thermoplastic polymers: a review. *Journal of Micromechanics and Microengineering*, 17:96–109, 2007.

12. S. Giselbrecht, T. Gietzelt, E. Gottwald, A. Guber, C. Trautmann, R. Truckenmüller, and K. F. Weibezahn. Microthermoforming as a novel technique for manufacturing scaffolds in tissue engineering (cell chips). In *IEE Proc. Nanobiotechnology*, Volume 151, pp. 151–157, 2004.

13. S. Giselbrecht, T. Gietzelt, E. Gottwald, C. Trautmann, R. Truckenmüller, K. F. Weibezahn, and A. Welle. 3d tissue culture substrates produced by microthermoforming of pre-processed polymer films. *Biomed. Microdevices*, 8:191–199, 2006.

14. G.Menges, W. Michaeli, and P. Mohren. *Spritzgiess-Werkzeuge*. Hanser Publishers, 5th edition, 1999.

15. L. J. Guo. Recent progress in nanoimprint technology and its applications. *Journal of Physics D: Applied Physics*, 37:R123–R141, 2004.

16. M. Heckele and W. K. Schomburg. Review on micro molding of thermoplastic polymers. *Journal of Micromechanics and Microengineering*, 14:R1–R14, 2004.

17. B. Heidari, I. Maximov, and L. Montelius. Nanoimprint lithography at the 6 in. wafer scale. *J. Vac. Sci. Technol. B—Microelectronics and Nanometer Structures*, 18(6):3557–3560, 2000.

18. R. Hull, T. Chraska, Y. Liu, and D. Longo. Microcontact printing: new mastering and transfer techniques for high throughput, resolution and depth of focus. *Materials Science and Engineering C*, 19:383–392, 2002.

19. S. Kang. Replication technology for micro/nano optical components. *Japanese Journal of Applied Physics*, 43(8B):5706–5716, 2004.

20. H. Lee, S. Hong, K. Yang, and K. Choi. Fabrication of nano-sized resist patterns on flexible plastic film using thermal curing nano-imprint lithography. *Microelectronic Engineering*, 83:323–327, 2006.

21. V. Piotter, T. Gietzelt, and L. Merz. Micro powder-injection moulding of metals and ceramics. *Sadhana Acad Proc Eng Sci*, 28:299–306, 2003.

22. V. Piotter, K. Mueller, K. Plewa, R. Ruprecht, and J. Hausselt. Performance and simulation of thermoplastic micro injection molding. *Microsystem Technologies*, 8:387–390, 2002.

23. O. Rötting, W. Röpke, H. Becker, and C. Gärtner. Polymer microfabrication technologies. *Microsystem Technologies*, 8:32–36, 2002.

24. R. Ruprecht, T. Benzler, T. Hanemann, K. Mueller, J. Konys, V. Piotter, G. Schanz, L. Schmidt, A. Thies, H. Woellmer, and J. Hausselt. Various replication techniques for manufacturing three-dimensional metal microstructures. *Microsystem Technologies*, 4:28–31, 1997.

25. R. Ruprecht, T. Gietzelt, K. Müller, V. Piotter, and J. Haußelt. Injection molding of microstructured components from plastic, metals and ceramics. *Microsystem Technologies*, 8:351–358, 2002.

26. J. L. Throne. *Technology of Thermoforming*. Hanser Gardner Publications, 1996.

27. J. L. Throne. *Understanding Thermoforming*. Hanser Gardner Publications, 1999.

28. C. M. Sotomayor Torres. *Alternative Lithography—Unleashing the Potentials of Nanotechnology*. Kluwer Academic/Plenum Publishers, 2003.

29. R. Truckenmüller and S. Giselbrecht. Microthermoforming of flexible, not buried hollow microstructures for chip based life sciences applications. In *IEE Proc. Nanobiotechnology*, Volume 151, pp. 163–166, 2004.

30. R. Truckenmüller, Z. Rummler, Th. Schaller, and W. K. Schomburg. Low-cost production of single-use polymer capillary electrophoresis structures by microthermoforming. In *12th Micromechanics Europe Workshop (MME)*, Cork, Ireland, pp. 39–42, 2001.

31. R. Truckenmüller, Z. Rummler, Th. Schaller, and W. K. Schomburg. Low-cost thermoforming of micro fluidic analysis chips. *Journal of Micromechanics and Microengineering*, 12:375–379, 2002.

32. J. L Wilbur, A. Kumar, H. A. Biebuyck, E. Kim, and G. M. Whitesides. Microcontact printing of self-assembled monolayers: applications in microfabrication. *Nanotechnology*, 7:452–457, 1996.

33. Y. Xia and G. M. Whitesides. Soft lithography. *Angew. Chem. Int. Ed.*, 37:550–575, 1998.

34. S. Zankovych, T. Hoffmann, J. Seekamp, J.-U. Bruch, and C. M. Sotomayor Torres. Nanoimprint lithography: challenges and prospects. *Nanotechnology*, 12:91–95, 2001.

35. J. Zhao, R. H. Mayes, G. Chen, H. Xie, and P. S. Chan. Effects of process parameters on the micro molding process. *Polymer Engineering and Science*, 43(9):1542–1554, 2003.

3 Molding Materials for Hot Embossing

Molding processes, also hot embossing, cannot be described without the fundamentals of the molding materials. Especially topics like thermal behavior, flow behavior, shrinkage, and warpage and their mathematical descriptions are important issues for an analysis of molding and demolding. The objective of this chapter is therefore to give the reader a basic knowledge of the behavior of molding materials. Most applicable for hot embossing are thermoplastic polymers. Thus, the priority of this chapter is in the characterization of thermoplastic polymers, with the focus on topics relevant for molding. Nevertheless, alternative molding materials like glass, metals, or ceramics will also be discussed. In detail the following aspects will be described:

- A description of suitable materials for replication will be presented. A distinction between unconventional materials and polymer materials shows the bandwidth of molding materials suitable for hot embossing.
- The rheologic behavior of polymer melts determines the flow behavior in the residual layer and during mold filling, and it is therefore an essential issue in molding processes. The stationary flow based on shear and strain, the viscoelastic behavior of the melt, and the modeling are also topics like the molecular orientation and the stress relaxation.
- During cooling the polymer behavior is characterized by solidification. The process parameters in this step decide the orientation and crystallinity of molded parts. But also stress, shrinkage, and warpage of the molded parts are influenced by the process parameters.
- The behavior of solid polymers especially near the softening range is an important factor to understand the behavior of polymers during demolding. Aspects of viscoelasticity will also be discussed, like the stress strain behavior at different temperatures, the creeping of polymers, and the measurement by dynamic mechanical analysis. An important point is also the friction between mold and polymer. High values of friction force are one of the reasons for the damage of structures during demolding.
- The thermal behavior of thermoplastic polymers, based on the aggregate states, already defines molding windows. An analysis of the thermal behavior will give a first estimation about the suitable molding and demolding temperatures for amorphous and semicrystalline polymers and will allow further the specification of the bandwidth of molding temperatures—here, called molding windows.
- Finally, the measurement of the thermal behavior of polymers will be presented, with the aim to measure all necessary data for the

further analysis and modeling of the material behavior. This section of the chapter also gives an overview of the most important thermal properties of polymers and discusses the affiliated measurement methods.

3.1 Classification of Alternative Molding Materials

The objective of this section is to specify the wide range of molding materials for micro replication processes. In macroscopic molding, polymers are an established molding material. The advantage of polymers is the large bandwidth of different properties that allows one to find a suitable polymer for most applications. The great advantage is, further, the moderate temperature range and, if thermoplastic polymers are used, the reversible switch between the solid and viscous aggregate states. The knowledge of material behavior is essential for an analytic view on every replication process. Nevertheless, of the dissemination of polymer materials in the replication processes, alternative materials are also suitable for molding. These materials close a gap because these materials show properties that cannot be achieved by polymer materials. In this section some alternative molding materials and their relevance in micro replication processes are presented.

Independent of the molding material, here are the basic requirements that have to be fulfilled.

- The material has to be set into an aggregate state, where molding is possible (e.g., by heat or by chemicals).
- The viscosity of the melt should be in a range that allows molding with established technologies.
- After molding the melt has to be cured (e.g., by cooling, UV radiation, or chemical reactions).
- Reversible switching between the solid state and the melt is desired.
- Fillers or additional components of the material have to be significantly smaller than the size of the micro- or nanocavities.

3.1.1 Hot Embossing of Glass

Glass is a material commonly used in microsystem technology. For the manufacturing of microstructures, serial (e.g., laser structuring) and parallel manufacturing processes like etching are available. Also, cutting processes like grinding, milling, and drilling are common. These structuring methods allow the fabrication of structures in the range down to 10 µm [18]. Nevertheless, the structuring methods are limited regarding the surface quality and the damaging of structures during manufacturing. The molding, also hot embossing, of glass is therefore an alternative structuring method for the fabrication of microstructures in glass.

3.1.1.1 Materials

Glass is an amorphous material and fulfills the requirements described above. Nevertheless, compared to polymer, the material properties are different. Suitable glasses for molding are standard glasses like Pyrex (Pyrex 7740) or Borosilicate glass (e.g., D263, Schott).

Pyrex glass is characterized by a glass-transition temperature of 560°C and a softening temperature of 821°C. The typical thermal expansion coefficient is in a range of 3.25×10^{-6}, significantly below the typical expansion coefficient of polymers, which are characterized by values in the range of 1×10^{-5} up to 1×10^{-4}. The thermal conductivity is typically in a range of 0.59–1.19 W/m K [21], in a comparable range to the thermal conductivity of polymers (0.1–0.8 W/m K [11]).

Borosilicate glass (e.g., Schott D263) is characterized by a glass-transition temperature of 557°C and a softening temperature of 736°C, approximately 80 degrees below the softening temperature of Pyrex glass. D263 is more pliable than Pyrex [26].

3.1.1.2 Molding Parameters

The molding parameters for hot embossing of glass depend, like the molding of polymers, on the design of the mold insert, the aspect ratio of the structures, the cross-section of the cavities, and the molded area.

Schubert et al. [18] used for the molding of Pyrex (7740) glass a molding window between 700–760°C. The viscosity in this temperature range was determined between $10^{9.5}$ and $10^{8.5}$ dPa. The replication refers to a silicon mold with smallest structures in the range of 10 µm and a height of 50 µm.

Takahashi et al. [21] also replicated structures with a height of 50 µm in Pyrex glass. Aspect ratios between 1 and 1.66 were investigated. The systematic replication experiments referred to a molding temperature range between 640°C and 650°C, a pressure of 2.83 MPa, and holding times between 300 seconds and 1,200 seconds. Microstructures with 1 µm line and space patterns with a height of 1 µm were replicated at 595°C, a pressure of 0.45 MPa, and a holding time of 180 seconds [22].

Yasui et al. [26] replicated micro- and submicrostructures into borosilicate glass D263. Patterns on an area of $20 \, \mu m^2$ with a depth of 6.5 µm were replicated successfully in a temperature range of 590–600°C and a pressure range of 0.22–0.66 MPa. If the structure size decreased to a line and space pattern of 0.4 µm, the molding temperature increased to 650°C at a pressure of 6.37 MPa.

3.1.1.3 Mold Inserts

Regarding the high temperatures during molding of glass, the requirements compared to the requirements of molding polymers expands to the issues of high temperature resistance and thermochemical stability.

Suitable are high temperature–firm metals like nickel-based alloys or molybdenum alloys. Nickel-Wolfram alloy Incoloy [26] has a thermal expansion similar to borosilicate glass. But also ceramic material, like silicon carbide or aluminum nitride, can be used. Silicon is also a suitable material. If a coating is desired, titanium nitride, for example, can be deposited.

An alternative is glassy carbon [21,22] structured in the micron range, for example, by focused ion beam–machining (FIB) or dicing. Glassy carbon is suitable up to a high temperature of 1,400°C, comparable to the transition temperature of quartz glass. The operation limit is in a range of 2,000°C and the thermal conductivity is in a range of 9 W/m K, significantly below metals.

3.1.2 Metals and Ceramics

Regarding the typically high softening temperature range of metals, only selected materials are suitable for their molding. An alternative for the molding of metals is the use of metal powder, which can be structured by molding in combination with a binder and solidified by sintering processes.

3.1.2.1 Hot Embossing of Metals

Compared to the molding temperatures of glass, the required molding temperatures of metals often increase significantly. Therefore, the metals that can be molded at moderate temperatures are limited. Cao et al. systematically analyzed the hot embossing of lead [2] and aluminum [1].

Hot Embossing of Lead A major topic of molding high-temperature materials is the chemical and mechanical interaction of the molding material with the mold insert, in this case an electroplated nickel mold insert. The interaction can result in high adhesion or bonding and can damage the mold insert and the molded part during demolding. Cao et al. demonstrated that between lead and nickel mold inserts no chemical interactions occurs. Therefore, a surface modification of a nickel mold insert is not necessary.

Independent of the melting temperature of lead of 327°C and the soft material, behavior results in a wide range of molding temperatures between 100°C and 300°C. The replication of structure details increases with the increasing molding temperature. In experiments [2] nickel mold inserts with 585 nickel micro pots with 150 μm diameter, a depth of 400 μm, and a spacing of 650 μm were replicated in lead plates (99.9%) in the form of circular discs with a 3.5 mm diameter and a thickness of around 0.6 mm. Already at 118°C and an acting force of around 780 N on the molding area, an increase of temperature in the range of 120°C up to 270°C improves the geometric details of the replicated features. The systematic experiments and the analyzed molding results demonstrate that the contact stress during molding and demolding is well below the yield stress of electroplated nickel-mold inserts. The experiments demonstrate that sharp microscale features can be replicated without degradation of the nickel-mold insert after repeated molding runs.

Hot Embossing of Aluminum   In contrast to the molding of lead, the molding of aluminum is more challenging regarding the molding temperature and especially the interaction of nickel-mold inserts with aluminum. Electroplated pure nickel molds are not suitable for molding aluminum. Because of the strong driving force of nickel and aluminum to form intermetallic compounds, a barrier between both components is required. This barrier can be implemented by a temperature-resistant coating. Cao and Meng [1] used a ceramic coating to modify the mold insert surface. By vapor deposition a Ti-containing hydrocarbon (Ti-C:H) layer was deposited.

Already below the melting temperature of aluminum of 660°C, microstructures can be replicated by hot embossing. Similar to the molding of lead, Cao and Meng [1] replicated a squared array of cylindrical micro pots with a height of 400 μm and diameter of 200 μm on an area of 1.8×1.8 cm^2 on an aluminum (99.9%) substrate in the form of circular discs with a 3.5 mm diameter and a thickness of around 0.6 mm. To compare the influence of coated and uncoated mold inserts, two mold inserts with and without coating were replicated, with the result that the structures from the uncoated mold insert could not be demolded successfully. The molding temperature was set in a range between 433°C and 460°C, with the molding force up to 1,700 N. It was found that the maximum compressive stress of 18 M Pa at 450°C was sufficient to mold and demold the microstructures without damaging the nickel structures on the mold insert and the corresponding molded aluminum structures.

An important aspect is the longtime stability of the coating used. The Ti-C:H coatings are deposited at 250°C and may suffer hydrogen desorption at higher molding temperatures. The recommended molding temperature for aluminum will, therefore, degrade the coating with the number of molding steps. This degradation increases if the molding temperature increases in the range of 500°C or higher. The loss of hydrogen and the graphitization of the coating used reduce the durability and the lifetime of the coating. The molding of aluminum is, therefore, a function of the coatings and their longtime stability.

Metallic Glass   Independent of the molding of metals, the molding of metallic glass has been investigated. Pan et al. [13] investigated a thermoplastic forming process of bulk metallic glass ($Mg_{58}Cu_{31}Y_{11}$), in theory by a simulation with commercial simulation software, and verified the study by hot embossing experiments. Due to the moderate molding temperature of 140–150°C, the molding is quite different from the molding of conventional metals or glass. The hot embossing experiments referred to the replication of an Ni–Co electroplated mold insert with a micro lens array. The height of the structures was 14 µm and the diameter 330 µm. Systematic measurements and experiments demonstrated the hot embossing capability of this material class. It could be found that under molding conditions, the shrinkage of the molded parts (0.61%) was about 10 times higher than the measured shrinkage of PMMA (0.06%).

3.1.2.2 Micro Powder Molding

Micro powder (injection) molding (PIM) of metals or ceramics is an alternative to fabricate structures of these materials in large series, especially in the micron range. This method is well established in replication by injection molding and metal injection molding (MIM) or ceramic injection molding (CIM).

Fine metal or ceramic powder is mixed with a binder system into a feedstock and injected into a mold containing a microstructured mold insert. After demolding, the so-called green compact is processed further in the furnace, mostly under an oxidizing atmosphere, to remove the binder system, thus changing into a brown compact. In this step, the binder is removed and the first contact spots between the powder particles are produced. Subsequently, in a defined atmosphere, the brown compact is sintered into a solid microstructure of close-to-theoretical density [16].

Typically, materials for metal powder are, for example, carbonyl iron powder, stainless steel (316L), or alloyed hardenable steel (17-4PH). Typically, ceramic powders are aluminum oxide powder, ZrO_2-strengthened Al_2O_3, or zirconium oxide powder. The mean particle sizes are typically in a range of 0.3 µm up to 5 µm (e.g., iron powder with a mean particle size of 4–5 µm, aluminum oxide powder with a mean particle diameter of 0.6 µm, zirconium oxide powder with a mean particle diameter of 0.3–0.4 µm). The binders are commercially available and consist of a polyolefin/wax compound or a polyacetal-based system [14,17].

Depending on the particle shape and the particle size distribution, the feedstock typically contains 50 vol% up to 60 vol% of metal powder. Because of the high porosity, the linear shrinkage of a molded part after sintering is in a range of 15% up to 22%. The precision that can be achieved by these materials is about ±0.3% compared to the original shape [6]. The particle size also determines the quality of the molded part, especially the surface roughness. But also the filling of filigree cavities depends on the particle size. For micro features, the average particle size should be one or two magnitudes smaller than the structural details of a design. This can limit the minimal achievable structure size.

The replication of microstructures therefore requires very fine powders in the submicron range. Further, the temperature of the binder determines the molding and demolding behavior of the feedstock. To achieve a successful molding of microstructures, a variotherm molding cycle is required [6]. Especially the demolding behavior and the possible damage to the molded part correspond to the strength of the feedstock after injection and cooling. Fu et al. [5] analyzed the demolding behavior in the case of micro metal injection molding. This analytic approach can be transferred also to the demolding of polymers.

Nevertheless, these materials are successfully implemented for injection molding but are also suitable for hot embossing. By injection molding, a pillar array with pillars of 100 µm [10] or 60 µm [4] have been replicated.

3.2 Polymer Materials

3.2.1 Advantages and Typical Properties

Polymers are the most common materials in molding processes. The use of polymers in molding macroscopic and microscopic structures is based on several advantages and properties of polymers:

- Compared to glass, polymers have a low molding temperature. Molding temperatures are in the range between approximately 80°C (e.g., polystyrole) up to 340°C (e.g., PEEK) for high-temperature polymers. The reduction of molding temperature results in a decrease of heating and cooling time and finally effects shorter process cycle times.
- A large variety of polymers with different thermal, optical, and chemical properties allows one to find the most appropriate polymer material for a specific application. Some polymers are characterized by biocompatibility and are therefore suitable for medical or biological applications.
- Compared to metals or ceramics, the specific weight of polymers is much lower. The density of polymers is in the range of 0.8–2.2 g/cm^3. In combination with reinforcing fillers, polymers can be used as lightweight building material.
- The bulk modulus and the mechanical strength of polymers are spread over a wide range. Therefore, the mechanical behavior of each polymer can also be different, from rubberlike behavior up to a behavior in the range of metals (e.g., aluminum).
- The heat conduction of polymers is in the range of 0.1–0.8 W/m K. Compared to metals, the range is approximately three dimensions lower than metals.

- Polymers are insulators. For good conductivity, additives like graphite or soot can be used.
- Polymers are partially transparent. Some amorphous polymers (e.g., PMMA, PC) show optical properties similar to glass.
- Some polymers show a high resistance to acids, alkali solutions, and saline solutions. They are predestined for use in chemical environments.
- Polymers can be combined during polymerization to create new materials with supplementary properties. So-called copolymers are polymeric materials with two or more monomer types in the same chain (e.g., ABS).
- Polymer blends belong to another family of polymers. Mixing or blending of two or more polymers allows the enhancement of the physical properties of each individual component (e.g., PP-PC, PVC-ABS). For example, a high-impact strength can be combined with a reasonable modulus by blending. Another example is the combination of high-impact strength with higher heat-deflection temperature or the combination of a reasonable modulus with a higher heat-deflection temperature.
- Additives can be mixed with a polymer to improve, e.g., mechanical, optical, electrical, or acoustic behavior.
- Fillers can be used to reinforce the polymer and improve mechanical performance or to take up space to reduce the amount of polymer. Fillers can also be used to improve the electricity of a polymer by dispersing into the polymer. Using filled polymers in micro replication, the size of both the fillers and the microcavities has to be taken into account.
- Because of the large distances of molecules, polymers are often permeable for gases or some liquids. This can be a disadvantage for some applications.

The items named above show the diversity of polymers, so that for each application a polymer with suitable properties should be found. But not every polymer is suitable for molding with each process; especially molding of microstructures is characterized by advanced requirements on polymers. To select the applicable polymers for molding and to classify basic classes of polymers, knowledge of their molecular structure is essential.

3.2.2 Molecular Architecture

The molecular architecture of polymers is based on macromolecules generated synthetically or through natural processes. Examples for natural structures of macromolecules are natural rubber, wood, or cotton. The capability of carbon atoms to build molecular chains is fundamental for most natural and synthetic polymers. Polymers consist of basic units called monomers. For the synthetic generation of polymers, monomers are linked via addition polymerization or condensation polymerization to a molecular chain—the macromolecules. The molecular chains can be linear or branched. In one macromolecule, typically at least several hundred monomers are linked together. Depending on the nature and the number of linked monomers, typical molecular weights are in the range of 10^4–10^7 g/mol.

Important for the characteristic of every polymer is the arrangement of the molecular chains in space. The arrangement decides the polymer's suitability for molding. Two different kinds of arrangements can be distinguished: physical arrangements and chemical arrangements. Chemical arrangements are irreversible because the molecular chains are cross-linked: they are chemically bonded. Physical arrangements are reversible: the molecular chains are only intertwined.

With regard to the different arrangements of macromolecules in space, four different kinds of polymers with different characteristics can be classified (Fig. 3.1).

Thermoplastics consist of linearly or branched physically bonded macromolecules. At room temperature they are hard, and some thermoplastics show a glassy and brittle behavior. With an increase of temperature, thermoplastics soften up to the melting range. With a decrease of temperature, the polymer melt cures to amorphous or semicrystalline structures. This behavior is reversible: thermoplastics have the ability to remelt after they have solidified. The reversibility is caused in the physical arrangement of the macromolecules. With higher temperatures a sliding between the intertwined macromolecules is possible. Because of the arrangement, two different classes of thermoplastics are distinguished. Amorphous polymers show intertwined macromolecules, and semicrystalline polymers, in addition on crystalline areas, a parallel arrangement of linear macromolecules (Section 3.5).

Thermosets show, in addition to physical arrangements, chemical arrangements. The macromolecules are cross-linked to other macromolecules like a network and ensure that thermosets cannot melt. The typical maximum density of

Arrrangement of macro molecular chains

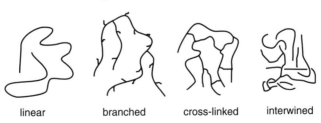

linear   branched   cross-linked   interwined

Figure 3.1 Physical and chemical arrangement of molecular chains of macromolecules. The arrangement characterizes the kind of polymer: thermoplastic, elastomer, or thermoset.

chemically bonded contacts is approximately one chemical contact in 20 atoms of the main chain. In additional to the chemically bonded parts, physically bonded parts are effective. With an increase of temperature the physically bonded contacts can soften so that stiffness can be reduced in the range of two dimensions. Nevertheless, because of the chemically bonded parts the softening state cannot be achieved, even with higher temperatures when the physically bonded parts are not effective.

Elastomers show next to physical arrangements coarse-meshed chemical arrangements. The typical density of chemical contacts for coarse-meshed elastomers is one contact per 1,000 atoms. Additional intertwined macromolecules and physically bonded parts are effective. The physical arrangements allow a sliding of the molecular chains at room temperature, which results in a rubber-like behavior at room temperature. An increase of temperature will not effect a change in the softening range because of the chemically bonded parts.

Thermoplastic elastomers are, with regard to molecular structure, similar to elastomers. Under the effect of heat the softening range can be achieved. The reason for this behavior is caused by different rates of meltable and non-meltable parts in the molecular chains. With differences in arrangement of these parts, a wide range of structures and properties beginning from easy melting up to chemically bonded is available.

With regard to the requirements for micro replication of polymers, only thermoplastic polymers and, with restriction, thermoplastic elastomers can be used. Common for replication are thermoplastics. Schulz et al. [19] investigated the molecular weight distribution with the respect to their imprint behavior. It was confirmed that higher molecular weight materials are beneficial to support successful imprint at reduced temperatures by utilizing shear rate effects. They recommended polymers with a polydispersity index of 2 and a maximum value of the molecular weight of 10^6 kg/mol.

3.3 Polymer Melts

For replication of polymers, knowledge of the behavior of polymer melts is essential, because in most replication processes polymers are processed in the melting state. However, the molecular architecture of polymers results in a complex flow behavior of polymer melts. The field of science that studies the flow behavior during flow-induced deformation is also called rheology [12]. Regarding the flow behavior of polymer melts, the relationship between stress and strain is fundamental. Two kinds of deformation can be distinguished:

- Deformation based on shear. Shear rate–dominated processes are, e.g., injection molding, injection compression molding, or selected process variations of hot embossing.

- Deformation based on elongation. Processes that are dominated by elongation or a combination of shear and elongation are thermoforming or blow molding.

An essential property of polymer melts is their viscoelastic behavior. Viscoelastic means that the behavior is between that of a viscous fluid and an elastic solid. Polymers show the ability to decrease stress time dependently. This effect can be observed by a non-stationary deformation of the melt, which is typical for polymer processing. Another important point is the deformation and orientation of macromolecules during flow. Orientations can be frozen during cooling and are reasonable for properties of the molded parts in the solid state. Generation of molecule orientations and reduction by relaxation are also part of this section.

3.3.1 Shear Rheologic Behavior

3.3.1.1 Stationary Flow

In most replication processes polymer melts undergo shear deformation. Shear results from adhesion on the sidewalls of the flow channels during low flow velocities. Because of the high viscosity of polymer melts, the Reynolds number is small, so that in typical replication process conditions a laminar flow can be assumed.

To describe shear deformation the well-known Newtonian model can be used (Fig. 3.2). The model consists of a fluid between two plates. One of the plates can be moved horizontally with constant velocity. Because of the motion of one plate against the other plate, the fluidic layers can glide on each other. The melt will be sheared. The mathematical description can be found by the reflection on a rectangular fluidic element. Shear stress deforms the rectangular element with the shear velocity.

$$\frac{dv}{dy} = \frac{d}{dy}\left(\frac{ds}{dt}\right) = \frac{d}{dt}\left(\frac{ds}{dy}\right) = \left(\frac{d\gamma}{dt}\right) = \dot{\gamma} \tag{3.1}$$

This model of shear deformation can also used to describe the movement of only one sidewall. This case can be found during polymer flow in screws of injection molding machines. Another case of deformation is the shear in a pipe based only on a pressure difference.

The description of the shear viscosity based on stationary conditions means a fluidic element undergoes the same deformation during flow. The shear velocity is constant for every fluidic element. The elastic behavior does not have to be taken into account. Stationary conditions can be found by constant flow in a channel with a constant cross-section. To describe the viscous properties of a

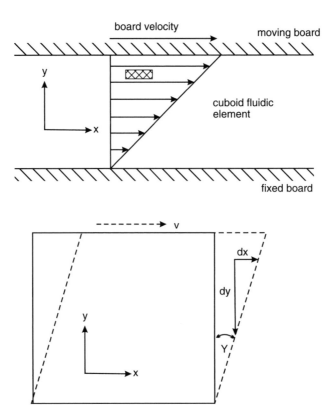

Figure 3.2 Newton model based on two plates to explain the shear deformation of a fluid.

polymer melt, only the viscosity has to be taken into account. The viscosity causes a resistance of the melt against flow so that a constant force is needed to obtain a constant flow. Based on the force, shear stress is induced in the polymer melt.

The relationship between shear and the necessary force is described by the viscosity η,

$$\eta = \frac{\tau}{\dot{\gamma}} \tag{3.2}$$

where η is the viscosity, τ is the shear stress, and $\dot{\gamma}$ is the shear velocity.

Compared to an elastic body, in this case the velocity of deformation is part of the equation, not the deformation itself. Compared to the shear behavior of Newtonian fluids, polymer melts are different from a classic Newtonian fluid. The viscosity shows dependencies of the molecular structure, the temperature, the density of chemical arrangements, and with low priority, the pressure. Nevertheless, a distinctive dependency can be seen by the influence of the shear velocity.

3.3.1.2 Shear Viscosity/Shear Thinning

Many fluids show a constant viscosity, for example, water or oil. Shear stress and shear velocity are proportional. For polymer melts, this relationship is only valid for small velocities. For higher velocities, the linear dependency/the Newtonean flow is not valid. Polymer melts show a shear thinning or structurally viscous behavior, which means if the shear velocity increases the corresponding shear stress increases in a degressive behavior. The reason for this behavior is based possibly on the dissolution of intertwined molecular chains. For the description of this behavior, two diagrams are established: the flow curve and the viscosity curve (Fig. 3.3).

The viscosity of a polymer melt can be influenced by pressure and temperature. An increase in temperature effects a decrease in viscosity. Conversely, an increase in pressure effects also an increase in viscosity. Figure 3.4 demonstrates that by an increase in pressure or temperature, the viscosity curve will be moved along the y-axis. The shape of the curve remains identical.

3.3.1.3 Mathematical Description of Shear Thinning Behavior

The mathematical description of the viscosity is fundamental for calculation and simulation of the flow in replication processes. For approximation of the shear thinning behavior, different empirical, semi-empirical, and theoretical models are used. The models allow the prediction of the shear thinning behavior of a polymer melt over a large range of shear velocity. Some of the most used models are the:

- exponential model from Ostwald and de Waele—simple but not suitable for small shear velocities
- Bird-Carreau-Yasuda model

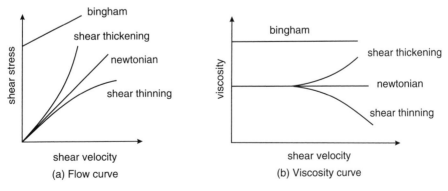

Figure 3.3 Structure-viscous behavior of different kinds of fluids. Polymer melts show shear thinning behavior. With an increase of shear velocity, the viscosity shows a degressive increase in shear stress.

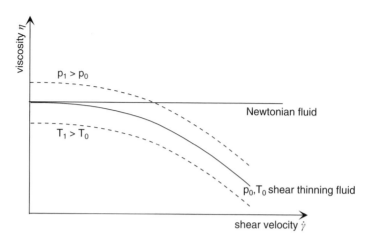

Figure 3.4 Influence of temperature and pressure on the viscosity of a polymer melt. If the temperature increases, the viscosity will decrease. If the pressure increases, the viscosity will also increase. Independent of the influence of temperature and pressure, the shape of the curve will remain constant.

- Herschel-Bulkley-WLF model used for underfill encapsulation
- Carreau model
- Cross-WLF model [8]

For the simulation of micro replication processes, especially for hot embossing, it is important to describe the shear thinning behavior for small shear velocities. The exponential model from Ostwald has the advantage of a simple model, but it cannot describe the behavior for small velocities precisely. The models from Carreau and Cross, with three parameters, are well suited to reflect the shear thinning behavior at small shear velocities. The Cross-WLF model is described below (Fig. 3.5).

The Cross-WLF model is marked by two tangents and their point of intersection. This point defines the transition of the Newtonian behavior and the shear-dependent behavior. The gradient of the line in the shear-dependent area is defined by the reciprocal flow exponent n, which is the inverse of the gradient of this line,

$$\eta(\dot{\gamma}) = \frac{\eta_0}{1 + \left(\frac{\eta_0 \dot{\gamma}}{\tau}\right)^{(1-n)}} \tag{3.3}$$

with

$$n = \frac{1}{m} = \frac{\Delta(\log \tau)}{\Delta(\log \dot{\gamma})} \tag{3.4}$$

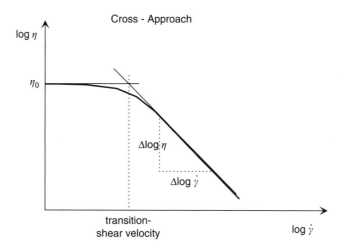

Figure 3.5 Cross-model to describe shear thinning behavior. Characteristic is the splitting into two sections, a Newtonian behavior and, after the transition point, the shear thinning behavior. The transition point is defined by the intersection of both tangents.

where η_0 is the viscosity at zero rate of shear, $\dot{\gamma}$ is the shear velocity, τ^* is the shear stress at the intersection of the two lines defining the change between Newtonian fluid and shear thinning behavior, and n is the reciprocal flow exponent.

The abbreviation for WLF refers to a model from Williams, Landel, and Ferry that describes the influences of temperature on thermoplastic polymers [24]. This model can be used for the temperature dependency of the shear viscosity and the relaxation behavior of polymer melts. The range of validity trends to amorphous polymers; for semicrystalline polymers, the model from Arrhenius is recommended. This elementary time-temperature equivalence of polymers is fundamental and describes, besides the viscosity behavior, also the stress relaxation behavior (Section 3.4.3).

3.3.1.4 Mathematical Description of Temperature Function

The influence of temperature on the viscosity can be illustrated by a logarithmic graph of the viscosity curve (Fig. 3.6). Both the shear velocity and the viscosity are plotted in logarithmic dimensions. It can be shown that, for most polymers, the different viscosity curves measured at different temperatures can be transferred into one master curve [11]. This transformation is possible because, for each temperature, the shape of the curve remains identical; only the position in the diagram is a function of temperature. To obtain a master curve the viscosity curves for each temperature have to shift along a line with a gradient of –1. This line defines a line with constant the shear stress $\eta \times \dot{\gamma}$. Each curve is shifted over a distance from $\log(\eta_0(T))$ along the ordinate and

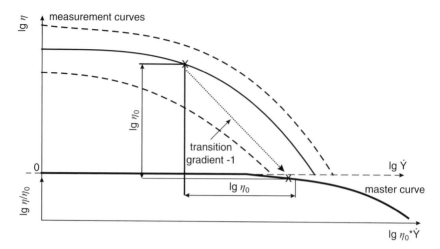

Figure 3.6 Time-temperature shift of viscosity curves. The measured viscosity behavior dependent on pressure and temperature can be shifted to a so-called master curve. Therefore, the values are printed in logarithmic coordinate systems and are shifted by a transition gradient of –1.

against the coordinate. The shift results in a master curve with the relative viscosity of η/η_0 over $\eta_0 \times \dot{\gamma}$. The master curve represents a characteristic function for the selected polymer. The temperature as reference temperature can be selected in the temperature range of the melt.

$$\frac{\eta(\dot{\gamma}, T)}{\eta_0(T)} = f(\eta_0(T) \cdot \dot{\gamma}) \tag{3.5}$$

To find now the viscosity function for a defined temperature, it is necessary to calculate the shift from a given master curve or from a viscosity curve measured at different temperatures. The shift is defined as time-temperature shift factor a_T. The factor describes the shift of the reference viscosity curve at temperature T_0 to obtain the viscosity curve for a selected temperature,

$$a_T = \frac{\eta_0(T)}{\eta_0(T_0)} \tag{3.6}$$

or

$$\log a_T = \log \frac{\eta_0(T)}{\eta_0(T_0)} \tag{3.7}$$

The principle of the shift described above is called the time-temperature shift. It can be used not only for the calculation of viscosity functions, but also for the calculation of the relaxation behavior (Section 3.4.3).

3.3.1.5 Mathematical Description of the Time-Temperature Shift

For the calculation of the shift factor, many approaches can be used. Two of the most important approaches—the Arrhenius approach and the WLF equations—are presented in detail below.

Arrhenius Approach The Arrhenius approach is well suited for the characterization of the temperature dependency of the viscosity of semicrystalline polymers. The approach refers to thermal-activated processes of molecules,

$$\log a_T = \log \frac{\eta_0(T)}{\eta_0(T_0)} = \frac{E_0}{R} \cdot \left(\frac{1}{T} - \frac{1}{T_0} \right) \tag{3.8}$$

where E_0 is the specific activation energy of flow (J/mol) and R is the universal gas constant ($R = 8.314$ J/(mol K)).

WLF Approach The approach from Williams, Landel, and Ferry (WLF) [24] is based on the theory of free volume between the molecules. This approach was first developed to describe the temperature dependency of relaxation behavior, and it can also be used for the description of temperature dependency of the viscosity.

$$\log a_T = \log \left(\frac{\eta(T)}{\eta(T_s)} \right) = -\frac{C_1 \cdot (T - T_s)}{C_2 + (T - T_s)} \tag{3.9}$$

The equation refers by constant shear stress the viscosity η for a selected temperature to a viscosity at a defined temperature. For a wide range of polymers, the constants C_1 and C_2 can be approximated. Menges et al. [11] showed that, for a temperature of 50°C over the softening temperature, the values of the constants are $C_1 = -8.86$ and $C_2 = 101.6$.

3.3.2 Flow in Capillaries

In replication processes, a second typical flow based on pressure difference occurs. For filling of molds or pressure drop in runner systems, it is significant to obtain information about the interaction of pressure and flow. For simple geometric cross-sections like circular, quadratic cross-sections or circular gaps, it is possible to obtain an analytic result. As an example, the equations for a circular cross-section will be derived below. Compared to a Newtonian fluid, the calculation for a shear thinning fluid based on the theory of the representative viscosity will be shown.

3.3.2.1 Newtonian Fluid

For a Newtonian fluid, the relationship between pressure drop over the length of a capillary and the shear stress is based on a balance of force on a fluidic element. The distribution of shear stress over the cross-section is given by

$$\tau(r) = \frac{\Delta p}{2L} \cdot r \qquad (3.10)$$

where L is the length of capillary, r is the coordinate beginning from the center of the capillary, $\tau(r)$ is the shear stress, and p is the pressure.

The shear stress is independent of the fluid. In contrast to the shear stress, the shear velocity is a function of the volume flow

$$\dot{\gamma}(r) = \frac{4\dot{V}}{\pi R^4} \cdot r \qquad (3.11)$$

With $\eta = \tau/\dot{\gamma}$, the pressure–volume flow equation results in

$$\Delta p = \frac{8\dot{V}L\eta_{newton}}{\pi R^4} \qquad (3.12)$$

The equations named above are valid for Newtonian fluids based on a linear dependency between shear velocity and radius of the capillary. The flow behavior of a shear thinning fluid is completely different. Compared to the linear velocity distribution of a Newtonian fluid, a parabolic velocity distribution is characteristic for shear thinning fluids. To calculate the relationship between pressure drop and volume flow for a shear thinning fluid, an approach from Schuemmer based on the concept of the representative viscosity can be used [11].

3.3.2.2 Shear Thinning Fluid

The pressure drop in a capillary flow of a shear thinning fluid can be estimated by the approach of the representative viscosity (Fig. 3.7).

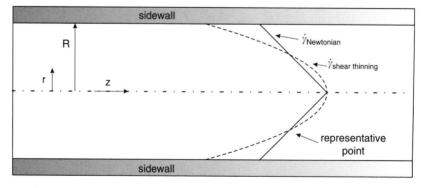

Figure 3.7 Estimation of the pressure drop in capillaries by the approach of the representative point. At this point, there is an intersection between the shear velocity profiles of a Newtonian fluid and a shear thinning fluid.

For an equal pressure drop and equal volume flow, the streamline of shear velocity of a shear thinning polymer shows different behavior. Compared to a Newtonian fluid, the shear velocity of a shear thinning fluid is characterized by lower values in the center and it shows higher values at the border of the flow channel. Because of this, there exists a point where the shear velocities are equal. This point is named the representative point. Viscosity and shear velocity at this point can be denominated as representative. For a circular capillary, the representative point e_0 is given by

$$e_0 = \frac{r_{rep}}{R} \approx \frac{\pi}{4} \tag{3.13}$$

The calculation of the shear velocity at the representative point is based on the equation of the Newtonian fluid:

$$\dot{\gamma}_{rep} = \dot{\gamma}_{newtonian}(r_{rep}) = \frac{4\dot{V}}{\pi R^3} \cdot e_0 \tag{3.14}$$

where $\dot{\gamma}$ is the shear velocity and $\dot{V}$ is the volume flow.

The relationship between pressure drop Δp and volume flow is similar to the calculation of a Newtonian fluid but based on the viscosity of the polymer melt at the representative point:

$$\Delta p = \frac{8\dot{V}L\eta(\gamma_{rep})}{\pi R^4} \tag{3.15}$$

The equations are valid for thermal and mechanical homogeneous polymer melts at stationary flow. These equations can also be used for a first-draft estimation of the necessary pressure to fill cavities (Section 6.4.2). Nevertheless, during hot embossing the cavity filling is also characterized by non-stationary flow behavior, especially at the beginning of the velocity-controlled molding cycle. Independent from this, the concept allows a first estimation about the polymer flow in capillaries. The concept of the representative point can finally also be adapted to other geometries (Table 3.1, [11]).

3.3.3 Viscoelastic Behavior of Polymer Melts

3.3.3.1 Definition of Viscoelasticity

For polymers, three different superimposing deformations can be distinguished [3]:

- elastic deformation (spontaneous and reversible)
- viscoelastic deformation (time dependent and reversible)
- viscous deformation (time dependent and irreversible)

Table 3.1 Estimation of Pressure Drop during Flow of Shear Thinning Polymers

Geometry	Representative Shear Velocity $\dot{\gamma}_{rep}$	Pressure Drop $\Delta p/l$
Capillary	$\dfrac{4\dot{V}}{\pi R^3} \cdot 0.815$	$\dfrac{8\eta_{rep}\dot{V}}{\pi R^4}$
Annular gap	$\dfrac{\dot{V}}{(R_a^2 - R_i^2)\bar{R}}$	$\dfrac{8\eta_{rep}\dot{V}}{\pi(R_a^2 - R_i^2)R^2}$
Rectangular gap	$\dfrac{6\dot{V}}{BH^2} \cdot 0.722$	$\dfrac{12\eta_{rep}\dot{V}}{BH^3}$

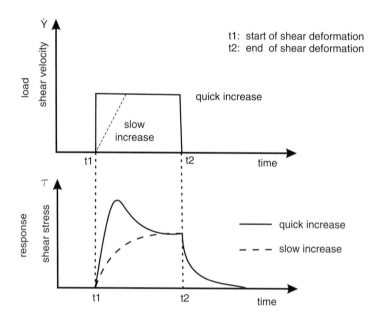

Figure 3.8 Time-dependent response of a viscoelastic medium to a time-dependent load. A fast increase in the shear velocity results in peaks of shear stress. A slow increase in the load results in a slow increase in the shear stress. A fast decrease in the load results in time-dependent relaxation of the shear stress.

Most polymers show an elastic and viscous behavior. This viscoelastic behavior is valid for solids as well as for melts. This viscoelastic behavior in a polymer melt becomes noticeable if the state of deformation changes over time. A pure viscous media during a deformation, the shear stress is proportional to the velocity of deformation. Viscoelastic behavior is, in contrast to this, characterized by a time-dependent answer. A change in the deformation state results in a time-delayed answer of the shear stress. Also, the velocity of deformation influences the time-dependent shear stress. A fast increase in shear velocity results in high shear stress; a slow increase results in low shear stress (Fig. 3.8).

3.3.3.2 Mechanical Models to Describe Viscoelasticity

Viscoelastic behavior can be described with mechanical models constructed of elastic springs obeying Hook's law and viscous dashpots obeying Newton's law of viscosity. By an arrangement of the element's spring, dashpot, and an element for friction, nearly all polymers can be modeled. The arrangement of these elements is based on a serial or parallel connection or a combination of these [20]. A simple combination is the serial connection of a spring and a dashpot. This arrangement is known as the Maxwell model (Fig. 3.9). For this model, the mathematical equations are shown below. For other models, for example, a parallel connection of spring and a dashpot like the Kelvin or Voight model, the reader will find more information in the literature [9,7,12,23].

The mathematical description of the viscoelastic behavior can be derived for a single Maxwell element, in this case a simple form of a viscoelastic material model. As a result of stress, the serial connection of spring and dashpot undergoes an elongation that results in strain on both elements. The strain ϵ can be split into two parts: an elastic strain ϵ_{el} and a viscous strain ϵ_{vis}

$$\epsilon = \epsilon_{el} + \epsilon_{vis} \tag{3.16}$$

The first derivative results in

$$\dot{\epsilon} = \dot{\epsilon}_{el} + \dot{\epsilon}_{vis} \tag{3.17}$$

Considering Hook's law and Newtonian flow

$$\dot{\epsilon}_{el} = \frac{\dot{\sigma}}{E} \tag{3.18}$$

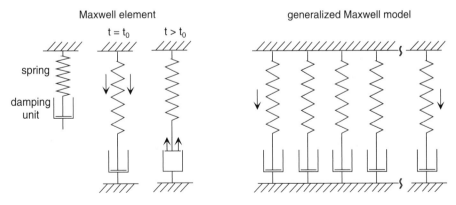

Figure 3.9 Maxwell element and generalized Maxwell model for the modeling of viscoelastic behavior. A single Maxwell element consists of a serial connection of an elastic spring and a damping unit (dashpot). A single element cannot describe the behavior of a real polymer. Therefore, a generalized Maxwell model is used, a parallel connection of n Maxwell elements.

$$\dot{\epsilon}_{vis} = \frac{\sigma}{\eta} \tag{3.19}$$

results in a differential equation

$$\dot{\sigma} + \frac{E}{\eta} \cdot \sigma - E \cdot \dot{\epsilon} = 0 \tag{3.20}$$

This approach is also valid for shear stress and shear velocities. To describe the relaxation, the time-dependent decay of stress under a constant deformation, the strain remains constant. $\epsilon = const.$, respectively $\dot{\epsilon} = 0$. Under this condition, a simple differential equation to describe the behavior of stress can be derived:

$$\dot{\sigma} + \frac{E}{\eta} \cdot \sigma = 0 \tag{3.21}$$

The solution of this differential equation by integration results in:

$$\sigma = \sigma_0 \cdot e^{-\frac{t \cdot E}{\eta}} = \sigma_0 \cdot e^{-\frac{t}{\lambda}} \tag{3.22}$$

This equation describes the decay of stress with time. The exponent $-t \cdot (E/\eta)$ results in a value of 1 for $t = \eta/E$. This time will be called relaxation time λ:

$$\lambda = \frac{\eta}{E} \tag{3.23}$$

The relaxation time defines the time in which the stress decays down to $e^{-1} = 0.37$ of the initial value. For an approximation of the relaxation, it is sufficient to know that for four times of the relaxation time the stress is decayed to 1.8% of the initial value.

In analogy to relaxation, the equation for retardation can also be derived. If the load on a Maxwell element is removed, the system reacts with a time-dependent strain, a kind of creeping. The time-dependent strain results in

$$\epsilon(t) = \frac{\sigma}{E} \cdot \left(1 - e^{-\frac{t \cdot E}{\eta}}\right) \tag{3.24}$$

Real polymers cannot be described with a single Maxwell element. Because of the complexity of the material behavior, more than one time constant is required. In this case, a parallel connection of several Maxwell elements is suitable. This arrangement is called a generalized Maxwell model. For a parallel connection of Maxwell elements, the relaxation behavior can be described by the sum of the parameters of each single element.

$$\sum_{i=1}^{n} E_i \cdot e^{\left(\frac{-t}{\lambda_i}\right)} \tag{3.25}$$

All these equations are only valid for small deformations in one dimension. For larger deformations, the elements are characterized by a nonlinear behavior of spring and dashpot, resulting in complex models.

3.3.4 Measurement of Viscosity

To characterize a polymer melt for replication processes, the knowledge of its viscosity is essential. The measurement of the viscosity under different temperatures and shear velocities allows one to obtain several points of the characteristic flow curves of a polymer melt. Three typical measuring systems are suitable: the measurement of the melt index, the capillary rheometer, and the rotation rheometer [12]. The following sections will only describe the basics of these measurements.

3.3.4.1 Melt Index Measurement

The flow of a polymer can be characterized by the melt flow index (MFI). This value defines the mass of the melt that, in the span of 10 minutes, will be pressed through a defined nozzle by a piston. The piston itself will be effected by a load with a defined mass. The dimensions of nozzle, piston, and load of piston as well as the dimensions of the cylinder are standardized. Alternative to the MFI index, referring to the mass of the melt, the MFV index, the volume of the melt, can be used. The measured values have to refer to the temperature and the load of the piston. For example, the MFI specification 190/2.16 means that the melt index refers to 190°C and a piston load of 2.16 kg.

One disadvantage of this measurement system is the fact that only one point of the viscosity curve can be determined. Therefore, this measurement system is well suited for comparative measurements, fo example, for quality control. Another disadvantage is the missing consideration of the shear thinning behavior and the elastic properties of the melt.

3.3.4.2 Capillary Rheometer

Compared to the melt index measurement, the elastic properties of the melt have to be taken into account. The construction is similar to the melt index measurement system. The only difference is that the piston will be moved with a constant velocity so that, in the integrated capillary, a constant-volume flow can be achieved. With the measurement of the pressure drop in the capillary, the volume flow, and the dimensions of the capillary, all information is given to calculate one point of a viscosity curve. With the variation of temperature and piston velocity, different flow curves can be determined.

3.3.4.3 Rotation Rheometer

In contrast to the flow characterized by pressure in capillary rheometer, rotation rheometer is based on shear flow like the Newtonian model of two plates. The translational motion of the plate is substituted by a rotary motion of a circular plate. For the determination of viscosity curves, the geometry of the plates and the gap between the plates have to be taken into account. For the determination of the shear force, the measured torque, and for the determination of the shear velocity, the rotation speed, have to be analyzed. An advantage of rotation rheometer is the diversity of measurements. The viscous flow behavior can also be determined, like the elastic properties of a polymer melt. In combination with the high sensitivity of the measurement systems, the oscillating behavior of the polymer melt can also be measured. This allows one to determine the elastic behavior.

Rotation rheometers are based on rotation of circular plates or other geometries. Common are two flat plates or, most often used, a combination of a flat plate and a cone. The advantage of a combination of cone and plate is the achievement of a homogeneous shear velocity over the whole gap. With only two plates, the shear velocity is a function of the radius and has to be taken into account via a mathematical correction.

The flexibility of the rotation rheometer regarding the time-dependent control of torque and rotation speed allows one to perform basic rheologic experiments [12].

- One of the two plates will escalate to a constant angular speed. The torque will be measured. These experiments are used to measure the viscosity curves.
- The creeping experiment analyzes the behavior of the melt during a jump to a constant shear force with respect to a constant torque.
- The relaxation experiment analyzes the shear stress during a jump to a defined deformation.
- The oscillation experiment analyzes the behavior of the melt during an oscillated shear velocity. For small amplitudes, the behavior is linearly viscoelastic, which means that the torque with respect to the shear stress also shows an oscillated behavior. Shear velocity and shear stress are out of phase. The phase difference is in a range between 90° for an elastic body and 0° for a viscous fluid. The difference of phase is also a function of the frequency. With this experiment, it is possible to characterize the time-dependent viscoelastic properties of a polymer fluid.

3.3.5 Strain-Rheologic Behavior

During replication processes the melt undergoes shear deformation, as described above, and also deformation based on strain. For example, strain deformation can be found in flow channels with a change in cross-section, where the melt undergoes an acceleration or a deceleration. Typical replication processes in which strain deformation is the dominant deformation are thermoforming and blow molding. Two kinds of strain deformation can be distinguished: axial deformation and biaxial deformation. The case of axial deformation is valid by the flow of melt in flow channels and can be described in analogy to the shear viscosity.

$$\mu = \frac{\sigma}{\dot{\epsilon}} \qquad (3.26)$$

For low shear velocities, the relationship between strain viscosity μ, or so-called Trouton viscosity, and shear velocity η can be described with

$$\mu = 3 \cdot \eta \qquad (3.27)$$

For higher shear velocities, this relationship is not valid. The curves between shear viscosity and strain viscosity show different behavior. For hot embossing with typical flow velocities and shear velocities, this equation can be used for the approximation of stationary strain viscosity.

3.4 Molecular Orientation and Relaxation

Based on flow, the macromolecules of polymers undergo shear and orientation in the flow direction. The orientation of the molecules can remain by cooling in a short time—the orientation can be frozen. Orientations are responsible for the material behavior in the solid state, so the knowledge of their generation and their reduction via relaxation and retardation is essential.

3.4.1 Orientation

During flow or during embossing, the macromolecules of the polymers are oriented. This orientation can be defined as an imposed formation. This formation is based on the energy generated by the molding process. In this formation, the oriented molecules are characterized by a lower entropy and would like to go back to the initial, not oriented, state. This behavior is also known as entropy elastic behavior. The orientation and the return to the not-oriented state is a function of temperature and time. Under a short induction period of force, a polymer shows elastic behavior. During the short time the force is effective, the molecules are oriented. After the release of force, the orientation of the molecules goes back to the initial state. A different behavior can be seen under a longtime effect of force and higher temperature. The molecules also undergo an orientation, but under the effect of the force and the higher motility of the molecules related to the higher temperature, the orientation will degrade. The result is a new state without orientation. In molding processes like injection molding, the melt does not have enough time for relaxation; the orientation will be frozen. If such a molded part is heated up, the orientations of the molecules go back to the initial state, which results in a deformation or warpage of the part. This effect is based on the entropic elastic behavior of polymers.

3.4.2 Relaxation and Retardation

The degradation of orientation is a function of temperature and is based on two mechanisms: relaxation and retardation.

The mechanism of relaxation describes the reduction of orientation under constant deformation, which results in a decay of stress in a molded part. Under constant deformation, like a molded part in a mold, the motility of the molecules is limited. The molecules can only glide against each other, so this mechanism needs more time than the mechanism of retardation.

The mechanism of retardation describes the reduction of orientation without any imposed forces or deformations. Therefore, the motility of the molecules is higher than the motility in the mechanism of relaxation. This results in a faster reduction of orientation (approximately 10^4 times faster than relaxation), because groups of molecules can be arranged to the initial state. The result of the motility is an undesired deformation of the molded parts. But the mechanism of retardation can also be used in specific ways. For example, common heat shrinking of foils is based on the retardation of orientations. The foils are stretched during processing and, with higher temperatures, the shrinkage of these foils can be used in many applications.

The effect of the mechanism of retardation can also be seen by the use of polymer foils as semifinished products for hot embossing. During heating, the foils undergo a deformation so that, without any contact force, the distribution of the melt over the microstructured area is undefined.

3.4.3 Mathematical Characterization of Relaxation and Retardation

The time behavior of relaxation and retardation of polymer melts shows a similar dependency of temperature to that of the viscosity. Therefore, for a description of relaxation and retardation behavior, the principle of the time-temperature shift from William et al. [24] can be used. Measurements show that the shape of the curves for relaxation and retardation is similar. Therefore, the measured curves at different temperatures can be shifted along the time axis to

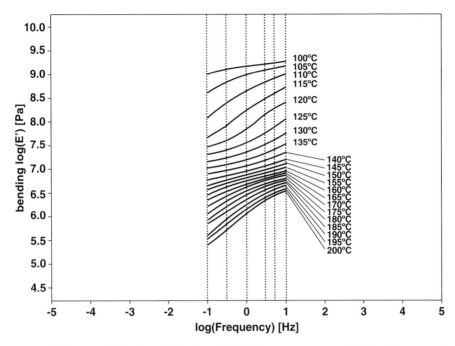

Figure 3.10 Measured elastic modulus in a defined frequency range at different temperatures. The measured curves at different temperatures can be shifted along the x-axis to a so-called master curve. The master curve refers to a selected reference temperature and allows one to determine the behavior over a wide range of frequencies.

a master curve. The shift factor a_T is influenced by temperature and can be calculated by

$$a_T = -\frac{C_1 \cdot (T - T_{ref})}{C_2 + (T - T_{ref})} \tag{3.28}$$

where T_{ref} is the reference temperature and C_1 and C_2 are constants (depending on material and reference temperature).

The shift factor represents a relation between the master curve and the relaxation curves for different temperatures.

In analogy to the relaxation behavior, Young's modulus of the polymer shows the same time-temperature behavior. Systematic DMA measurements allow the determination of the temperature-dependent elastic modulus in defined ranges of frequency. Based on the equivalence between time and temperature, the individually measured curves can be shifted along the time axis to one master curve (Fig. 3.10). In this example of PMMA, the shift refers to the reference temperature of 110°C. The resultant constants C_1 and C_2 are only valid for this temperature.

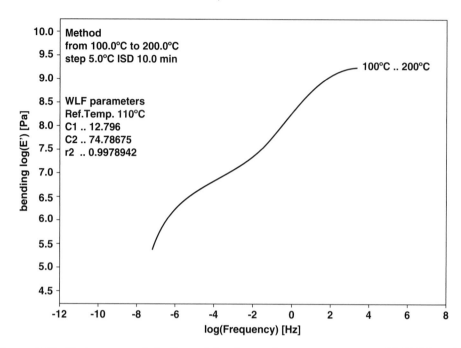

Figure 3.11 Master curve of elastic modulus for a reference temperature of 110°C based on the shift of single curves measured at different temperatures.

With this master curve and the associated shift function, a calculation of the elastic modulus dependent on time and temperature is possible (Fig. 3.11). The shape of the curve will remain constant; the curve will only be shifted along the time axis. This kind of curve can be used to describe the material behavior in simulation models for polymer replication processes.

3.5 Solidification

Carrying off heat from a polymer melt, the molecular chains lose their motility if the viscosity of the melt increases up to the solidification at the transition point where the melt is frozen. The transition point refers to a temperature, but the process of solidification trends to a transition range over several degrees. With regard to the arrangement of the macromolecules, two different kinds of morphologic structure can be distinguished: the amorphous structure and the semicrystalline structure.

3.5.1 Amorphous and Semicrystalline Structures

An amorphous structure is characterized by an orderless arrangement of the macromolecules. The molecular chains are intertwined and build a network with detachable, physical connectivity points. The volume of an amorphous polymer changes with temperature linearly, but with different gradients in the melting range and in the solid state. The morphologic structure is responsible for an isotropic behavior of melted and solid bodies. Common amorphous polymers are, for example, PMMA, PC, or PSU.

The semicrystalline structure is characterized by a crystalline part and an amorphous part. The crystalline structure is the result of the solidification of molecular chains with regular arrangements. Already above the glass transition range, groups of macromolecules with regular architecture can arrange themselves with high density—they crystallize. Because of the structure of macromolecules, the rate of crystallization during solidification is in the range of 30–70%. Because of the ramifications of the macromolecules, a complete crystallization is not possible. The crystallization generates crystallization heat and so, before the temperature of the melt can decrease, the crystallization heat has to dissipate. The cooling rate is one significant parameter of the crystallization. A high cooling rate results in a lower crystallization temperature: the frozen free volume rises and the ductility increases. Ductility results also from the amorphous part. Above the glass-transition temperature and under the crystallization temperature, the semicrystalline polymers show a solid behavior because of the strength of the crystalline part. The ductility of the polymer in this range results from the unsolidified amorphous part. Common semicrystalline polymers are, for example, PE, PP, PEEK, or LCP.

3.5.2 p-v-T Diagram

The solidification of a polymer melt and the resultant changes of aggregate states of amorphous or semicrystalline polymers can be represented by the p-v-T diagram (Figs. 3.12 and 3.13). This kind of diagram describes the relationship between pressure p, specific volume v, and temperature T. Polymers are compressible in the melt and in the solid state. The specific volume is therefore a function of pressure, so that in a p-v-T diagram a set of curves of isobars is shown. The behavior of the curves is different for amorphous and semicrystalline polymers.

Significant for amorphous polymers is the linear behavior of the curve in the melting range and in the solid state. The point of intersection is defined as the glass transition point, the related temperature as glass-transition temperature T_g. Semicrystalline polymers show a similar behavior of the curves in the melting range. Their difference from amorphous polymers can be seen by the behavior of the curve in the solid state. Opposed to the amorphous polymers, semicrystalline polymers show a parabolic behavior of the isobars in the solid state. The point of intersection between the linear curve and the parabolic curve is defined as the melting point or crystallization temperature.

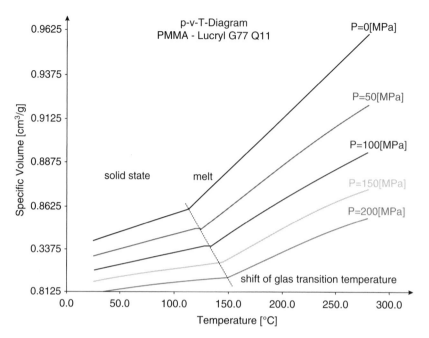

Figure 3.12 p-v-T diagram for an amorphous polymer. Characteristic are the differences of the gradient of the curves in the solid state and in the melt. Characteristic for polymers is the increase in the glass-transition temperature with an increase in pressure.

By means of the p-v-T diagram the following essential aspects can be explained.

- The glass-transition temperature for amorphous polymers and the crystallization temperature of semicrystalline polymers are a function of pressure. With higher pressure, the glass-transition temperature and the crystallization temperature are shifted linearly to higher temperatures.
- The volumetric shrinkage of a polymer replication process can be determined. The difference between the specific volume at the moment of solidification and the specific volume at room temperature describes the volumetric shrinkage of a molded part under the chosen process parameter.
- Because of the parabolic behavior of the isobars in the solid state, the volume shrinkage of semicrystalline polymers is higher than the shrinkage of amorphous polymers.
- The p-v-T diagram is the basis for the calculation of derived temperature and pressure-dependent physical values like density, volume-based thermal coefficient, and isothermal compressibility:
  — density:

$$\rho(p, T) = \frac{1}{v(p, T)} \tag{3.29}$$

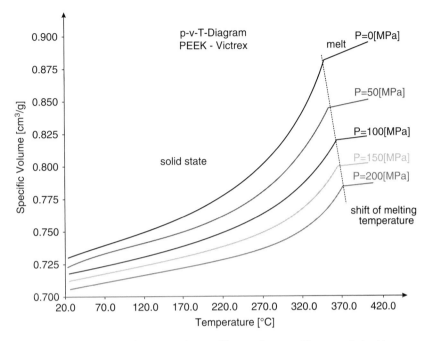

Figure 3.13 p-v-T diagram for a semicrystalline polymer. Compared to the amorphous polymer, the shape of the curves in the solid state is not linear; the volume shrinkage increases with temperature. The melting temperature is a function of pressure and will shift toward higher temperatures if the pressure increases.

— thermal coefficient:

$$\alpha = -\frac{1}{v}\left(\frac{\delta v}{\delta T}\right)_p \tag{3.30}$$

— compressibility:

$$\chi = -\frac{1}{v}\left(\frac{\delta v}{\delta p}\right)_T \tag{3.31}$$

The p-v-T diagram can be useful for an estimation of the volume shrinkage of molded parts during a polymer replication process. Therefore, the characteristic parameters—like molding and demolding temperatures, related pressures, and solidification temperature—have to be entered as points in the diagram. For the estimation of shrinkage, the process of hot embossing can also be entered in the p-v-T diagram (Section 4.4.3). For process simulation, the p-v-T behavior can be described by several approaches. Two common approaches are the coefficient approach, based on empirical studies, and the Tait approach, which is modified for use in commercial simulation software like Moldflow.

3.6 Solid Polymers

The behavior of polymers in the solid state is essential for the analysis of the demolding step of every replication process. Typically, the temperature range during demolding is near or below the glass-transition temperature or the crystallization temperature of a polymer. Solid polymers are also characterized by viscoelastic behavior. To describe viscoelastic behavior in the solid state, the models used to describe behavior in the melt state are valid.

3.6.1 Linear Viscoelasticity in the Solid State

Viscoelastic behavior in the solid state can also be described with the Maxwell element, a serial connection of a dashpot, and a spring. For a single element, the behavior of the element can be described by the differential equation

$$\frac{d\sigma}{dt} + \frac{E}{\eta} \cdot \sigma = E \cdot \dot{\epsilon}_0 \tag{3.32}$$

For a constant velocity of strain, the solution of this equation, the relationship between stress σ and strain ϵ, is given by

$$\sigma = \eta \cdot \dot{\epsilon}_0 \left(1 - e^{-\frac{E}{\eta}t}\right) = \eta \cdot \dot{\epsilon}_0 \left(1 - e^{-\frac{E}{\eta_0}E}\right) \tag{3.33}$$

where η is the viscosity of the dashpot and E is the elastic modulus of the spring.

The relaxation time is a function of the viscosity and the modulus of the spring—the ratio of viscosity and modulus. By a short impulse the Maxwell element shows a behavior like an elastic spring, because the velocity of strain is higher than the inverse of the relaxation time. Conversely, when the deformation is very slow, the model shows a viscous behavior. The maximal stress is given by the multiplication of viscosity and strain velocity.

The model of a single Maxwell element is not accurate enough to describe the behavior of a real polymer in the solid state. Like the models to describe behavior in the melt state, a parallel connection of single Maxwell elements is used. The stress of the system can be described by the sum of the stress of each single Maxwell element.

$$\sigma(t) = \sum_{i=1}^{n} \sigma_i \tag{3.34}$$

Every element undergoes the same deformation and the same velocity of deformation. So the stress-strain relationship for a parallel connection can be given by

$$\sigma(t) = \dot{\epsilon} \cdot \sum_{i=1}^{n} \eta_i \left(1 - e^{\left(-\frac{E_i}{\eta_i}t\right)}\right) \tag{3.35}$$

The Maxwell model is based on a linear dependency of deformation and stress. The spring and the dashpot undergo the same deformation at the same time. The linear viscoelastic model described here is based on the constant relation between stress and strain. Because of the linearity, the superposition principle of Boltzman can be used to analyze a complex profile of deformation. Therefore, the profile will be cut into several linear segments and can be analyzed with the described model. Superposition allows the addition of the single results to a total result. So the sum of the deformation of every segment can be used to calculate the total deformation as a result of the performance of a complex stress profile. The principle of superposition can also be used to calculate the deformation of a multiaxial stress profile regarding direct stress. The effect of every direct stress can be calculated and added to the total effect.

In the field of viscoelasticity, an approximation can be made if the equations of the science of elasticity are used. This procedure is called the correspondence principle. The stress σ, the strain ϵ, and the elastic modulus E can be replaced by the time-dependent functions $\sigma(t)$, $\epsilon(t)$, and $E(t)$. For the multiaxial state of tension, the time-dependent Poisson's ratio $\nu(t)$ has to be taken into account.

The Maxwell model described above is an illustrative model to characterize linear viscoelastic behavior. But this model is only valid for small deformations. For large deformations the mechanical behavior of polymers shows, besides time dependency, a nonlinear dependency between stress and strain. To consider this nonlinearity, the viscosity function of the dashpots used is also nonlinear. For example, the fluid in the dashpot can, in the case of an increase in stress, be modeled by a superproportional increase in the flow rate. The disadvantage in modeling nonlinear behavior is the mathematical complexity of the equations and the lack of an analytic solution. Therefore, often numeric algorithms are used.

3.6.2 Creeping and Relaxation

Along with the time dependency of load, the kind of load has to be considered. Two kinds of loads are essential: a constant load and a constant deformation. Therefore, the elastic modulus is defined in relation to these kinds of loads.

The case of constant load can be demonstrated by the creep experiment. A viscoelastic body undergoes a constant load at $t = t_0$. The elastic part allows a spontaneous deformation. On the one hand, if there is only elastic behavior, this deformation will be constant. On the other hand, for a viscoelastic body, the deformation will increase with the time of the load. This behavior is called creep. The time-dependent elastic modulus under constant load is defined as the creep modulus $E_c(t)$.

$$E_c(t) = \frac{\sigma_0}{\epsilon(t)} \qquad (3.36)$$

The creep experiment is characterized by a constant load and an increasing deformation as a result of the load, so that the creep modulus will decrease over time. This material behavior should be taken into account when molded components are used under a constant load.

The second case of load is the case of a constant deformation, for example, in a screwed connection. Similar to the creep experiment, a relaxation experiment is defined. The viscoelastic body is deformed at $t = t_0$ by a constant value. The resulting stress decreases over time. This relaxation behavior can be defined by the relaxation modulus $E_R(t)$

$$E_R(t) = \frac{\sigma_t}{\epsilon_0} \tag{3.37}$$

The relaxation modulus shows, like the creep modulus, a decreasing behavior over time. Comparing both, it can be shown that $E_R(t) < E_c(t)$.

3.6.3 Mechanical Properties

For the calibration of the material model described above, or for mechanical dimensioning, material parameters of the selected polymer have to be found. Because of viscoelastic behavior, it is often only possible to obtain relative parameters. The values have a comparative character and should not be used for the structural dimensioning of components. To get absolute values, some testing methods have to be modified, like a strain-controlled tensile test. For replication processes, the analysis of behavior during demolding is essential. Typical deformations during demolding are characterized by deformation based on tensile load. Therefore, tensile experiments allow one to characterize the behavior of the polymer during demolding, especially when the geometries, tensile velocities, and temperatures used in the tensile experiments are similar to the conditions during the demolding state.

3.6.3.1 Tensile Experiments

Tensile tests are established testing methods in material science. Compared to the tensile behavior of metals, polymers are characterized by a different behavior caused by their viscoelastic material properties. The influence of temperature and time can be shown in the tensile curves (Fig. 3.14). The influence of time is represented by the tensile velocity, which influences the deformation of the geometry. The properties of mechanical strength used in tensile experiments are listed below:

- The yield stress σ_s is defined as the tensile stress at the first point of the stress-strain curve, where the gradient of the curve is zero.
- The yield strain ϵ_s is the corresponding strain to the yield stress.

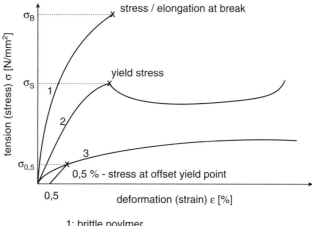

1: brittle poylmer
2: ductile polymer with yield strength
3: ductile polymer without yield strength

Figure 3.14 Tensile behavior of different kinds of polymers. Brittle polymers show a characteristic curve with a maximum stress and strain at break. Ductile polymers show no point of break. In this case, the yield stress σ_s is defined as the tensile stress at the first point of the stress strain curve, where the gradient of the curve is zero. Finally, for ductile polymers without visible yield stress, the stress at the offset yield point 0.5–2% is defined.

- The tensile strength σ_B is defined as the stress at maximum force in the tensile experiment.
- The stress at offset yield point $\sigma_x\%$ is defined as the tensile stress at the point where the stress-strain curve has a difference in a range of 0.5–2% related to the linear part of the curve.
- The stress at break σ_R defines the stress at the point where the specimen for the tensile test breaks.

The stress at offset yield point $\sigma_x\%$ can be used for the dimensioning of polymer parts when the stress-strain curve shows no significant yield stress. The behavior can be seen by tensile experiments near the softening temperature of the polymer. The characterization of the tensile behavior of polymers in this temperature range is essential in analyzing the demolding behavior during replication processes.

3.6.3.2 Strength of Polymer Material during Demolding

To analyze the demolding of structures, the temperature-dependent stress-strain characteristics of the material at demolding temperature and at demolding velocities have to be determined. For this purpose, tension specimens were produced by hot embossing with a geometry according to DIN EN ISO527-2, with a view to obtaining process-specific material properties. The tensile tests were carried out using a universal test machine in the temperature range typical of demolding, from 60°C to 100°C. Apart from the temperature, typical demolding rates were considered by the tension rate being varied in the range of 0.5 mm/min to 10 mm/min. Based on the tension curves obtained, the yield stress was defined as the first stress value, at which strain is increased without increasing stress. The corresponding yield strain was in the range of about 4% for all temperatures studied. Figure 3.15 shows the stress-strain curves of PMMA for a constant tensile velocity of 1 mm/min with a variation of temperatures. Figure 3.16 show the curves measured at 90°C, with a variation of demolding rates.

3.6.3.3 Dynamic Mechanical Analysis

From tensile experiments the dynamics of a viscoelastic material cannot be determined completely. Therefore, for the determination of the time- and temperature-dependent material behavior, another measurement system is required,

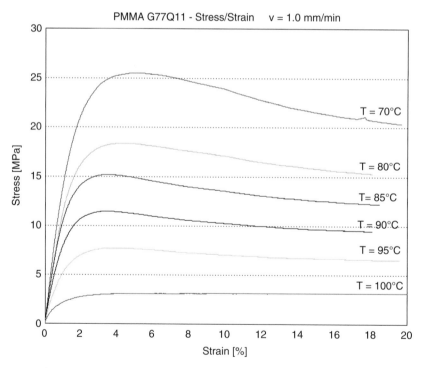

Figure 3.15 Stress-strain curves of PMMA measured as a function of temperature at a constant tension rate of 1 mm/min. The strength of the plastic decreases drastically with increasing temperatures. For defect-free demolding, an adapted demolding temperature has to be selected that is as close as possible to the glass-transition temperature, but ensures sufficient strength to prevent the component from being destroyed.

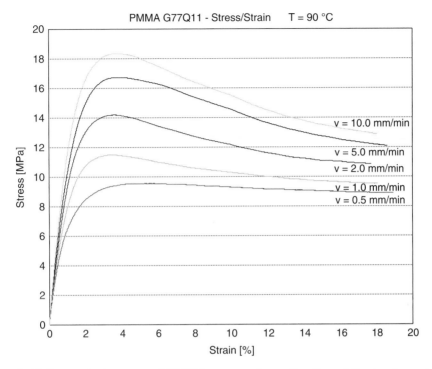

Figure 3.16 Stress-strain curves of PMMA measured as a function of the tension rate at a constant temperature of 90°C. The strength of the plastic is increased with increasing tension rate due to the viscoelastic behavior.

dynamic mechanical analysis (DMA). This measuring method is based on a sinusoidal load applied to a polymer sample. The range of the load should be in the linearly elastic range of the sample and has to be set significantly below the critical load. The applied sinusoidal load results in a sinusoidal response of stress or deformation, depending on if the load is applied by stress or by deformation. Besides the measurement of load and strain, the response is analyzed concerning the amplitude of the signal and the time difference of phase between load and response characterized by the phase angle δ. These results allow the determination of characteristic time-dependent material properties (Fig. 3.17):

- The complex elastic modulus $E^* = \sigma_A/\epsilon_A$ is the ratio between the amplitude of stress and strain.
- The storage modulus E' describes the stiffness of a viscoelastic polymer and is proportional to the stored elastic part of work. The storage modulus can be compared to the elastic modulus found in databases.
- The loss modulus E'' is proportional to the dissipative work. The loss modulus describes the energy of oscillation that disappears with heat.

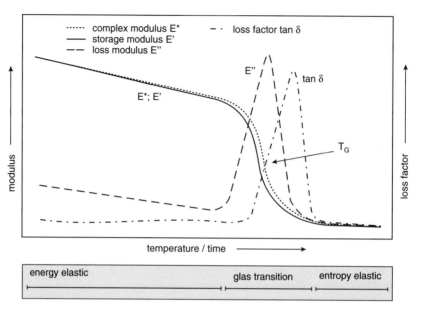

Figure 3.17 The mechanical properties of polymers can be described by the complex elastic modulus E^*, a function of the elastic storage modulus E', and the dissipative loss modulus E''. The loss factor defines the ratio between E'' and E'' [3].

- The phase angle δ characterizes the difference of phase between stress and related deformation during zero-crossing.
  - Elastic samples are characterized by a phase angle of zero. Without any delay, a response is given to an oscillated load.
  - Viscous samples are characterized by a phase difference of 90° $= \pi/2$. The load is completely lost with heat.
  - Viscoelastic samples like polymers show a phase angle in the range of $0 < \delta < \pi/2$. The response of an oscillated load is time delayed. An increase in the phase angle defines a higher damping of the oscillation.
- The loss factor $\tan \delta$ defines the ratio between loss modulus E'' and storage modulus E'. The loss factor can be interpreted as the ratio of the loss of energy to the recuperative part of the energy and indicates the mechanical damping or the internal friction of a viscoelastic polymer. Similar to the phase angle, a loss factor of zero defines an elastic material and a high loss factor defines a viscous material.

3.7 Friction

3.7.1 Friction between Mold and Polymer

Friction between polymer and mold is an essential point of view during demolding of microstructures. High demolding forces correspond to high static-friction forces and are responsible for the damaging of microstructures, especially if the cross-section of the structures decreases and the aspect ratio increases. Friction forces can be described by Coulomb's law with a separation in static friction, effective at the beginning of demolding, and dynamic friction, effective during sliding. The friction forces are characterized by the friction coefficient and the normal forces acting on the contact area. The normal forces are a result of the difference in shrinkage between polymer and metal mold insert. Polymers are characterized by higher shrinkage, resulting in high contact stress between mold insert and polymer at the vertical sidewalls (Fig. 3.18). In the nanoscopic range, the friction behavior may change. Riedo and Brune [15] investigated the nanoscopic sliding on different coatings. They found that the nanoscopic friction coefficient is directly linked to the Young modulus. Independent of these results, the description of friction during demolding can be done by the conventional Coulomb's law.

Real surfaces, also microstructured molds, are determined by a surface profile. During molding, the polymer is pressed by the molding pressure form-fitted into the cavities and the sidewall profiles of the mold (Fig. 3.19). During the demolding step, the cured molded part begins to move relative to the mold. Because of the differences in stiffness of a metal mold and a molded polymer, the surface profile of the mold will scratch along the sidewalls of the molded part. Furthermore, the surface of the polymer can be ablated by the relative movement and fit to the surface profile of the mold, which results in a decrease in friction force. Because of this mechanism, the friction force is a function of the material properties, especially of the temperature.

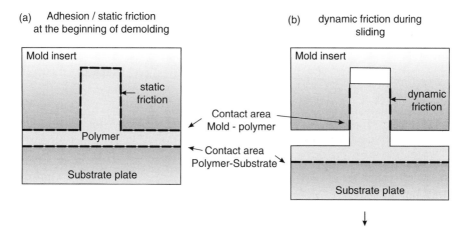

Figure 3.18 Different kinds of friction during demolding. At the beginning of demolding, the friction between mold and polymer is characterized by static friction. If the break-away torque is overcome, the friction is characterized by sliding based on dynamic friction. Independent of the friction between polymer and mold, the friction between polymer and substrate plate is characterized by adhesion and static friction, which should be on a higher level than the static friction between mold and polymer.

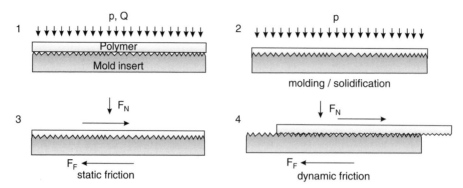

Figure 3.19 Friction between mold and polymer under the aspect of molding processes. In the case of molding processes, the static friction increases because of the molding step. The polymer melt flows into the surface roughness and will cause a perfect form-fitting contact between mold surface and polymer. After cooling, the difference in shrinkage will cause high normal forces, which results in combination with the form-fitting contact in high static- and dynamic-friction forces.

Especially for process simulation and process optimization, the determination of friction coefficients is desired. Under real hot embossing conditions with microstructured molds, only integrated friction forces (tensile forces) can be measured. A determination of friction force fails because of the unknown normal forces at each microstructure. Therefore, a measurement method has to be developed to give an estimate of the friction coefficients under hot embossing process conditions.

3.7.2 Measurement of Friction between Mold and Polymer

A central task is the development of a measurement method, by means of which static and dynamic friction can be determined during demolding under typical hot embossing conditions. Static and dynamic friction forces in microstructured tools, however, can only be measured as the sum of demolding forces. Detailed analysis fails, as it is impossible to determine normal forces acting on the individual sidewalls of the structured tool, which largely depend on the shrinkage of the component and, thus, on a multitude of process parameters and influencing factors. Moreover, force measurement systems of high spatial resolution, which might be integrated in the microstructured tool, are lacking. For this reason, static and dynamic friction forces are measured using a macroscopic measurement system (Fig. 3.20, [25]). The approach to determining friction coefficients according to DIN/ISO is not suited, as the boundary conditions defined in these standards do not reflect those of the hot embossing process. A measurement process has been developed, which includes an embossing process prior to the measurement of static and dynamic friction forces. Apart from the

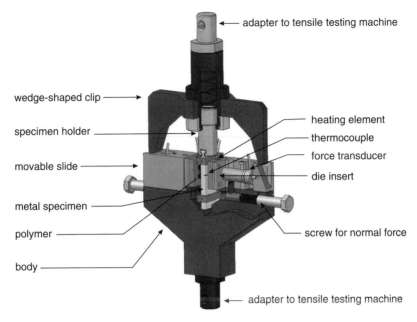

Figure 3.20 Test arrangement to determine adhesion and friction under typical hot embossing conditions. This test arrangement can be used in a tensile testing machine. The advantage of the arrangement is an integrated molding cycle before the measurement of friction is started.

surface roughness of the specimen and the defined adjustable normal force, the influence of the process parameters of hot embossing— that is, embossing pressure, embossing temperature, demolding temperature, and demolding rate—on the demolding force is considered.

As far as the material and surface roughness are concerned, the specimens used represent typical values of microstructured tools. Both surface roughnesses of tools produced by electroplating and of tools manufactured by micromechanical processes are considered. The advantage of this measurement method is that the history of the plastic, namely, the embossing process, is taken into account. Furthermore, the individual parameters influencing the demolding force can be assessed systematically.

3.7.2.1 Components of the Measurement System

The measurement system may be integrated in any tensile testing machine. The embossing force and tension movement required for friction measurement are applied via the tensile testing machine. The embossing force is transmitted from the tensile testing machine to a wedge-shaped clip and, further, to two movable slides with die inserts. Between them, a metal specimen is located that is connected with the tensile testing machine via a spring system. The normal force between the metal specimen and polymer required for measurement is applied by a spring system independent of the tensile testing machine. To control the normal force and embossing force, each slide is equipped with a force transducer. To reach the embossing temperature required, heating elements are installed in the slides.

3.7.2.2 Functioning

At the beginning of the experiment, a thin polymer foil is positioned between the test metal specimen and each of the slides. This foil is fixed by the slides. The die surface is provided with undercuts, such that the polymer is fixed after embossing and only a relative movement between specimen and polymer is possible. Embossing is started by the tensile testing machine being loaded by pressure. The force is transmitted to the metal specimen and the polymer plates via both wedge-shaped clips. With increasing temperature of the dies, the polymer and specimen are heated far above the glass-transition temperature of the polymer. Embossing takes place. As the plastic flows during embossing, the force has to be readjusted. After embossing, the specimen and polymer are cooled down to demolding temperature and measurement of the friction forces is started. The embossing force is reduced by reducing the pressure forces until a defined normal force can be applied by the spring system of the slides.

Friction forces are measured by a velocity-controlled tensile test. Velocity corresponds to the typical demolding rates of the hot embossing process. Construction of the measurement system allows for a displacement of about 1 mm or 2 mm, before tensile forces are determined as a measure of friction forces. In this way, it is ensured that the system is in a stationary state and that the break-away force is determined at constant tensile velocity.

3.7.3 Friction Force Curves

Two characteristic measurement curves are shown in Fig. 3.21. Breaking away of the structures at the beginning of demolding is of decisive importance to demolding the microstructures, as the highest forces are encountered at this moment. This point is characterized by static friction. In contrast to dynamic friction, static friction is defined clearly by the peak of the measurement curve. After the plastic has broken away from the tool, further demolding is characterized by sliding along the vertical sidewalls. Dynamic friction is also reflected by the stick-slip effect.

Based on the friction forces measured, static and dynamic friction coefficients are determined assuming Coulomb's friction

$$\mu_{stat} = \frac{F_{stat}}{F_N} \tag{3.38}$$

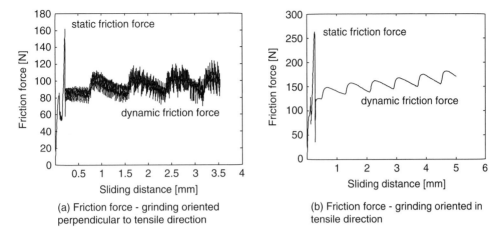

(a) Friction force - grinding oriented perpendicular to tensile direction

(b) Friction force - grinding oriented in tensile direction

Figure 3.21 Measurement of typical static- and dynamic-friction forces. The peak of the curves determines the static-friction force; the dynamic-friction force can be estimated by averaging. The surface roughness of approximately 200 nm is oriented perpendicularly to the direction of tensile and in the tensile direction.

$$\mu_{dyn} = \frac{F_{dyn}}{F_N} \quad (3.39)$$

where F_{stat} is the static friction force measured, F_{dyn} is the mean dynamic friction force measured, and F_N is the normal force between mold and polymer.

Due to the integrated molding cycle, static and dynamic friction forces can be measured under typical demolding conditions. With the measurement arrangement, friction coefficients can be calculated for different material combinations and surface roughness, and under various molding conditions.

3.8 Thermal Aggregate States of Polymers

During a molding cycle a thermoplastic polymer passes several temperature-dependent aggregate states. Switching to another aggregate state changes the material properties significantly. Because of the molecular structure of amorphous and semicrystalline polymers, different aggregate states and different transition ranges can be determined. The thermal behavior of thermoplastic polymers can be described, illustrated by the time-dependent shear modulus over temperature.

3.8.1 Thermal Behavior of Amorphous Polymers

The time-dependent behavior of the shear modulus is illustrated for an amorphous polymer in Fig. 3.22.

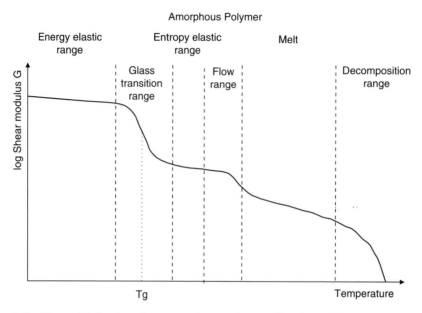

Figure 3.22 Thermal behavior of an amorphous polymer. The thermal behavior is characterized by a change of aggregate states. Beginning in the glassy state, where the polymer shows an
elastic behavior, the transition range follows. In this state the softening range is achieved, and in
the following entropy-elastic state the polymer can be formed. With an increase in temperature
the shear modulus decreases, and after the flow range is passed the state of a polymer melt is
achieved. A further increase in temperature will decompose the molecular structure.

For an amorphous polymer, four substantial aggregate states can be
distinguished:

- energy-elastic state or glassy state. The polymer is in the solid state
  and shows an energy-elastic behavior.
- rubber-elastic state or entropy-elastic state
- melting state characterized by a viscous behavior
- decomposition range. The molecular chains are destroyed.

The switch between the aggregate states is characterized by so-called transition ranges.

- The glass-transition range defines the transition between the glassy
  state and the entropy-elastic state.
- The flow-transition range defines the transition between the
  entropy-elastic state and the melting state.

In the energy-elastic state, the polymers are in the solid state, and they show a
brittle behavior like glass, especially with a decrease in temperature. With an
increase in temperature, the polymer comes into the elastic-entropy state.

In this range, the polymer shows ductile properties, a behavior like that of rubber. The transition between the glass state and the entropy-elastic state is defined by the glass-transition range or softening range. Range means that no specific point can be defined where the aggregate state changes. The transition is a continuous process over a temperature range, and the point with a maximum change of enthalpy defines the glass-transition temperature (Section 3.9.2). With a further increase in temperature, the flow-transition range of the polymer is reached. The transition between the entropy-elastic state and the melting state occurs continuously and can be extended over a temperature range of approximately 15 K. In this temperature range, physically bonded arrangements of the macromolecules are not effective and the macromolecules can glide off each other. The shear modulus decreases significantly. With a further increase in temperature, the polymer comes into the melting state. This range is characterized by a large temperature interval in which the shear modulus decreases with a further increase in temperature. Finally, if the increase in temperature continues, the decomposition range is reached, in which the macromolecules are destroyed by thermal energy.

3.8.2 Thermal Behavior of Semicrystalline Polymers

The thermal behavior of semicrystalline polymers is marked by the existence of an amorphous and a crystalline part. Because of this, the shear modulus over temperature is characterized by aggregate states similar to pure amorphous polymers but also by transition ranges that are typical for crystalline materials (Fig. 3.23).

Characteristic is a second transition range and a melting range instead of the flow range, so two transition areas can be defined.

- The glass-transition range of the amorphous part defines the transition between the glass state and the entropy-elastic state.
- The melting range of the crystallites defines the transition between the entropy-elastic state and the melting state.

In the energy-elastic state, the semicrystalline polymers show a mechanical behavior similar to that of an amorphous polymer. But in contrast to an amorphous polymer, in the glass-transition range only the amorphous part is influenced by temperature. Only the amorphous molecular chains glide off; the crystalline part is not influenced. This results in an only moderate decrease in the shear modulus. The material behavior is like a solid body with a reduced elastic modulus. With a further increase in temperature, the melting-transition range is reached. In this range, which is characterized by a small temperature gap, the shear modulus decreases significantly, so that the behavior of the polymer changes in a small temperature range of approximately 2 K or 3 K. In this

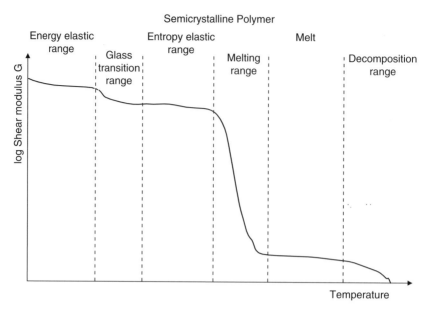

Figure 3.23 Thermal behavior of semicrystalline polymers. Semicrystalline polymers are marked by an amorphous and a crystalline part. The thermal behavior is also characterized by a glass-transition range and additionally by a melting range in which the crystals will be melted. This melting occurs in a small temperature range and results in a significant change of the shear modulus. The following melting state is typically marked by a low viscosity.

temperature range, the behavior of a semicrystalline polymer changes from a nearly solid body to a fluid with low viscosity. In the following melting state, the change of the viscosity is not distinctive. With a further increase in temperature, a thermal decomposition of the macromolecules also occurs.

3.8.3 Thermal Molding Windows

Based on the diagrams described above, typical molding windows for embossing amorphous and semicrystalline polymers can be specified. Amorphous polymers show theoretically a wide temperature range for hot embossing, beginning at the glass-transition temperature and ending before the decomposition range. In practice, the molding window depends on several other parameters and has to be set in dependency on the design, the molding area, and the technique of the embossing machine. Because of this, the molding window is smaller than the theoretical window and can be set approximately in the range of 20 K up to 100 K above the transition temperature. In contrast to the amorphous polymers, semicrystalline polymers show only a small gap of temperature, suitable for hot embossing. As described above, the decrease in the shear modulus occurs in a small temperature gap in the melting range. At the beginning of this range, the stiffness of the polymer is too high for molding. The risk of damage to a microstructured mold is high. At the end of the range, marked by a low shear modulus, semicrystalline polymers typically show a behavior similar to a fluid with low viscosity, which is not suitable for hot embossing because the pressure needed for embossing with free-flow fronts cannot be achieved. So the molding window can be found approximately in the middle of the small gap of the melting range. In practice, an exact temperature has to be set during molding. Therefore, the temperature range of amorphous polymers is much larger than the range of semicrystalline polymers, which in practice makes it easier to mold amorphous polymers by hot embossing (Fig. 3.24).

3.8.4 Commercially Available Polymers

For hot embossing of microstructures, typically commercially available thermoplastic polymers are suitable. Regarding the size of the microcavities, it should be taken into account that fillers should be in a range significantly smaller than the cavity size; otherwise an incomplete filling or a separation between fillers (like glass fibers) and polymer occurs. In most cases pure polymers are recommended. The commercially available polymers are manifold. Material properties are modified in a wide range, using additives, for example, to improve the demolding behavior or the impact strength. Nevertheless, the characteristic of the basic polymer is still present. Regarding the range of the glass-transition temperature or the melting temperature, the basic thermoplastic polymers can be classified into three groups (Fig. 3.25):

- Bulk plastics produced in mass. For hot embossing replication, suitable are, e.g., polyethylene (PE), polypropylene (PP), or styrene polymers (PS).
- Technical or engineering plastics are characterized by improved mechanical, thermal, and electrical properties. Representative of this group of polymers are polyamide (PA), polyoxymethylene (POM), polycarbonate (PC), polyethyleneterephthalate (PET), and copolymers of styrene (ABS, ASA, SAN).
- High-performance plastics refer to the technical plastics with advanced properties, especially in thermal resistance. Representative are, for example, polysulfone (PSU), polyetheretherketone (PEEK) or liquid crystal polymers (LCP), and fluoro polymers (PTFE).

Typical micro applications are based on a small volume of polymer, so the price of the material is not the determining factor in the cost of an application. Because of this, high-performance plastics can be used if needed. For hot embossing, the polymers commonly are processed in semifinished products like polymer foils or thin polymer sheets. The desired thickness of polymer

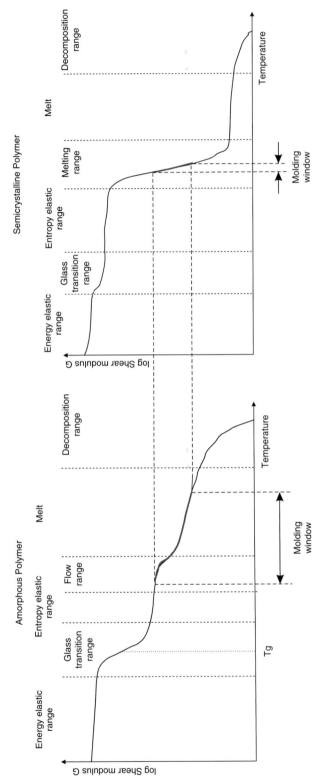

Figure 3.24 Thermal molding windows for hot embossing. The molding windows correspond to the viscosity range that allows the achievement of the necessary pressure during hot embossing. In the case of amorphous polymers, the viscosity of the melt can be influenced in a wide range by the molding temperature, which allows the use of a relatively large temperature molding window. Semicrystalline polymers are characterized by a significant change of the viscosity in a small temperature gap, which results in a small molding window, in special cases only a few degrees. The viscosity of some polymer melts is too low for an embossing with free-flow fronts, which underlines the importance of precise setting of the molding temperature.

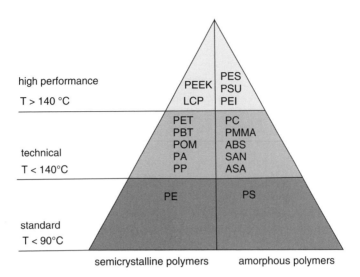

Figure 3.25 Classification of thermoplastic polymers. Polymers are classified into three groups depending on thermal resistance: bulk plastics, technical plastics, and high-performance plastics.

foils depends on the depth of the microcavities and the desired thickness of the residual layer. Typical thicknesses of commercially available foils are in the range of approximately 50 μm up to 1000 μm. Nevertheless, not every polymer foil or sheet is commercially available in the desired thickness. In this case, the polymer foils can be fabricated, for example, by large-area hot embossing using granular material.

3.9 Thermal Properties of Polymers

The thermal properties of polymers play an important role during heating and cooling of the polymer. For process simulation, the knowledge of the heat capacity, the thermal conductivity, or the temperature-dependent density of amorphous and semicrystalline polymers is required.

3.9.1 Thermal Material Data

3.9.1.1 Density

The temperature-dependent density of a polymer can be calculated from the data of the p-v-T diagram as the inverse of the specific volume,

$$\rho_T = \rho_{T_0} \frac{1}{1 + \alpha(T - T_0)} \qquad (3.40)$$

where ρ_{T_0} is the density at reference temperature T_0 and α is the linear coefficient of thermal expansion.

The equation is only valid for the linear part of the p-v-T curves. This equation is therefore valid only for amorphous polymers in the solid state up to the glass-transition temperature and also above the glass-transition temperature. For semicrystalline polymers, this equation is only valid above the melting point.

3.9.1.2 Heat Capacity

The enthalpy and heat capacity of the polymer is required for the dimensioning of cooling and heating systems and for material models used in simulation software. The energy needed for heating or cooling a molded part can be determined by

$$\dot{Q} = \dot{m} \cdot \Delta h \qquad (3.41)$$

The enthalpy Δh can be calculated by

$$\Delta h = \int_{T_1}^{T_2} c_p(T) dT \qquad (3.42)$$

where $\dot{Q}$ is the heat flow, $\dot{m}$ is the mass flow, Δh is the specific enthalpy, T_1–T_2 is the temperature range, and c_p is the specific heat capacity.

The specific heat capacity c_p can be determined by caloric measurements. Compared to an amorphous polymer, for which the heat capacity shows no significant change over temperature, the heat capacity of a semicrystalline polymer shows a discontinuity at the melting point. This discontinuity characterizes the necessary heat for melting the crystals ([11], Fig. 3.26).

3.9.1.3 Heat Conduction

During molding, especially during cooling and heating, temperature distribution in the molded part can be observed. This results in a heat flow between areas with higher temperatures and areas with lower temperatures. Under stationary conditions, this heat flow is proportional to the gradient of temperature. The proportional factor is called the coefficient of thermal conduction λ. Compared to metals for which the heat conductivity is based on free electrons, the energy will transmit by the oscillation of the chains of macromolecules. The theory of heat flow in polymers is based on the transmission of elastic waves

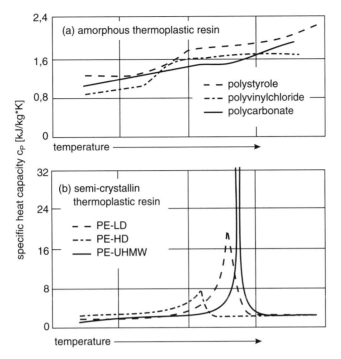

Figure 3.26 Heat capacity of amorphous and semicrystalline polymers. For amorphous polymers, the heat capacity changes between the solid state and the melting state. Semicrystalline polymers show a maximum at the melting point [11].

along molecular chains and was analyzed by Debye. Based on this theory, heat conduction is proportional to the heat capacity c_p, the density ρ, the velocity of the elastic waves u, and the free length of path l of the waves.

$$\lambda \approx c_p \cdot \rho \cdot u \cdot l \tag{3.43}$$

With the theory of Debye, heat conduction is influenced by several factors [11]:

- With an increase in temperature, oscillation of the macromolecules increases, which results in an increase in heat capacity. Heat conduction increases.
- With an increase in temperature, specific volume increases, which results in reduced transmission of elastic waves. Heat conduction decreases.
- The structural constitution of the polymer shows more influence on heat conduction than the temperature. Because of this, the heat conduction in semicrystalline polymers is higher than the heat conduction in amorphous polymers.

- Generally heat conduction can be increased with higher pressure. Higher pressure results in higher density and in an increase of the linkage force, which results in easier transmission of elastic waves.
- The percentage of crystals also defines the quality of heat conduction. With an increase in the crystalline part, heat conduction increases.
- Heat conduction shows a dependency on the orientation of molecules, which can result in anisotropic heat conduction over the molded part.

3.9.1.4 Thermal Diffusivity

Thermal diffusivity characterizes the time-dependent behavior of non-stationary heat conduction processes and relates to the Fourier differential equations. Thermal diffusivity is defined by the relation of the heat conduction coefficient λ and the heat storage capability $\rho \cdot c_p$.

$$a = \frac{\lambda}{\rho \cdot c_p} \tag{3.44}$$

3.9.2 Measurement of Calorimetric Data

Knowledge of thermal material behavior is required to predict the process parameter in molding processes. Material data are also fundamental for any simulation of polymer replication processes. To measure the different specific data of polymers, different specialized measurement processes are developed. Determination of the glass-transition temperature and characterization of the phase transition can be achieved by the measurement of calorimetric behavior. Two related measurement systems are suitable: differential thermal analysis (DTA) and differential scanning calorimetry (DSC). The principle of these measurements can be shown using the method of DSC.

3.9.2.1 Differential Thermal Analysis

DTA is based on the precise measurement of the temperature of a polymer specimen and a reference material in a heater (Fig. 3.27). Measurement occurs by thermoelectric couples integrated in the symmetrically arranged specimen and the reference material. The thermoelectric couples are connected in this way so that no thermoelectric voltage occurs when the polymer specimen and the reference material show the same temperature. Measurement starts with a linear increase in temperature. If during heating a phase transition occurs, the

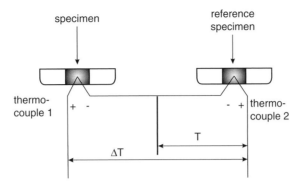

Figure 3.27 Schematic view of the principle of differential thermal analysis (DTA). An unknown and a reference specimen are connected with thermocouples, isolated from ambient influences and heated up in a heater. During heating, differences in temperatures are measured by the interconnected thermocouples. The difference of voltage relates to the difference of temperature and is used to identify the calorimetric behavior of the unknown specimen.

measured thermoelectric voltage can be used to characterize the temperature of reaction, the heat of reaction, and the kind of reaction (endothermic or exothermic).

3.9.2.2 Differential Scanning Calorimetry

In contrast to the DTA measurement system, in the DSC device thermoelectric couples are not placed directly in the specimen. Instead, they are embedded in the specimen holder (Fig. 3.28). The thermoelectric couples make contact with the containers from the outside. In contrast to DTA, in which the specimens are in the range of 10 g, the DSC test needs samples in the range of only several mg. Because of this arrangement, the sensitivity of a DSC measurement is lower than with DTA.

A schematic curve obtainable from DSC measurements is shown in Fig. 3.29.

Regarding the schematic view of a DSC measurement curve, the following information can be extracted.

- The glass-transition temperature of amorphous polymers can be determined by a step in the specific heat. In the DSC diagram, this effect is visualized by a shift of the curve. The shift is proportional to the step of the specific heat. The glass-transition point is in the range of the inflection point of the curve and can be determined graphically. Therefore, three tangents are fit to the step of the curve. The half of the distance between the intersection of the tangents defines the glass-transition temperature.

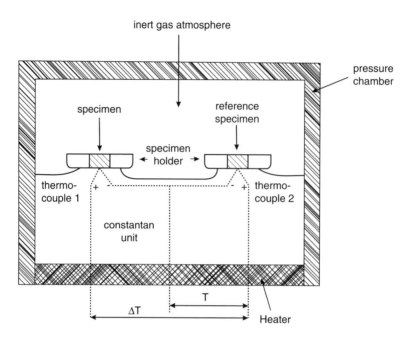

Figure 3.28 Principle of differential scanning calorimetry (DSC). Inside a pressure chamber the unknown and the reference specimen are put inside the specimen holder. The inert gas–filled and isolated chamber is heated up and the differences in temperature of the specimen holder is measured by thermocouples.

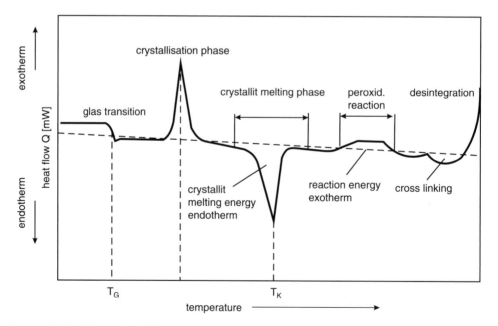

Figure 3.29 Schematic DSC measurement curve. By a DSC measurement, glass-transition temperature, melting temperature, specific heat, and the part of crystals of a polymer can be determined.

- The melting point and the melting heat are defined by the area between the trend line and the base line. The area is the amount of the heat needed for transition.

- The crystalline part χ can be determined from the ratio of the heat of fusion of a polymer sample ΔH_{sc} and the enthalpy of a 100% crystalline sample ΔH_c.

- The specific heat c_p can be determined during a DSC temperature sweep. The sample pan and the reference pan are maintained at the same temperature. This allows the measurement of the differential energy required to maintain identical temperatures. The sample with the higher heat capacity will absorb a larger amount of heat, which results in a shift of the DSC curve in the y-direction. This shift is proportional to the difference between the heat capacity of the sample and the reference specimen.

References

1. D. M. Cao and W. J. Meng. Microscale compression molding of Al with surface engineered LiGA inserts. *Microsystem Technologies*, 10:662–670, 2004.

2. D. M. Cao, W. J. Meng, and K. W. Kelly. High-temperature instrumented microscale compression molding of Pb. *Microsystem Technologies*, 10:323–328, 2004.

3. W. Ehrenstein. *Polymer-Werkstoffe—Struktur, Eigenschaften, Anwendung*. Hanser Publishers, 1999.

4. G. Fu, N. H. Loh, and S. B. Tor. A variotherm mold for micro metal injection molding. *Microsystem Technologies*, 11:1267–1271, 2005.

5. G. Fu, N. H. Loh, S. B. Tor, B. Y. Tay, Y. Murakoshi, and R. Maeda. Analysis of demolding in micro metal injection molding. *Microsystem Technologies*, 12:554–564, 2006.

6. T. Gietzelt, O. Jacobi, V. Piotter, R. Ruprecht, and J. Hausselt. Development of a micro annular gear pump by micro powder injection molding. *Journal of Materials Science*, 39:2113–2119, 2004.

7. R. N. Haward and R. J. Young. *The Physics of Glassy Polymers*. Chapman & Hall, 1997.

8. G. T. Helleloid. On the computation of viscosity-shear rate temperature master curves for polymeric liquids. *Morehead Electronic Journal of Applicable Mathematics*, 2001.

9. I. M. Ward. *Mechanical properties of solid polymers*. John Wiley & sons, 1985.

10. S. G. Li, G. Fu, I. Reading, S. B. Tor, N. H. Loh, P. Chaturvedi, S. F. Yoon, and K. Youcef-Toumi. Dimensional variation in production of high-aspect ratio micro pillars array by micro powder injection molding. *Applied Physics A*, 89:721–728, 2007.

11. G. Menges, E. Haberstroh, W. Michaeli, and E. Schmachtenberg. *Werkstoffkunde Kunststoffe*. Hanser Publishers, 2002.

12. M. Pahl, W. Gleissle, and H.-M. Laun. *Praktische Rheologie der Kunststoffe und Elastomere*. VDI-Verlag GmbH, 1995.

13. C. T. Pan, T. T. Wu, Y. C. Chang, and J. C. Huang. Experiment and simulation of hot embossing of a bulk metallic glass with low pressure and temperature. *Journal of Micromechanics and Microengineering*, 18: 025010, 2008.

14. V. Piotter, T. Gietzelt, and L. Merz. Micro powder-injection moulding of metals and ceramics. *Sadhana Acad Proc Eng Sci*, 28:299–306, 2003.

15. E. Riedo and H. Brune. Young modulus dependence of nanoscopic friction coefficient in hard coatings. *Applied Physics Letters*, 83(10):1986–1988, 2003.

16. R. Ruprecht, T. Benzler, T. Hanemann, K. Mueller, J. Konys, V. Piotter, G. Schanz, L. Schmidt, A. Thies, H. Woellmer, and J. Hausselt. Various replication techniques for manufacturing three-dimensional metal microstructures. *Microsystem Technologies*, 4: 28–31, 1997.

17. R. Ruprecht, T. Gietzelt, K. Müller, V. Piotter, and J. Hausselt. Injection molding of microstructured components from plastic, metals and ceramics. *Microsystem Technologies*, 8:351–358, 2002.

18. A. Schubert, J. Edelmann, and T. Burkhardt. Micro structuring of borosilicate glass by high-temperature micro-forming. *Microsystem Technologies*, 12:790–795, 2006.

19. H. Schulz, M. Wissen, N. Bogdanski, H.-C. Scheer, K. Matthes, and Ch. Friedrich. Choice of the molecular weight of an imprint polymer for hot embossing lithography. *Microelectronic Engineering*, 78–79:625–632, 2005.

20. G. Strobl. *The Physics of Polymers*. Springer-Verlag, 1997.

21. M. Takahashi, Y. Murakoshi, R. Maeda, and K. Hasegawa. Large area micro hot embossing of Pyrex glass with GC mold machined by dicing. *Microsystem Technologies*, 13:379–384, 2007.

22. M. Takahashi, K. Sugimoto, and R. Maeda. Nanoimprint of glass materials with glassy carbon molds fabricated by focused-ion-beam etching. *Japanese Journal of Applied Physics*, 44(7B):5600–5605, 2005.

23. T. Osswald and G. Menges. *Materials Science of Polymers for Engineers*. Hanser Publishers, 1995.

24. M. L. Williams, R. F. Landel, and J. D. Ferry. The temperature dependence of relaxation mechanisms in amorphous polymers and other glass-forming liquids. *Journal Am. Chem. Soc.*, 77:3701–3707, 1955.

25. M. Worgull, J.-F. Hétu, K. K. Kabanemi, and M. Heckele. Hot embossing of microstructures: characterization of friction during demolding. *Microsystem Technologies*, 14:767–773, 2008.

26. M. Yasui, M. Takahashi, S. Kaneko, T. Tsuchida, and Y. Hirabayashi. Micro press molding of borosilicate glass using plated Ni-W molds. *Japanese journal of Applied Physics*, 46: 6378–6381, 2007.

4 Molded Parts

Replication processes cannot be analyzed without their results—in this case the molded parts. These replicated parts with their microstructures give feedback to the process parameters and boundary conditions of the replication process. Therefore neither component—the process nor the replicated parts—can be analyzed independently.

What does it mean to characterize a molded part, especially under the aspect of microstructuring? This chapter will define characteristic properties of replicated parts, especially those replicated by hot embossing, and will discuss the technology that is available to measure the characteristics of each molded part in the micro or nano range. Here the reproducible accuracy of each measurement and the resolution of typical measurement systems are limitation factors for determination of these characteristics. Further, with the background of the characteristics and their determination, quality criteria can be defined. They provide feedback on the molding process, the technology, the polymer materials used, and the quality of the microstructured mold insert. Here the application and potential further process steps of the molded parts will decide which level of quality is necessary or which quality criteria will be weighted by different factors. For example, for optical components the surface roughness is an important criteria of quality; in contrast, for a housing of a micro system the surface roughness is typically not critical.

The analysis of replicated parts will focus on hot embossed parts, especially on single-sided replicated parts. Referring to this, first typical components of a replicated part shall be defined.

4.1 Components of Hot Embossed Parts

What are the typical components of a molded part fabricated by hot embossing? Referring to the process (Chapter 5), a characteristic of molded parts is the residual layer as a carrier layer of microstructures. The microstructures are arranged on this carrier layer in the microstructured area (Fig. 4.1). The size of this area depends on the design and the available technology for fabrication of the related mold inserts and can increase to diameters of several inches. The applications for large molded areas are, for example, Fresnel lenses for overhead projectors. The carrier layer or so-called residual layer or remaining layer can, on one hand, be part of the design—for example, for housings of microstructures or as a carrier layer for micro-fluidic channels. On the other hand, for some applications the residual layer has to be removed to separate the micro structures for further processing. Here the molding of thin residual layers is an

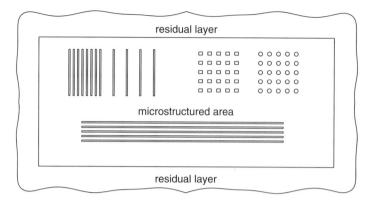

Figure 4.1 Characteristic components of a molded part: a microstructured area on a carrier layer, so-called residual layer, or remaining layer.

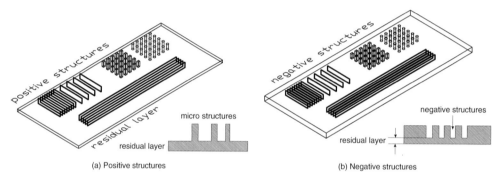

(a) Positive structures   (b) Negative structures

Figure 4.2 Classification of molded structures. Positive structures are characterized by a higher level than the remaining layer, for example, freestanding structures like cylinders or walls. Negative structures are characterized by a ground level below the top of the residual layer, typical for these structures are holes, trenches, or fluid channels.

important aspect, performed by the selection of the thickness of the semi-finished products and the right set-up of the process parameter molding force, molding temperature, and holding time.

Molded microstructures can be classified in relation to the classification of structures of the mold. In general, two kinds of structures can be distinguished (Fig. 4.2). So-called positive structures define the freestanding structures like cylinders or walls. Opposed to the positive structures, negative structures define structures like holes or trenches.

To classify a design, the density of structures in the microstructured area can be an appropriate definition. The density can be defined by the lateral areas on different ground levels, the area at zero level (here the top of the residual layer) and the area at the top level of the structures. This level can be positive or negative depending on the kind of structures. The ratio between these two levels will give an estimate of the density of structure compared to the molding area and can help classify a design in relation to the estimated demolding forces.

$$\text{Structure density} = \frac{\text{Area of top level of structures}}{\text{Area of residual layer}} \qquad (4.1)$$

4.2 Characterization of Molded Parts

Which properties can be defined characterizing molded parts replicated by hot embossing? It is obvious that in general the shape of the molded part has to be replicated as best as possible compared to the shape of the inverse mold insert. Nevertheless, looking in detail the shape of the molded part can be analyzed under different aspects.

- The lateral dimensions of single structures specified by characteristic length of the structures should be identical to the lateral dimensions of the corresponding structure on the mold insert. This requirement should be also valid for the whole molded part with much larger lateral dimensions. Here the influence of shrinkage, especially the differences in thermal expansion between metal mold inserts and molded parts, effects a difference in lateral dimensions. These differences increase with an increase of the microstructured area. The shrinkage is finally responsible for a difference in the lateral geometries regarding single structures and the arrangement of structures or groups of structures in relation to each other.
- Besides the lateral dimensions, the height of the molded structures should be identical compared to the depth of the related microcavities of the mold. Here the influence of incomplete mold filling or remaining resistance in microcavities can reduce the final height of molded structures.
- If the mold insert is characterized by structures with vertical sidewalls, the molded parts should replicate this verticality in the same way. Here also the influence of shrinkage, internal stress of replicated structures, and deformations during demolding can deform vertical sidewalls.
- The surface roughness of sidewalls in the microstructured mold insert defines the surface roughness of molded parts. Here a local deformation of the surface of a mold insert can scratch the surface of the replicated structure over the whole demolding distance.

Besides the geometric aspects characterized by the mold insert, the geometry of the microstructures can also be influenced by the process parameters, the available technical equipment, and the polymer material.

- The thickness of the residual layer can be mainly controlled by the process parameter molding force and molding temperature. Here a constant thickness over the whole replicated part is desired and, for some further process steps, essential. For example, the separation of microstructures by the removal of the residual layer by ion etching or other abrasive methods like grinding requires a constant thickness of the residual layer. Here the measurement of the thickness of the residual layer at different points, beginning in the center up to the margin regions, can determine the thickness distribution of a replicated part.

- Corresponding to the constant thickness of the residual layer the evenness of this layer as a carrier layer of the molded part can be essential for the successful application of further process steps.

- The influence of shrinkage can be partially controlled by the process parameter and the technical solutions that allow the achievement of a homogeneous pressure distribution during molding and cooling. Inconsistent pressure distribution inside molded parts will result in differences of shrinkage inside the molded parts and finally result in warpage of the whole part, especially the residual layer. This warpage can be measured by the determination of the final shape.

- Sink marks are visual, typically immeasurable, criteria to characterize the pressure and the material flow during molding and cooling states. Sink marks are the result of inadequate pressure and material flow during the cooling state of a replication process. These sink marks will finally also qualify the accuracy of the replicated shape and surface.

- Residual stress in the molded part is also responsible for the warpage of the whole part. The measurement of internal stress, especially in microstructures, is difficult. Here the warpage as a result of internal stress can be determined. The stress inside a molded part can be visualized in a qualitative way by optical methods, for example with polarized light.

In practice, not every criterion has to be taken into account for a verification of the demanded quality. Depending on the application some of the criteria are negligible. For example, the criteria of surface roughness will be important for optical components; for housing of a microsystem this criteria will have lower priority. One important point regarding the criteria is the technique and the complexity to measure the interesting values. The next section will present the most important measurement systems.

4.3 Measurement Systems for Characterization

The measurement of the characteristic values of the criteria named above is especially ambitious in the micro or nano range. The available measurement systems, their resolution, and repeatable accuracy determine the available quality of the characterization of molded parts. Different kinds of measurement techniques are suitable to determine characteristic values like lateral dimensions or surface roughness. However, the difficulties increase with the decrease of structure size. Especially in the nano range the number of suitable measurement techniques decreases. A main focus for characterizing will be the measurement of lateral dimensions, the height of structures, and the surface roughness [1,3,7].

The suitable measurement methods can be split into two groups: contact and non-contact measurement systems. Contact or tactile measurement systems refer to a point of contact between a sensor and the microstructure. Non-contact measurement systems are based mostly on optical measurement systems and allow the characterization of points, areas, and volumes. Non-contact measurement techniques can be also distinguished so that a further differentiation of surface measurement techniques can be given by:

- Optical measurement systems
  - Microscopy, autofocus measurement systems
  - Electromagnetic measurement systems, for example, scanning electron microscope (SEM)
- Tactile measurement systems
  - Contact-based measurement systems, for example, brush analyzer or contact stylus instrument
  - Measurement systems based on a probe or sensor, for example, scanning tunneling microscope (STM)

4.3.1 Tactile Measurement Systems

Tactile measurement systems detect the contour of a molded part along a line. The stylus touches the workpiece at one point with a low touch force. In a second step the stylus moves along the surface. The vertical amplitude of the stylus as a function of the covered distance characterizes the surface profile along a selected line. Systematic measuring of different lines will give a workpiece a surface profile. Typical resolution of tactile measurement systems is in the range of 0.5 μm.

4.3.1.1 Measurement Systems Based on a Probe

Representative for a probe-based measurement system is the atomic force microscope (AFM). The measurement principle is based on an atomic thin pin that is moved along a grid over the surface of a workpiece. Therefore the workpiece can be moved in all directions via a piezoelectric system. The pin is fixed at the end of a cantilever flat spring. The amplitude of the flat spring can be measured by a laser measurement system. With this measurement system the topology of a surface can be determined with a vertical and lateral resolution of 1 nm.

The measurement principle refers to atomic interaction between the atoms of the pike and the atoms of the surface. If the distance of the pike is in the range of one nanometer, attracting and repulsing forces appear. These forces can be measured with the elongation of the flat spring. Based on this effect, different measurement modi can be used. Two characteristic modi are the contact modus, based on a mechanical contact of the pin and the workpiece, and the non-contact modus, based on dynamic oscillation of the flat spring.

In the contact modus the pin touches the workpiece with a contact force in the range of 10^{-8} down to 10^{-10} N. This range is about four dimensions lower than typical tactile measurement systems, which means that filigree structures will not be damaged during movement of the pin. The motion of the workpiece in the contact modus allows characterization of friction behavior. The non-contact modus is suitable to measure very delicate structures. The measurement refers to an oscillating flat spring ($\omega \approx 100\text{--}500$ kHz), whose frequency changes with the distance to the surface of the workpiece. This modification of frequency can be measured, for example, by the change of phase or the change of amplitude. The resolution of this measuring modus is several nanometers lower than the measurement in the contact modus.

The advantages of this AFM measurement system can be summarized:

- High local resolution—resolution down to atomic dimensions
- Samples can be used with less or no preparation
- Measurement of interaction forces
- Characterization of topology and additional information about elastic modulus, friction, adhesion

The disadvantage is the relatively small area than can be measured: typical areas are in the range of $100\,\mu\text{m} \times 100\,\mu\text{m}$, which can result in time-consuming analysis of large-area molded parts. The focus of this measurement system is therefore only to analyze selected areas of a molded part.

Molded parts are also characterized by larger dimensions in the range of millimeters, for example, the distance of structures in a micro-optical bench. To measure larger distances coordinate measurement machines are established. The measurement system refers also to a probe, but with larger dimensions. Typical diameters of the circular touch probes range down to $25\,\mu\text{m}$. The advantage is the large area that can be analyzed. The resolution of such measurement systems underlines a continuous improvement and can be determined in the range of $1\text{--}2\,\mu\text{m}$. The challenge of measurement is the characterization of the surface quality in holes or trenches. The diameters of the probes are here the limiting factors.

Besides the AFM and coordinate measurement machines, related measurement techniques based on a probe are common—for example, a scanning tunneling microscope, which refers to a tunneling current between a conduction atomic sharp pin and a conducting workpiece. Another example is the scanning near-field optical microscope (SNOM), which avoids the boundaries based on the deflection of light. An optical probe scans in a short distance the surface of the workpiece. The resolution of this measurement system is characterized by the properties and size of the probe and is not a function of the wave length. The typical resolution is in the range of 50 nm.

4.3.2 Optical Measurement Systems

The measurement of lateral dimensions of molded parts can be determined by measurement microscopes, focusing on selected edges of molded structures. The magnification and the accuracy of the mechanical equipment determine the resolution. These kinds of measurement systems are well suited for the measurement of microstructures in a range down to approximately 1–10 µm depending on the measurement system. The determination of lateral geometries will be supported by scanning hardware and analysis software.

The determination of the height of molded structures can be done by focusing on the different levels of the structures, here on the ground level, the residual layer, and the top level of the structures. The accuracy here depends also on the magnification and the precision of the optical elements and especially on the accuracy of the mechanical measurement unit. An alternative in determining the height of structures is non-focal measurement systems. With these, a small laser point is projected onto the surface and the reflected light is detected. The intensity of the reflected light is used to determine the focus of the beam. The lens of the microscope is focused until the maximum intensity is detected.

Finally, for many tasks optical measurement microscopes are well suited to qualify the molded parts and to give initial feedback on the parameters of the molding process.

4.3.3 Electromagnetic Measurement Systems

The scanning electron microscope (SEM) is named representatively for the large group of electromagnetic measurement systems. Instead of visible light an electron beam is focused by magnetic lenses on a beam with a diameter in the range of a few nanometers. This beam is focused on the surface of the workpiece, with the effect that secondary electrons are emitted. The intensity of the secondary electrons is related to the height of the structure and will be collected, amplified, and transformed to an electric signal used for visualization on a screen. To get a picture of the workpiece, the electron beam scans the workpiece line by line. Resolution in the range of nanometers can be achieved by this technology. The SEM is mostly used to qualify the surface and the contours visually; detailed quantitative measurements can be achieved with an adequate analysis software. The SEM is well suited for the visual control of defects or deformation of molded structures. The measurement of lateral geometries is also suitable, but the determination of the height of the structures is difficult. Here the tactile systems like AFM are recommended.

4.4 Quality of Molded Parts

What defines the quality of a replicated part? The criteria and their related measurement techniques named above characterize the geometry and the surface roughness as a result of the molding process. The results of these measurements are fundamental to analyzing the molded part regarding the flow behavior of the polymer melt, the isotropy of shrinkage, the warpage of thin parts, or the deformation because of internal stress. All these aspects result in a deviation from the ideal shape, here the inverse shape of the microstructured mold insert, and provide feedback on the molding process, the molding parameter, and the molding technology. The quality of a molded part therefore will include, besides the measurable geometry of molded parts, also the visible and immeasurable aspects like sink marks or stress distribution and the reasons that are responsible for deviations of the optimal shape.

4.4.1 Classification of Failures

Deviations from the optimal shape can be classified into different groups of failures. Hsueh et al. [2] analyze the failures of delamination and buckling of molded lines. They identify temperature, relaxation effects, pressure distribution, and friction as significant parameters. Nevertheless, the reasons and the effect on the shape of microstructured molded parts are manifold, beginning with a small deviation of freestanding structures from the vertical line by the effect of shrinkage up to the separation of structures from the residual layer during demolding. A classification of failures can be done by separating them into failures regarding the filling of micro cavities, failures occurring during cooling, and finally failures generated during demolding.

- The incomplete filling of microcavities refers to inadequate pressure for a given viscosity at a selected temperature. Characteristic for this failure are round profiles of structures instead of sharp contours. This failure becomes more important with a decrease of structure size, especially for structures with small cross sections and high aspect ratios (Fig. 4.3(a)).

- Weld lines result from a connection of two flow fronts, for example in a flow channel filled from both sides. At the intersection the orientation of the molecules can be different from the regions with

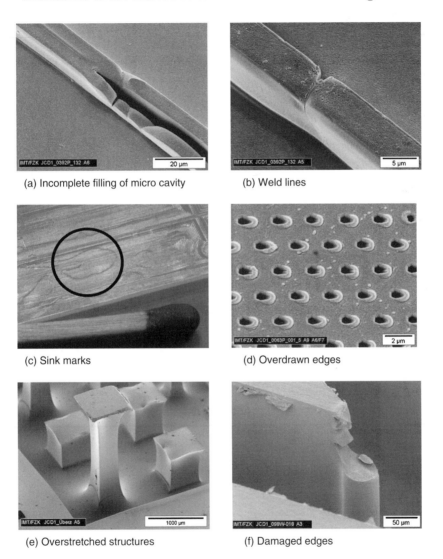

(a) Incomplete filling of micro cavity   (b) Weld lines

(c) Sink marks   (d) Overdrawn edges

(e) Overstretched structures   (f) Damaged edges

Figure 4.3 Classification of failures of molded structures. The failures can be split into three groups: failures referring to the filling of microcavities, failures referring to cooling, and finally failures referring to demolding. Besides the failures during filling, incomplete filling and weld lines, and during cooling, sink marks as a result of deficient dwell pressure during cooling, the most failures occur during demolding: overdrawn edges, damaged edges of structures, overstretched structures, and with an increase of demolding forces also from the residual layer, separated structures. Warpage of structures is a result of differences of shrinkage and can deform structures in a way that they cannot be used in the desired applications.

 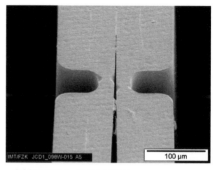

(g) Separated structures   (h) Warpage of structures

Figure 4.3 Classification of failures of molded structures. The failures can be split into three groups: failures referring to the filling of microcavities, failures referring to cooling, and finally failures referring to demolding. Besides the failures during filling, incomplete filling and weld lines, and during cooling, sink marks as a result of deficient dwell pressure during cooling, the most failures occur during demolding: overdrawn edges, damaged edges of structures, overstretched structures, and with an increase of demolding forces also from the residual layer, separated structures. Warpage of structures is a result of differences of shrinkage and can deform structures in a way that they cannot be used in the desired applications (*continued*).

homogeneous orientations of molecules. Weld lines therefore can be a kind of predetermined breaking point (Fig. 4.3(b)).

- Sink marks can occur during cooling of molded parts. If the dwell pressure during cooling is not high enough to compensate for the temperature-dependent shrinkage by a continuous flow of polymer, the contact pressure between molded part and mold insert will partially be lost. This lack of contact results in a bad surface quality like a non-uniform surface and rounded edges (Fig. 4.3(c)).

- Overdrawn edges or walls of microstructures occur during the demolding step and are characteristic failures referring to the lateral shrinkage of the residual layer. Because of the differences in shrinkage between molded parts and mold, shear forces occur, deforming the top of the structures during demolding (Fig. 4.3(d), Chapter 6).

- Overstretched structures result from high adhesion and friction between structure and mold during demolding. Especially filigree freestanding structures are periled to be overstretched. Adhesion and friction forces have to transfer over the cross section of the structure and induce tensile stress inside the structures. In the case of a higher induced stress compared to the yield point of the material at demolding temperature, the structures will be deformed. The direction of the deformation corresponds to the demolding direction, which results in overstretched structures with a characteristic reduction in cross section (Fig. 4.3(e)). Here the demolding temperature is an important factor because the yield point of the polymer and the demolding forces are a function of temperature.

- Damaged edges are also a typical failure occurring during demolding. Because of the difference in shrinkage, the contact stress between mold insert and the top of the microstructures increases during the last micrometers of the demolding path. At this state the yield point of the polymer can be achieved, which can result in deformation or, in the worst case, in damaging the top of the structures up to a separation of parts of the structure. This kind of failure is an enhancement of the failure of overdrawn edges, because the reason is similar (Fig. 4.3(f)).
- The separation of freestanding structures from the residual layer during demolding is the worst case of failures (Fig. 4.3(g)). This case can be interpreted as an increase of the case of overstretched structures. The induced stress is higher than the maximum stress at yield point, which results in a separation of the structures from the residual layer. Similar to the overstretched structures, filigree structures with small cross sections and high aspect ratios are periled. The risk of separation increases significantly if the sidewalls of the microstructured mold insert are characterized by undercuts (Chapter 6).
- Warpage of single structures refers to differences in shrinkage within the molded parts. Warpage can occur by local differences in cooling or at the intersection of structures with different thicknesses, for example filigree structures and the residual layer. Also, internal stress that is not yet relaxed can be a reason for warpage of single, especially freestanding, microstructures with a high aspect ratio (Fig. 4.3(h)).

4.4.2 Surface Quality of Molded Parts

The quality of the surface of a molded part corresponds obviously to the surface quality of the mold insert. Here the surface roughness and the evenness of the mold insert are the main characteristics. The surface of molded parts can also be influenced by sink marks as an effect of shrinkage (Section 4.4.3). Nevertheless, the best surface quality that can be achieved by replication is at best identical to the surface quality of the mold insert. If an exact replication of the microstructures is achieved, in general also the surface roughness of the mold insert is replicated to the molded structures. But one factor can increase the surface roughness: the demolding process. During demolding, for example, an area with higher surface roughness or a defect on the mold insert can increase the surface roughness of the molded part in a larger area by the relative motion between structures and mold insert during demolding. Here already a small defect can scratch the molded structures over the demolding distance, here the height of the structure. The surface roughness of mold inserts depends on the fabrication processes and can be achieved in a range between 10 nm and

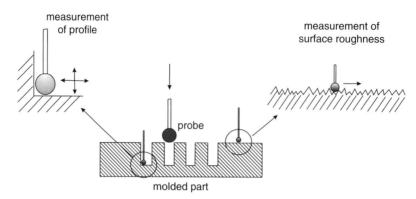

Figure 4.4 Limitation of tactile measurement for the determination of roughness and profiles in horizontal and vertical direction, e.g., small microcavities. The limitation results from the diameter of the probe compared to the profile of the surface roughness. To follow the contours of sharp corners the diameter of the probe is the limiting factor.

below for X-ray lithographic mold inserts, up to several hundred nanometers for the fabrication process of electro-discharge machining (Chapter 9). An increase of surface roughness, especially by mechanical fabrication processes, will also increase the risk of scratching the polymer sidewalls during demolding. For example, unavoidable grooves in the nano range, typical for the fabrication of mold inserts by milling, can increase the surface roughness if the direction of theses grooves is perpendicular to the demolding direction.

The characterization of surface roughness is well established for horizontal surfaces; for example, atomic force microscopes and tactile measurement systems are suitable. Nevertheless, by the replication of microstructures the surface roughness on the vertical sidewalls is at the forefront of interest. Here the established technologies can reach their limits, especially when the surface roughness in micro cavities with small cross sections should be determined.

The resolution of tactile measurement techniques is also determined by the size of the probe. Depending on the diameter of the probe, sidewalls in microcavities or molded grooves or holes can be characterized. Nevertheless, the diameter of the probe limits the resolution of the measurement (Fig. 4.4). The contours of the surface roughness cannot be measured exactly, because the probe size cannot follow the surface profile precisely. Here mostly average values are available, which can be sufficient. Compared to tactile systems, AFM measurements allow the measurement of surface roughness in higher resolution but only in a comparatively small area in the range of several hundred square micrometers. Nevertheless, this technology is in general not suitable in measuring the surface roughness on vertical sidewalls. Here there is a lack of suitable measurement technique that will allow the measurement of the roughness of vertical sidewalls in microcavities, for example in holes with a diameter below 20 μm.

4.4.3 Shrinkage and Warpage

The thermal behavior of thermoplastic unfilled polymers is like that of most materials characterized by thermal expansion. This thermal expansion is not constant over the temperature and it can be determined by the p-v-T-diagram of the polymer (Section 3.5.2). Semicrystalline polymers especially show a significant change of thermal expansion by temperature. Compared to typical mold inserts fabricated from metal, the thermal expansion of polymers is significantly higher than the thermal expansion of metals. This difference of thermal expansion especially results during cooling in high-contact stress between molded part and mold insert and finally in demolding forces. The influence of this difference in thermal expansion is therefore essential for every molding process, especially if the feature size decreases. For example, the molding of submicron structures with high aspect ratios can show overdrawn edges, a typical sign of the influence of shrinkage.

Shrinkage, defined as the relative difference of characteristic length between mold insert and molded part, considers besides thermal expansion as material property also the influence of the replication process, like the kind of replication process, the technology, and finally also the process parameters. Besides the geometry of the molded part, especially the temperature and the dwell pressure during cooling will influence the shrinkage of a molded part. Here it is important to obtain a homogeneous pressure distribution inside the molded part during cooling; otherwise differences in shrinkage can occur that will result in warpage of the molded part. The prediction of shrinkage and their anisotropy is part of process simulations in injection molding [5] and in injection compression molding [8]. Fan et al. [4] investigate the warpage of optical media like CDs fabricated by compression molding. Simulation and experimental results indicate that the processing conditions like melt and mold temperature and packing pressure have effects on warpage. Also gravity has a significant influence on the magnitude and curvature behavior of disc warpage.

4.4.3.1 Shrinkage of Embossed Parts

As mentioned above, replicated parts molded by hot embossing are characterized by the microstructures and the residual layer. Compared to the volume of the structures, the volume of the residual layer can be the characteristic element, determining the shrinkage behavior of the whole replicated part. This shrinkage behavior can be compared with the shrinkage of a circular disc, which will shrink during cooling in the direction of the center of the disk. These behaviors can be transferred also to rectangular residual layers, where typical lateral shrinkage is oriented toward the center of the mold (Fig. 4.5). To reduce shrinkage, therefore, the thickness and the area of the residual layer should be reduced to a necessary minimum, especially when filigree freestanding microstructures are being replicated.

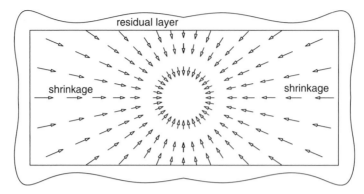

Figure 4.5 Direction of shrinkage of an embossed part. The lateral shrinkage of an embossed part is characterized by the residual layer and is directed to the center of the molded part.

How can the shrinkage be determined? Shrinkage can be mathematically defined as the difference between any length l_m on the mold insert compared to the same length on the replicated part l_{mp} related to the corresponding length on the mold insert.

$$S = \frac{l_m - l_{mp}}{l_m} \tag{4.2}$$

This definition is not exclusive in characterizing shrinkage. For example, the mold insert also changes the shape by thermal expansion and can be elastically deformed by the load during molding. Further, the time-dependent behavior of the polymer, the relaxation and retardation of the molded part, has to be taken into account. The sum of these influences on the shrinkage results in a schematic diagram characterizing the shrinkage of a molded part over time and defines different states of shrinkage (Fig. 4.6) [6].

These states of shrinkage can be split off into the shrinkage at demolding time, when the temperature of the molded part is at the demolding temperature, after cooling to room temperature, and finally after a defined time of several hours. Relaxation and retardation processes will further influence the shrinkage over a long time period, but these processes will not change the shrinkage significantly. Practically all measurements of shrinkage always refer to the mold insert at room temperature. Regarding the time- and temperature-dependent behavior of the shrinkage, it is recommended to measure the shrinkage under the same boundary conditions, for example after a certain time after demolding.

How can shrinkage be estimated before molding? This aspect becomes more influential when the structure size decreases, especially by the replication of microstructures, arranged on a large molding area. Here the accuracy of the lateral dimensions can be an important factor, for example in micro-optical

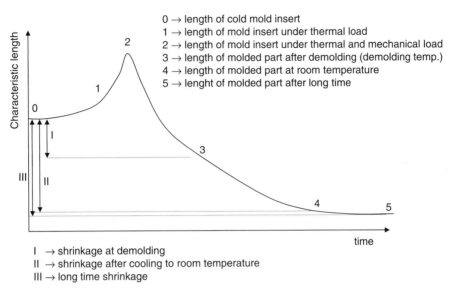

Figure 4.6 Different states (I, II, III) of shrinkage of a molded part. The shrinkage of a characteristic length is plot versus time under the consideration of the thermal and mechanical load during molding.

benches. The estimation of the shrinkage can be essential in correcting the design of the mold insert, including the effect of shrinkage of the molded part over a long time period after replication. As mentioned above, shrinkage will be influenced by the material property thermal expansion and the molding process. To estimate volume shrinkage the replication process can be linked to the material behavior of the polymer. For this, the process parameter, molding pressure, molding temperature, and demolding temperature can be entered in the p-v-T diagram of the polymer. Figure 4.7 shows the p-v-T diagram for PMMA, with the linked characteristic temperatures and pressures of a hot embossing cycle.

With this method the influence of the process parameters of molding temperature, molding pressure, and demolding temperature on the specific volume of the molded part can be estimated. This theoretical method has still some disadvantages for practical use. This method does not consider any designs and the linked pressure distribution inside a molded part. These kinds of diagrams are theoretically valid for only one point inside a replicated part. Further, the result refers to the difference in the specific volume, which is linked to the volume shrinkage. But at the forefront of interest is the shrinkage in lateral dimension. Therefore, the volume shrinkage should be transferred to the shrinkage in each direction, in lateral dimensions and height. This transfer is very difficult because of a lot of influence factors that are responsible for an anisotropy of shrinkage. In general, amorphous polymers show outside a mold an isotropic shrinkage behavior, but with pressure distribution during molding the shrinkage can be

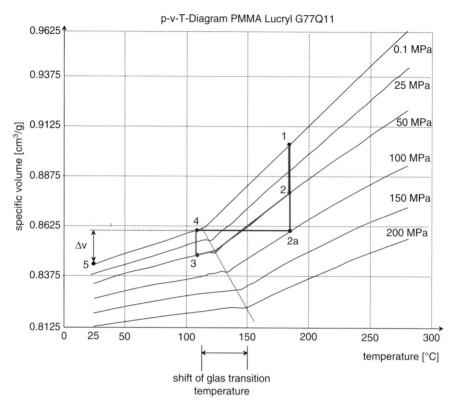

Figure 4.7 Determination of volume shrinkage under the use of the p-v-T diagram of a thermoplastic polymer (PMMA). The characteristic temperatures and pressures of the hot embossing process will be entered in the diagram and finally the differences of the specific volumes can be estimated.

influenced locally. Further, shrinkage is limited during cooling by the mold insert, which finally also can result in anisotropy.

Especially when semicrystalline polymers are molded the polymer itself shows anisotropic behavior because of the orientation of molecules during molding. This behavior is enhanced if fillers in the polymers are used. A further reason for anisotropic shrinkage is non-uniform cooling of the replicated part, in the case of hot embossing the residual layer as the carrier layer of microstructures. Depending on the cooling system, temperature differences of several degrees in the mold or substrate plate during the transient cooling state can result in different volume contraction of center regions or margin regions of a molded part. Because of the mechanical connection between these regions, an interaction between these areas will support the anisotropic behavior of the molded part.

These aspects show the difficulties previously in determining lateral shrinkage. Depending on the design, the polymer, the process, and the process parameters different shrinkages will occur. Therefore in literature mostly ranges of shrinkage for a polymer class will be given. The individual shrinkage can be determined only by measurement of characteristic length of the molded part. Nevertheless, with results from previous molded parts it is possible to estimate the shrinkage and correct the lateral dimensions of the microstructured designs in the mold insert.

Finally a coarse estimation of the shrinkage based on the thermal expansion without any adhesion of the molded part on the substrate plate can be given by the difference of thermal expansion coefficients:

$$\Delta l = l_0[(\alpha_{polymer}(T, p) - \alpha_{mold})\Delta T] \tag{4.3}$$

where l is the characteristic length and α is the thermal expansion coefficient. This equation does not consider any process-specific influence factors and illustrates only the difference in thermal expansion between mold and polymer.

4.4.3.2 Warpage of Embossed Parts

As described above, the reasons for the anisotropy of shrinkage are manifold. The measurement of anisotropy is difficult; here, the result of anisotropy of shrinkage, warpage, can be determined visually. Hot embossed parts are characterized by the microstructures and the residual layer, which makes it possible to split the effects of anisotropy into two different groups: (1) the warpage of the whole part, mainly characterized by the residual layer, and (2) local effects of warpage, characterized mainly by the design of microstructures. In the second case, for example, different thicknesses of connected structures or differences by heat transfer during cooling can effect local regions of warpage. Nevertheless, the first effect, the warpage of the residual layer, will influence the whole molded part. Here, the pressure distribution inside the part during cooling and differences in heat transfer, which result in non-uniform temperature distribution during cooling, will be responsible for warpage. Especially the typical parabolic pressure distribution of a squeeze flow (Section 6.4.2) will effect non-uniform shrinkage over the whole molded part, which also influences the microstructures on the residual layer. The result of this anisotropy is a concave or bowl-shaped warpage, which is an effect of higher shrinkage in the boundary regions of the molded disc compared to lower shrinkage in the center of the molded disc. This case of warpage is typical for embossed parts, especially for large areas. Opposite to this kind of warpage, the saddle-shaped warpage of the disc results from anisotropic shrinkage with higher shrinkage in the center of the molded part (Fig. 4.8). Especially in large-area hot embossing the effect of the bowl-shaped residual layer can be observed (Fig. 4.9).

The anisotropy in shrinkage can be reduced by some technical modifications that allow homogenization of the pressure during molding and eliminate the typical parabolic pressure distribution (see Section 9.5.3).

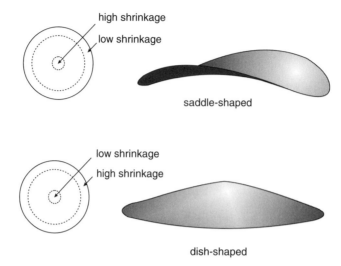

Figure 4.8 Typical warpage of embossed discs. In the first case, the bowl- or concave-shaped disc, the shrinkage in the center is below the shrinkage in the circular margin regions. The second case is opposite to the first case, a higher shrinkage in the center of the disc, which results in a saddle-shaped molded disk. These kinds of shrinkage can also influence the accuracy of the arrangement of the microstructures to each other.

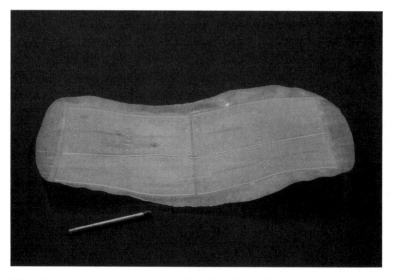

Figure 4.9 Bowl-shaped extreme warpage of an 8-inch molded part.

4.4.4 Stress Inside Molded Parts

Every molded part is characterized by stress. Depending on the molding process and the replicated design, this stress can vary over a wide range. Inside a molded part there exist, depending on the design, stress distribution with local concentration of stress especially at sharp corners. What are the reasons for this internal stress? During molding the polymer melt undergoes shear stress, which can be frozen by a short cooling time. In the case of short cooling times, below the relaxation time of the polymer at the molding temperature, the shear stress does not have enough time for relaxation, which results in frozen stress inside the part. For example, during injection molding high flow velocities and short cooling times are typical, which results in molded parts with high stress. Typical shear stress in the range of 10,000 s^{-1} and higher are common values. Compared to injection molding hot embossing is therefore characterized by low flow velocities and short flow distances, which results in low shear stress. Typical values are here in the range below 100 s^{-1}. Further, this moderate shear stress also has more time for relaxation because of the moderate cooling time in the range of minutes. Therefore, embossed parts are well suited for applications where low residual stress is needed, for example for optical lenses.

The distribution of the stress corresponds, besides the design, to the direction of the flow during mold filling. Hot embossed parts are characterized, after filling of the cavities, by a squeeze flow, a radial flow in the residual layer. Therefore, the direction of the inner stress of the residual layer is typically oriented in a radial direction (Fig. 4.10). A concentration of stress can be seen in the edges of the embossed part referring to sharp corners in the design and the radial shrinkage of the residual layer.

To see the differences in stress distribution between an embossed part and a part fabricated by injection molding, an example of a molded part is shown in Fig. 4.11. In this case of a flat device the stress is oriented to the flow behavior

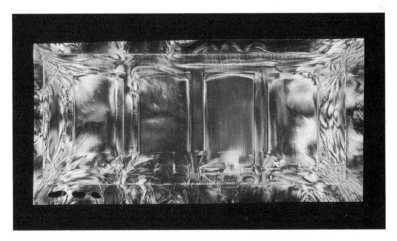

Figure 4.10 Residual stress of an embossed part with four microspectrometers, visualized by polarized light. The size of this example is about 30 × 70 mm^2; the thickness of the part is around 500 µm. Characteristic is a stress concentration at the edges, effected by the radial flow in the residual layer and shrinkage oriented toward the center.

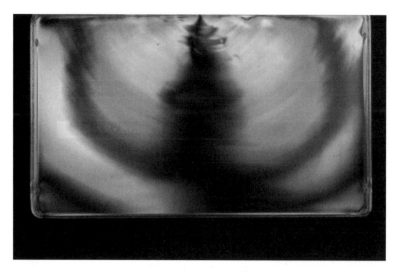

Figure 4.11 Residual stress of a part ($100 \times 70 \times 1$ mm^3) fabricated by injection molding. The stress corresponds to the flow behavior during mold filling, beginning from the injection point to the margin regions of the part.

during mold filling, here the flow from the injection point in the middle of the top side to the margin regions. Here, the analysis of stress will provide feedback about the filling behavior of the mold.

The internal stress can be further responsible for the deformation of the part, especially when the temperature of the molded part reaches values near the softening range of the polymer. With the influence of temperature the macromolecules of the polymer can slide against each other and will degrade the internal stress. This kind of retardation of the internal stress during heating results in a deformation of the shape. This effect can be demonstrated on a flat plate in the dimensions of a bank card with a thickness of 1.2 mm (Fig. 4.12). The injection point is in the middle of the large side. The card is heated up in the range of transition temperature with the effect that the internal stress deforms the card starting from the injection point.

4.4.5 Controlling Quality of Molded Parts

The aspects described above can be influenced by the process parameters and the technology of the tool and the machine. Splitting off the influence in these two groups the capability of influencing the quality of molded parts will be described in a general way. Depending on the design, the influence of each factor can vary.

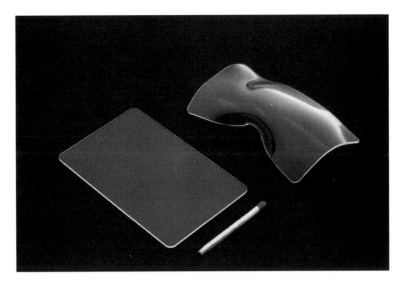

Figure 4.12 Deformation by retardation of a bank card fabricated by injection molding. The bank card is heated up to the softening range of the polymer (PMMA). The deformation corresponds to the internal stress caused by the frozen shear stress during molding.

4.4.5.1 Control of Quality by Process Parameters

- Shrinkage can be controlled by adequate dwell pressure during cooling. Here higher dwell pressure will reduce the shrinkage of the molded part. An estimate can be found by the use of the p-v-T diagram for the selected polymer.
- The cooling rate is also a factor that influences the shrinkage behavior, especially for semicrystalline polymers. A higher cooling rate increases the part of crystallites and results in a higher anisotropic shrinkage of the molded part.
- The thickness of the residual layer can be controlled in on sense by an adequate thickness of the polymer foil, but also by the press force, the molding temperature influencing the viscosity of the polymer melt, and the holding time. A high force at high temperature acting a long time will effect thin residual layers (see Section 6.1).
- Deformation of structures during demolding can be avoided by demolding near the glass transition temperature of the polymer (or melting temperature, if semicrystalline polymers are used). The difference in shrinkage between mold and polymer will effect higher contact stress if the demolding temperature decreases.

- To reduce internal stress the shear stress of the polymer during cavity filling should be minimized. This can be achieved by low molding velocities (for example, in the range of 1 mm/min).
- Internal stress can also be reduced by relaxation processes. Here, higher molding temperatures will effect shorter relaxation times, which can be decreased in the range of typical process times (Section 6.3.3).

4.4.5.2 Control of Quality by Molding Technology

The influence of the machine, the tool, and the mold on the quality of the molded part is complex and has to be individually specified. In general, some basic factors can be named.

- Warpage can be reduced by attaining a homogeneous pressure distribution during molding. To obtain a homogenization of pressure distribution in the residual layer, free-flow fronts should be avoided during the act of dwell pressure. The use of a simple frame around the mold insert, for example, can help to attain a homogeneous pressure and will further prevent an anisotropy of shrinkage (Section 9.5.3).
- To reduce warpage and deformation during demolding, the thickness of the residual layer should be as even as possible over the whole area. Therefore, even substrate plates and a minimum of bending of the tool under load will help to minimize non-homogeneous thickness distribution of the residual layer. Otherwise, bending of the crossbars and tools under load and uneven substrate plates result in a variation in thickness, typically in a bi-convex shape (Fig. 4.13).
- The bending of the machine and the tool under load can also result in a convex form of the residual layer. In extreme cases this shape of the residual layer can be one reason for higher demolding forces and deformation of structures. The influence of bending is illustrated in Fig. 4.14.

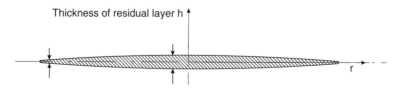

Figure 4.13 Typical shape of a residual layer as a result of bending or uneven substrate plates.

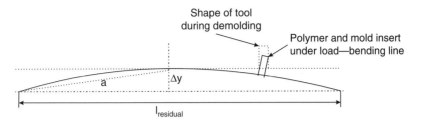

Figure 4.14 Principle of the deformation of structures in the case of bending of the microstructured mold. The molded part solidifies under the bending curve of the mold. During demolding an angle between mold and molded structure occurs, which can result in deformation of microstructures.

The cooling state is characterized by the act of the dwell pressure similar to the press force. During this state bending of the machine can deform the molded part and the thickness of the molded part along the bending line of the mold insert or substrate plate. If the temperature decreases, the polymer will solidify under this bending line. The convex shape of the molded part similar to the bending line of the mold will be frozen. During the following demolding step, the press force is not effective anymore and, because of the elasticity of the mold material, the shape of the mold will go back to the initial shape. In contrast, therefore, the solidified molded part retains the shape and this results in an angle between mold and structure during demolding. This angle can be one reason for deformation of structures during the demolding step. For a first approach the bending curve can be approximated by a straight line. The difference angle between mold and structure can be approximated by the bending of the mold and the length of the mold insert,

$$\tan \alpha = \Delta y / (l_{res}/2) \qquad (4.4)$$

where Δy is the bending of the mold and l_{res} is the length of the residual layer.

The influence of bending is not negligible. For example, for a mold insert with an area of 26 mm × 66 mm and a bending-under load of 30 μm, the angle is in the range of 0.5 degrees. A 100 μm high structure undergoes with this angle a shift of 0.9 μm, which results in additionally demolding forces (Section 6.4.4).

- Another influence factor is the homogeneous temperature distribution during heating and, more important, during cooling. During heating a non-homogeneous temperature distribution can cause a non-uniform melting of the polymer foil, which can result in a non-uniform flow of the polymer melt. A non-uniform viscosity of the polymer can cause problems during filling of microcavities and can be the reason for variations in thickness of the residual layer. During cooling a non-homogeneous temperature distribution becomes more important because of the risk of non-uniform cooling of the molded part. Non-uniform cooling can result in anisotropy of shrinkage, which is one reason for warping of a molded part. The technology of heating and cooling is therefore an important aspect in the development of molding tools (Chapter 8).

References

1. U. Brand and S. Büttgenbach. Dimensional metrology for components of microsystem technology. *Technisches Messen*, 69:542–549, 2002.
2. H.-Y. Lin, L.-S. Chen, W.-H. Wang, C.-H. Hsueh, and S. Lee. Analysis of mechanical failure in nanoimprint processes. *Materials Science and Engineering A*, 433:316–322, 2006.
3. G. Dai, F. Pohlenz, M. Xu, L. Koenders, H.-U. Danzebrink, and G. Wilkening. Accurate and traceable measurement of nano- and microstructures. *Measurement Science and Technology*, 17:545–552, 2006.
4. B. Fan, D. O. Kazmer, W. C. Bushko, R. P. Theriault, and A. J. Poslinski. Warpage prediction of optical media. *Journal of Polymer Science: Part B: Polymer Physics*, 41:859–872, 2003.
5. K. K. Kabanemi, H. Vaillancourt, H. Wang, and G. Salloum. Residual stresses, shrinkage, and warpage of complex injection molded products: numerical simulation and experimental validation. *Polymer Engineering and Science*, 38(1):21–37, 1998.
6. G. Menges, W. Michaeli, and P. Mohren. *Spritzgiesswerkzeuge*. Hanser Verlag, 1999.
7. A. Weckermann and R. Ernst. Future requirements on micro- and nanomeasurement technique—challenges and approaches. *Technisches Messen*, 67:334–342, 2000.
8. W. B. Young. On the residual stress and shrinkage in injection compression molding. *International Polymer Processing*, 18(3):313–320, 2003.

5 Hot Embossing Process

This chapter is intended to provide the reader with background information about the process steps and the related molding parameters. For this purpose, the components needed for hot embossing will be defined first. Based on the components, the process of one-sided molding will be outlined in principle. Each process step will be explained in detail, and all necessary definitions will be given. Furthermore, the molding parameters like process parameters, material parameters, and influencing factors will be presented. Their relevance to the quality of the process and the molded parts will be discussed. Finally, the manifold of the hot embossing process will be described by the presentation of different process variations.

5.1 Hot Embossing Principles

Three different forming principles of embossing can be distinguished (Fig. 5.1):

- Cylinder-to-cylinder embossing: This principle of embossing is characterized by two rotating cylinders with a polymer film in between. One of the cylinders is microstructured, with the contact area between the polymer and the cylinder being only a line. The advantage of this principle is a lower press force compared to flat-to-flat area embossing. Cylinder-to-cylinder embossing also allows for a high cylinder rotation speed and is therefore suited for the series production of large batches. The principle of rotating cylinders, however, may only be applied to produce microstructures with low aspect ratios and a low height, in the range of a few μm or in the nm range. As a result of the rotation principle, demolding of the embossed structures is not normal to the structures, which causes deformation of the structures. This deformation is a function of the height of the structures and the diameters of the cylinders. For molding microstructures with high aspect ratios and heights of several hundred micrometers, the diameter of the cylinders increases, which leads to problems of manufacturing. The fabrication of such microstructured cylinders and the necessary molding machine are only profitable for large series of microstructures with a very low height.
- Cylinder against a flat area: This principle is characterized by a rotating cylinder and a flat area, where the cylinder can unwind. It corresponds to the principle above, with the difference that the mold can be integrated in the plane area. The cylinder rolls along

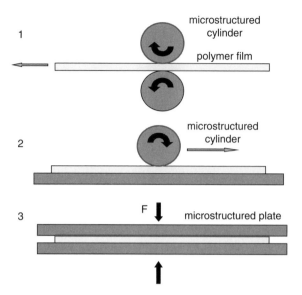

Figure 5.1 Three different hot embossing principles: (1) rotating cylinder against another rotating cylinder—roll-to-roll embossing; (2) rotation cylinder against a flat area—roller embossing; (3) flat area against another flat area.

the microstructured mold, and on the contact line, embossing of the structures takes place. In this case, the molding window should also be in the glass-transition range of the polymer foil. A disadvantage is the lack of a demolding system, which may lead to the damage of the structures during manual demolding. Consequently, this principle is only suitable for microstructures with low aspect ratios and low heights.

- Flat area against another flat area: This principle is characterized by an embossing of two flat areas, with a microstructured mold being integrated in one or both areas. Between both plates, a polymer film is positioned. As an advantage, the press force over the whole microstructured area allows the use of polymer melts, a prerequisite for molding high microstructures with high aspect ratios. This principle allows the switch from thermoforming, which uses the temperature range of glass transition, to molding a polymer in the temperature range of the melt. Another important point is the demolding of microstructures. In this case, structures can be demolded by a motion parallel to the structures. Hence, the risk of damaging the structures is lower than for the other principles. Due to these advantages and the fact that microstructured molds are typically produced on flat areas (e.g., wafers), this is an established principle of micro hot embossing technology.

5.2 Components for Hot Embossing

Independent from the manifold of the embossing processes described above, hot embossing is mainly based on the thermal forming of thermoplastic polymers by the use of an embossing technique. The details of the technique depend on the hot embossing machine and may vary. In general, however, the following components are essential for hot embossing (Fig. 5.2).

- a microstructured mold or mold insert that has to be replicated
- the tool to integrate the microstructured mold. This tool consists of the heating and cooling unit and, if needed, an alignment system for double-sided or positioned molding.
- the molding machine, in which the tool will be integrated. The machine consists of the press unit with a frame of high stiffness. This unit has to generate the molding force and the relative movement between mold and polymer.
- a vacuum chamber, in which the tool and the mold can be integrated

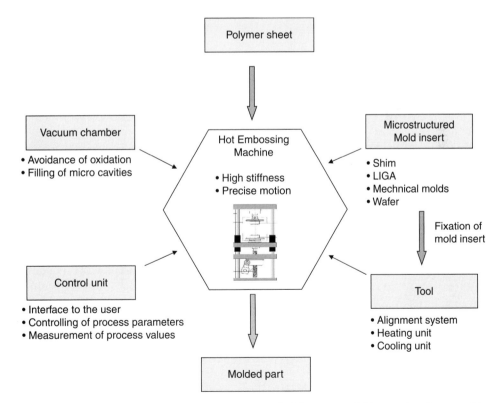

Figure 5.2 Schematic view of the components needed for successful hot embossing: machine, control unit, tool, mold insert, and a polymer film or sheet.

- the control unit with an interface to the user
- a measurement system documenting the process parameters for retracing the variations of process parameters
- in the case of one-sided molding, a rough plate with high adhesion, i.e., called the substrate plate. This plate is located opposite the mold.
- the molding material, in general, a semifinished product in the form of a thin film of thermoplastic polymer

5.3 Process Steps

The hot embossing process is divided into four major steps:

1. heating of the semifinished product to molding temperature
2. isothermal molding by embossing (displacement-controlled and force-controlled)
3. cooling of the molded part to demolding temperature, with the force being maintained
4. demolding of the component by opening the tool

The principle of a one-sided hot embossing cycle is represented schematically in Fig. 5.3. Between the tool and substrate a semifinished product, that is, a polymer foil, is positioned. The thickness of the foil exceeds the structural height of the tool. The surface area of the foil covers the structured part of the tool. The tool and substrate are heated up under vacuum to the polymer molding temperature. When the constant molding temperature is reached, the two steps of the molding cycle are initiated. In the first step, the tool and substrate are moved toward each other (in the range of 1 mm/min) until the pre-set maximum embossing force is achieved. In the second step, the embossing force achieved is held at a constant value for a defined holding time. To generate a constant force, the relative movement between the tool and substrate has to be controlled. The force is now kept constant over an additional time period (packing time, holding time). During this period, the plastic material flows in a radial direction and the residual layer is reduced under the act of constant force (packing pressure). At the same time, the tool and substrate move closer to each other, while the thickness of the residual layer decreases with packing time. During this molding process, temperature remains constant. This isothermal embossing under vacuum is required to fill the cavities of the tool completely. Air inclusions or cooling during mold filling may result in an incomplete molding of the microstructures, in particular at high aspect ratios. Upon the expiry of the packing time, cooling of the tool and substrate starts while the embossing force is maintained. Cooling is continued until the temperature of the molded part drops below the glass-transition temperature or melting point of the plastic. When the demolding

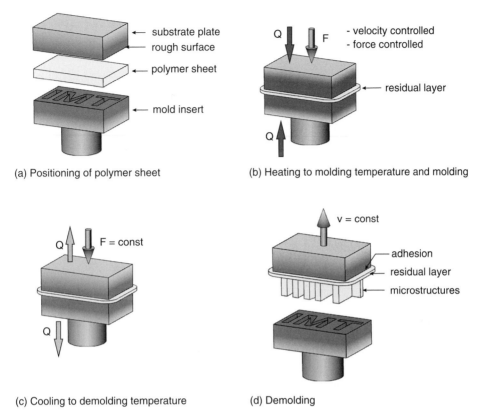

(a) Positioning of polymer sheet   (b) Heating to molding temperature and molding

(c) Cooling to demolding temperature   (d) Demolding

Figure 5.3 The hot embossing process: heating, molding, and demolding are the characteristic process steps. The hot embossing process is characterized by a residual layer that allows the vertical demolding because of an adhesion on the rough substrate plate.

temperature of the polymer is reached, the molded part is demolded from the tool by the relative movement between tool and substrate.

Demolding only works in connection with an increased adhesion of the molded part to the substrate plate. Due to this adhesion, the demolding movement is transferred homogeneously and vertically to the molded part. Demolding is the most critical process step of hot embossing. Depending on the selected process parameters and the quality of the tool, demolding forces may vary by several factors. In extreme cases, demolding is no longer possible: the structures are destroyed during demolding.

Apart from the one-sided molding described, the process is also used for double-sided positioned embossing. The principle of the process remains the same. Instead of the substrate, however, another tool is applied. To demold the molded part from one of the tool halves, special demolding mechanisms, such as ejector pins or pressurized-air demolding, are used. For a better understanding, the schematic representation of embossing in Fig. 5.3 is limited to the major process steps. Depending on the tool and polymer, the process and process parameters have to be adapted.

5.4 Molding Parameters

Molding parameters define all parameters and influencing factors that control the process or affect the quality of the molded part. In total, six groups of process parameters and factors that influence hot embossing can be distinguished by definition (Fig. 5.4). Together, they influence the quality of the microstructured molded parts.

Process parameters comprise those parameters that control the conduct of the hot embossing process. They include the molding temperature, embossing rate, embossing force, packing time, demolding temperature, and demolding rate. Another group consists of the parameters of the molding material. They largely determine the process parameters and the processing windows. In this chapter the aspects will only be named; the details are discussed in the chapter about the hot embossing technique and molding tools.

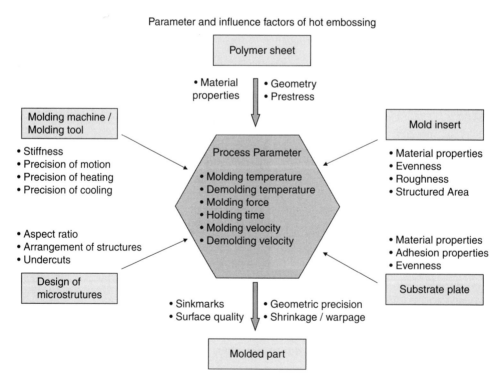

Figure 5.4 Schematic view of the process parameters and factors influencing the quality of the molded parts.

The six process and material parameters are complemented by a number of influencing factors. Four categories of influencing factors can be distinguished:

1. Tool: The evenness of the tool is of particular importance to ensure a homogeneous thickness of the molded part over the complete embossing surface. Depending on the type of tool patterning, surface quality may vary, which determines the forces needed for demolding.
2. Substrate: Like the tool, the substrate plate has to be extremely even over the complete surface area, also to ensure a homogeneous thickness of the molded part. The adhesion properties (e.g., fabricated by sand blasting or lapping of the surface) of the plate should not influence the evenness.
3. Machine: Influences of hot embossing machines are manifold. Heat- and force-induced strains and distortions, positioning accuracy of traverses, and the precision of control have to be taken into account.
4. Microstructures: The geometry, set-up, and surface roughness of microstructures on the mold insert influence the demolding behavior. In particular, the steepness of the sidewalls of the microstructures and the aspect ratios are decisive criteria of demoldability.

The parameters of the hot embossing process can be deduced for all process steps. Parameters of hot embossing comprise process parameters (controlled by the users), material parameters (physical aspects), and influencing factors mainly from the embossing machines.

5.4.1 Process Parameters

With regard to the machine technology, the following process parameters, controlled by the user, define a hot embossing cycle:

- molding temperature—heating rate
- touch force and molding force, molding velocity
- molding velocity
- holding time
- cooling time, cooling rate
- demolding temperature, demolding velocity

A general hot embossing cycle cannot be defined because of the large variety of process variations, their individual adaptation to the requirements, and the mold designs. Therefore for each process the values of each parameter have to be set up individually, depending on the design of the structures, the molding material, and the molding area. Systematic experiments help to determine the optimal parameters for the individual designs [9,11,23,35]. The time-dependent behavior of the process parameters explained below is characteristic of a typical hot embossing cycle (based on the embossing machine Jenoptik Mikrotechnik HEX03).

Figures 5.5 and 5.6 refer to the same molding experiment. Because of the different scaling required for a precise presentation of the measurement data, the data are split into two corresponding diagrams. Based on these diagrams, the process can be explained in detail:

- After placing the polymer foil onto the substrate plate, the vacuum chamber is closed first by the movement of the lower crossbar with constant velocity. The reference position is then reached. The mold and the polymer foil have no contact; there is a gap of a few millimeters.
- After closing the vacuum chamber, the evacuation of the chamber starts. The evacuation results in a negative peak in the diagram's force curve. When a selected pressure is reached (typical values are in the range of 1 mbar), the force is set to zero. The system is referenced. The evacuation is necessary to fill the microcavities without air bubbles and exhaust gas from the melted polymers.
- An advantageous step that is not necessary for every design is to apply a touch force between mold insert, polymer foil, and substrate plate. For this purpose, the lower crossbar moves with an adequate velocity in the range of 1 mm/min toward the mold insert until a selected contact force is reached. The contact force should be in the range of several 100 N only. In most cases, a contact force of

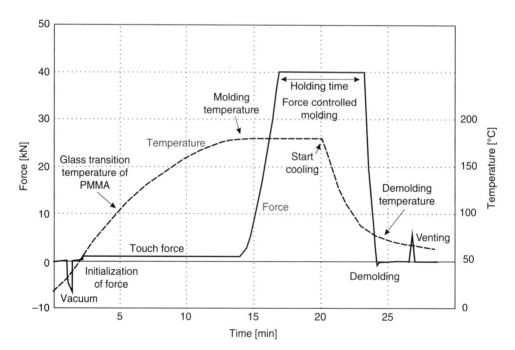

Figure 5.5 Typical behavior of the process parameters force and molding temperature of a hot embossing cycle. The behavior is exemplary and has to be adapted to every mold design.

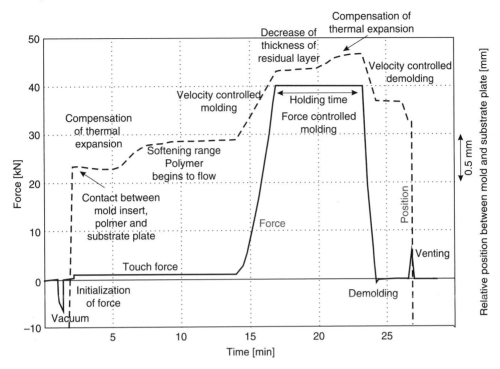

Figure 5.6 Detailed view of the relative position of microstructured mold insert and substrate plate with the polymer in between.

100 N or 200 N is sufficient. The advantage of the touch force is the fixation of the thin polymer foil in the microstructured area and the reduced delay in heating of the polymer foils. Another advantage is the balance of heat flow from the two integrated systems. As a result of the contact, an optimized heat transfer can be achieved between mold and substrate plate, which leads to a homogeneous temperature distribution in the polymer foil.

- While the touch force is acting, the heating of the mold and substrate plate begins. The temperature curve of the mold and substrate plate depends on the quality of the heating system, heat conduction of the materials, the amount and quality of heat transfer between the heating system and mold, and the heat capacity of the mold and tool. The time needed for reaching the desired molding temperature depends on these factors. Characteristic of many hot embossing machines are relatively long heating times compared to the embossing step.

- While the temperature rises to molding temperature, the polymer foils undergo changes in their states of aggregation, which are reflected by the motion of the traverse when the touch force is maintained at a constant value. In the first temperature range below the glass-transition range of the polymer, the motion of the crossbar is characterized by the thermal expansion of the mold substrate plate's heating and cooling unit. To maintain a constant touch force, the crossbar moves in the opposite direction to the mold. The behavior changes when the glass-transition temperature of the polymer is reached. At this temperature, the polymers begin to soften and the direction of movement of the crossbar changes toward the mold. With increasing temperature, the viscosity of the polymer decreases and the motion of the crossbar toward the mold continues.

- When the selected molding temperature is reached, the molding step starts with a continuous movement of the crossbar. A typical velocity is in the range of approximately 1 mm/min. The substrate and the polymer melt are pressed together until the selected force is obtained. The increase of the press force during motion with constant velocity is exponential. This behavior can be explained by an increase in flow through a decreasing gap. At constant velocity, the displaced polymer increases in every time step. In parallel, the gap, through which the displaced melt has to flow, decreases. This behavior is the reason a further increase of press force results in a small effect on the thickness of the residual layer only. The process switches from a velocity-controlled behavior to a force-controlled behavior, with the force defined by the user being measured. This force is kept at a constant value for a time that is also defined by the user. During this holding time, the polymer creeps under constant load, which results in a decrease in the residual layer and in the motion of the crossbar toward the mold. The holding time depends on the design and the desired thickness of the residual layer and can be set in the range of a few seconds up to a few minutes.

- After the holding time, the cooling cycle starts while the press force is maintained. The temperature decreases. When the demolding temperature is reached, the motion of the crossbar initiates the demolding step.

- The demolding step of a one-sided molding process uses the adhesion of the residual layer to the substrate plate, which should be higher than the adhesion of the microstructures in the mold. Demolding is initiated by a constant motion of the crossbar with the substrate plate. The velocity is very slow and set in the range of 1 mm/min. The corresponding force is reduced to negative values, representing the tensile forces. The point where the maximum negative value is reached represents the point of adhesion corresponding to the static-friction force. The microstructures have to transfer these forces via the residual layer to the substrate plate. At this point, the risk of damaging the molded structures is very high. After this break-away torque at the maximum negative

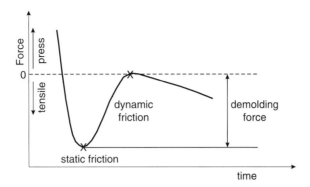

Figure 5.7 Characteristic shape of a measured demolding force. The minimal value characterizes the integrated static-friction force. This point is called break-away torque. After this critical point is passed, the friction force decreases because of the effect of sliding with a decreasing contact area and decreasing normal force.

value, the force increases and represents the sliding force of the microstructures along the sidewalls of the mold (Fig. 5.7). After demolding is finished, the force value corresponds to that at the reference point with the evacuated vacuum chamber.

- After demolding, the temperature decreases further to a value of approximately 80°C. This cooling is reasonable for avoiding oxidation when the chamber is opened at higher temperatures. To accelerate cooling, the vacuum chamber is vented with nitrogen, which results in an increase in and a peak of the measured force. Venting is also necessary to reduce the force needed to open the vacuum chamber.

5.4.2 Material Parameters

Besides the process parameters named above, knowledge of the material parameters and material properties is essential for successful molding.

- The process parameters, like molding temperature or demolding temperature, refer to the thermal properties of the molding material. For amorphous polymers, the glass-transition temperature, and for semicrystalline polymers, the melting temperature, are fundamental.
- Thermal contraction of the material is one of the reasons for demolding forces.
- The shrinkage and the warpage of a molded part are characterized by isotropic or anisotropic material behavior. Amorphous polymers are typically characterized by isotropic shrinkage, whereas semicrystalline polymers show anisotropic behavior. Therefore, the orientation of the crystallite structures determines the shrinkage and warpage of molded parts.

- The geometry of the polymer foil pertains to its thickness, area, and position relative to the mold.
- Pre-stress of the polymer foils caused by its manufacturing process may result in a change of the position of the polymer foils relative to the microstructured mold during heating. Therefore, the retardation of the polymer foil with increasing temperature results in this change of the position.

5.4.3 Influencing Factors

With the process parameters and the material parameters, an ideal hot embossing process can be described. In practice, the ideal process is approached by an embossing technology. This technology influences the ideal process and the molded parts. To describe these influencing factors, a classification into three groups is reasonable: influence of the machine, influence of the molding tool, and influence of the mold or mold inserts. These factors may influence the molding process and the quality of the molded part much more than will the process and material parameters. Consequently, these factors should not be neglected and will be described in detail in the following section.

5.4.3.1 Influence of the Hot Embossing Machine

The influence of the hot embossing machine depends on the technology used to apply the molding force. In general, the following aspects have to be taken into account:

- The stiffness of the whole construction has to be as high as possible. Especially the crossbars should be stiff enough to prevent bending under the typical embossing loads. Elastic bending of the whole system, including tool mold and substrate plate, may result in the damage of microstructures during demolding (Section 4.4.5).
- The motion of the crossbars of a hot embossing machine has to be parallel. The accuracy of the motion of the machine is mostly responsible for a parallel motion of the mold and substrate plate. A lack of parallelism results in a gradient in the residual layer and may be the reason structures get damaged during demolding. Parallelism is also required to obtain a homogeneously thin residual layer.
- The range of velocity and its reproducibility is an important factor in the molding and demolding step. The velocity has to be set in the range of 1 mm/min with high accuracy. The acceleration at the beginning of the motion has to be soft. Otherwise, the risk of damaging structures increases.

5.4.3.2 Influence of the Molding Tool

The influence of the molding tool with regard to heating and cooling and the integration of mold inserts may be described as follows:

- The accuracy of a homogeneous temperature distribution in the mold and substrate plate is essential for successful molding, especially in large microstructured areas. Hence, the heating system has to guarantee a homogeneous and stable molding and demolding temperature over the whole area. Nonuniform temperatures may result in differences of viscosity during molding and an inhomogeneous solidification during cooling.
- The gradient of heating and cooling influences the material behavior. Especially the crystallization of semicrystalline polymers may be influenced by the gradient of cooling. This gradient influences the isotropic behavior of the molded structures.
- A parallel and even integration of the mold insert with efficient heat conduction and heat transfer is a precondition for the desired parallel molding and demolding at a homogeneous temperature distribution.

5.4.3.3 Influence of the Mold Insert

Apart from the influencing factors of the embossing machine and the tool, the influence of the microstructured mold insert is essential for the quality of the molded part. (More detail is given on this in Chapter 4.)

- Most effective are undercuts of microcavities. These undercuts result in high demolding forces and increase the risk of damage to structures during demolding.
- The roughness of the mold surface is another factor responsible for high friction and demolding forces. It may cause the risk of damaging the structures to increase during demolding.
- The use of mold coatings and release agents can reduce friction, thereby reducing the risk of damaging the structures during demolding.
- The flatness of the mold is essential for molding structures on a thin residual layer.

5.5 Elementary Process Variations

As mentioned above, hot embossing is a very flexible technique, characterized by a diversity of process variations.

5.5.1 Position-Controlled Molding

The integration of an alignment system allows the adjustment of the substrate plate relative to the mold in a range of several micrometers. This permits the replication of structures in a precise position, for example, on prestructured substrates (Fig. 5.8).

5.5.2 Double-Sided Molding

Double-sided molding paves the way to a further large family of molded parts. For example, double-sided structured parts belong to microfluidic systems in which through-holes are essential features, although fraction lines with thin residual layers can be designed to separate molded parts. In general, with double-sided molding three-dimensional structuring can be enabled. Here a precondition for such designs is a correlation of two mold inserts that are positioned opposite each other and finally adjusted to each other. Depending on the design and the function of the structures, these adjustments should have a high precision down to several micrometers. For example, the molding of cone-shaped nozzles requires an accuracy much lower than the diameter of the cone (Fig. 5.9).

How does the alignment proceed? Here the use of markers or the orientation on characteristic structures of the design is recommended. After a coarse adjustment of both mold inserts, a first part should be molded. In a second step the differences in lateral dimension between the markers on the top side and the bottom side of the molded part have to be measured. Here, the transparency of most polymers is helpful, because of the determination of the misalignment of both

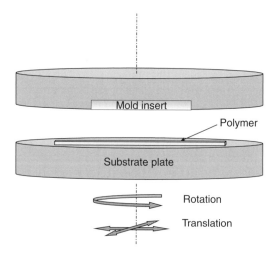

Figure 5.8 Position-controlled molding. The substrate plate can be aligned by a shift in two lateral directions and by a rotation around the center axis of the substrate plate.

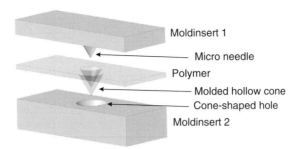

Figure 5.9 Cone-shaped nozzles as an example for double-sided molded structures. The molding of such nozzles requires, for example, an alignment below the diameter of the nozzle (Section 10.3).

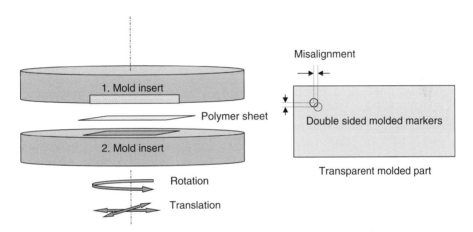

Figure 5.10 Principle of an alignment for double-sided molding. The alignment can be split into two transversal and one rotative direction. To adjust the position of one mold insert, the misalignment of both mold inserts has to be determined by the misalignment of several molded markers that are part of both mold inserts. Because of the transparency of the molded parts, the misalignment can be visualized and measured by measurement microscopy. The values obtained from the measurement can be used to correct the misalignment.

markers with an optical measurement system, focusing on the top and bottom of thin molded parts. The misalignment can be split into a transversal (x, y) direction and rotative (in the perpendicular axis of the part) misalignment. With this measured data the alignment of both mold inserts can be corrected by an alignment system integrated in a hot embossing tool (A detailed description of the technology of an alignment system can be found in Chapter 8). Here the alignment can be done separately in a transversal and rotative direction (Fig. 5.10). These alignment systems are optional components of commercial hot embossing machines (for example, Jenoptik Hex03) and a precondition for double-sided molding.

5.5.3 Molding of Through-Holes

In microsystem technology the need for molded parts with through-holes increases. Microfluidic components like micro pumps, micro valves, or fluidic components for lab-on-a-chip systems are especially common examples. The fabrication of through-holes is therefore an actual aspect in the fabrication of microsystem components. To reduce post-processing steps, through-holes should be fabricated during molding. Here, because of the molding principle, this requirement is difficult to achieve. As mentioned above, hot embossing is characterized by a squeeze flow of a polymer melt. Typical molded parts are therefore characterized by a residual layer. On the one hand, this residual layer is necessary to obtain the pressure for filling microcavities. On the other hand, for the fabrication of parts with through-holes, these residual layers should be completely displaced. The methods for fabricating these holes can be split into the post-processing methods to remove thin residual layers and methods that are characterized by selected substrate plates, allowing the fabrication of through-holes during molding. Finally a new development shall be discussed, a principle that allows the fabrication of through-holes by a two-step process, a molding cycle, and a cutting cycle, which gives its name to this process—hot cutting.

5.5.3.1 Post-processing Methods

A concept to realize through-holes is to optimize the molding process under the aspect of obtaining thin residual layers (Section 6.4.2). In a second process step, here post-processing, this residual layer can be removed by further processes.

Reactive Ion Etching Reactive ion etching is an established process in microsystem technology. This process can be used for structuring substrates through an etching mask but also for the displacement of residual layers of molded parts. This replacement can be done, on the one hand, from the top via an etching mask, and on the other hand, from the back side of the molded part. These processes have been adapted to remove residual layers from molded parts [21]. The adaptation of this technique for etching from the back side has also been used [22]. Nevertheless, this method is time-consuming because of the moderate etching rates. Therefore, the residual layer has to be molded as thin as possible.

Controlled Fracture of Residual Layer An alternative to the etching processes is the mechanical fracture of prestructured holes (Fig. 5.11). In this case, the bottom of molded holes has to be fractured manually, for example, by a needle. To achieve a nearly controlled fraction of the residual layer in molded holes, so-called predetermined breaking points are fabricated.

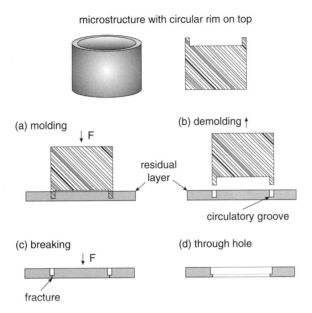

microstructure with circular rim on top

(a) molding

(b) demolding

residual layer

circulatory groove

(c) breaking

(d) through hole

fracture

Figure 5.11 Schematic view of the molding principle for through-holes fabricated by controlled fraction of the residual layer. The microstructures on a mold insert are structured with an additional circulatory rim on top, replicating a circulatory groove in the residual layer. The enclosed residual layer can be removed by controlled fraction.

The fracture refers to a kind of notch that creates high stress at the residual layer around a hole that makes it easier to remove the layer only in the cross-section of the hole. To obtain such predetermined breaking points, the top of the microstructures on the mold inserts has to be modified, for example, by the integration of a circulatory edge. This can be realized by mechanical machining, but it will cause problems by lithographic manufactured mold inserts (Fig. 5.12). Nevertheless, besides the further expense of structuring the mold inserts by a post-process, the holes are characterized by ridges that can reduce the quality of the holes fabricated by this method. And finally a second process step is needed that cannot be automated for every design. So this method is time-consuming.

An alternative method for molding through-holes is double-sided hot embossing with two interlocking mold inserts and a polymer film in between. Here the principle of a shear can be achieved by corresponding circulatory rims on the top of the structures on both mold inserts. For the principle described above, only a common substrate plate is required. In this case, the corresponding second mold insert has to be precisely aligned to the first mold insert to achieve an interlocking of both structures. To achieve through-holes the structures of the second mold insert are cone-shaped cylinders corresponding to cylindrical structures with a rim on top on the first mold insert (Fig. 5.13). With further modifications this principle can be used by the concept of "hot punching" (Section 5.5.5).

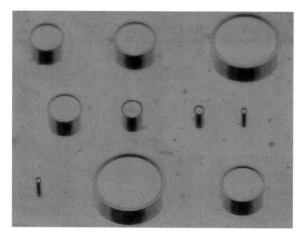

Figure 5.12 Mold insert for molding through-holes. The microstructures of the mold insert are structured with a circulatory rim on top.

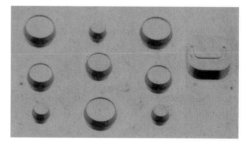

(a) Mold insert with microstructures with circulatory rims

(b) Corresponding mold insert with cone shaped structures

Figure 5.13 Mold inserts for molding through-holes by double-sided aligned hot embossing. The molding of holes refers to an interlocking of the top sides of the structures with a polymer film in between.

5.5.3.2 Molding on Selected Substrates

To avoid any post-processing of molded parts, through-holes should be fabricated during the embossing cycle. This can be achieved by the use of a combination of modified mold inserts and selected substrates [19]. The principle refers to a complete displacement of the residual layer in selected areas, achieved by embossing freestanding structures of mold inserts into modified substrates. This enhanced molding principle requires a sensitive set-up of the process parameters and a proper selection of the substrate materials, because of the relatively high load on the structures during molding. Here the risk of damaging filigree structures of the mold insert increases, especially when the diameter of the through-holes decreases.

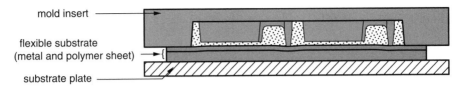

mold insert

flexible substrate
(metal and polymer sheet)

substrate plate

Figure 5.14 Schematic view of a configuration for the molding of through-holes under the use of a stack of foils.

Molding onto a Stack of Foils   By this approach, additional to the conventional metal substrate plate, a flexible layer is put on top. This layer consists typically of a polymer layer and a metal film. The flexibility of this combination can be influenced by the selection of the materials and film thickness. Figure 5.14 illustrates this principle. Because of the flexibility of the layer in selected areas, the polymer melt can be displaced completely from the top side of the mold insert. By this technique it is possible to keep the residual layer in large contact surfaces, for example, at the margin regions of the mold insert, and to displace the residual layer in small contact areas completely. During the cooling state the desired dwell pressure is generated by the flexible layer and not by the residual layer.

An advantage of these constructions is that the top side of the microstructures remains on the same level and no graded structures with two different levels are required. Therefore, for mold fabrication, all the established lithographic processes can be used. The disadvantage of the flexible layer is the deformation of the flexible layer under the load during molding, which disables the molding of parallel and even molded parts. Typical for replicated parts are, in this case, banked structures corresponding to the top-side structures of the mold insert. Figure 5.15(a) shows a cross-grinding of a molded example with cone-shaped through-holes. The dotted line represents the ideal parallel level of the replicated part. The top side of the molded nozzle shows the characteristic banked shape. The level of this kind of deformation corresponds to the flexibility of the polymer metal layer, the thickness of the layer, and the process parameter, especially the molding force. A more flexible layer or an increase in thickness results in an increase of the deformation. The use of metal foils with higher thickness compensates for the deformation but requires an increase in molding force and increases the risk of damaging structures on the mold insert. Therefore, for each design the optimal configuration of metal foil and polymer layer has to be determined by experiments. Figure 5.15(b) shows a molded through-hole embossed on a stack of steel (25 µm) and PTFE (250 µm).

Nevertheless, this approach is characterized by two main problems. First is the deformation of the flexible layer and the metal foil during the molding process. The mold inserts leaves markers on the metal foil after successful molding of through-holes, which makes it impossible to use the metal foil twice. Also, the polymer layer (e.g., PTFE or PVDF) tends to deform under high load by flow

(a) Cross-grinding   (b) Molded through hole

Figure 5.15 Cross-grinding of a molded example with cone-shaped through-holes. The dotted line represents the ideal parallel level of the replicated part. The top side of the molded nozzle shows the characteristic banked shape. The through-hole was embossed on a stack of steel (25 µm) and PTFE (250 µm) [19].

processes. The second problem is in guaranteeing the demolding by adhesion of the residual layer on the substrate plate. Because of the missing adhesion of the typically thin residual layer on the metal foil, the replicated part has to be demolded manually, which can damage the structures. Here the design of the mold has to be changed (e.g., the integration of sinks to achieve a thicker residual layer in the margin regions) to optimize the handling of the molded parts.

Molding onto Soft Metal Substrates   Another approach to mold through-holes is the use of a selected material combination of a hard mold insert and a comparatively soft metal substrate plate with a polymer foil in between. Ideal material combinations are characterized by a penetration of the top side of the mold insert into the substrate plate without damaging the structures of the mold insert. The deformation of the substrate plate has to be limited to the penetration of the top side of the mold insert; further deformations, especially around the penetration area, have to be prevented. Figure 5.16 illustrates the principle of this approach.

To impress the top side of the structures on the mold insert into the substrate plate, it is necessary that these structures on the mold insert stick up. These structures should be characterized by high stiffness to avoid any damage during impressing into the substrate plate. Further, the evenness of the mold insert should be as high as possible to avoid unequal impressing of single structures, which is caused by a high load on single structures. If this approach is used, the process parameter temperature and the thickness of the polymer film have to be optimized such that a minimum of molding force is required. During the first molding cycle the salient areas create an impression in the substrate plate (Fig. 5.17). In the following cycles the salient areas hit the already impressed areas in the substrate plate. Therefore, the precise guiding of mold

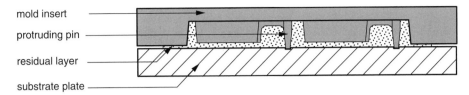

mold insert

protruding pin

residual layer

substrate plate

Figure 5.16 Schematic view of a configuration for the molding of through-holes under the use of different hardnesses of mold insert and substrate plate.

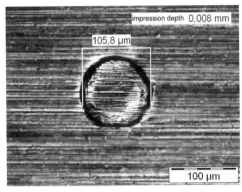

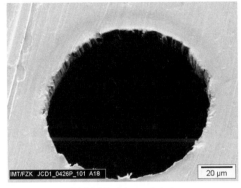

impression depth 0,008 mm

105,8 µm

100 µm

IMT/FZK JCD1_0426P_101 A18   20 µm

(a) Impressed structure on the surface of the substrate plate

(b) Molded through hole

Figure 5.17 In-polymer replicated through-holes. Because of the different hardnesses of mold insert and substrate plate, an impression into the substrate plate occurs. If a precise alignment is guaranteed, the mold insert always hits the same area of the substrate plate [19].

insert and substrate plate can be guaranteed by the precise guiding of hot embossing tools and hot embossing machines. Further, the temperature in every cycle should be set to the same values; otherwise, an alignment mismatch occurs by thermal expansion of mold and substrate plate, especially if different kinds of materials are used.

5.5.4 Multilayer Molding

Another illustrated approach for the fabrication of through-holes is the use of a polymer composite [22], a stack of two different polymer films (polymer 1 and polymer 2). The structured part with through-holes is molded on a layer of a second polymer (polymer 2) (Fig. 5.18). After demolding of the molded composite, the first polymer can be separated from the second polymer, in this case the carrier layer, by peeling. The advantage of this method is the maintenance of the adhesion on the substrate plate. This allows the demolding of the molded composite in a vertical direction by the precise movement of the hot embossing machine. The adhesion between the first polymer and the second polymer has

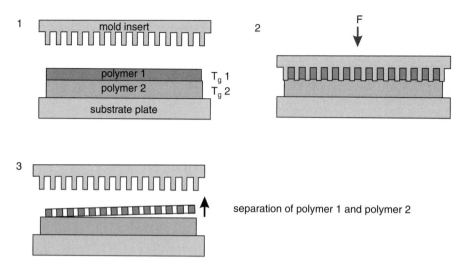

Figure 5.18 Principle of two-layer molding for molding through-holes. During molding, the mold insert penetrates both polymers. After demolding of the composite, the structured polymer layer on top can be separated from the second layer.

to be high enough to withstand the tensile forces during demolding. Here additionally a form-closed connection, effected by the penetration of the mold insert into the second polymer, will support the connection of both polymers. This approach is only suitable for selected material combinations, because of the required adhesion and further separation. A suitable material combination is cellulose acetate (AC) as the carrier layer and polyoxymethylene (POM) as the structured layer (Fig. 5.19).

5.5.5 Hot Punching

The principle of hot punching is a two-step method for molding through-holes (Fig. 5.20). In a first step, a selected mold insert is replicated in an amorphous polymer with high glass transition or a semicrystalline polymer with high melting temperature. The replication refers to a single molding on a rough substrate plate. Instead of peeling off the molded part from the substrate plate, the replicated part remains on the residual layer after demolding. The molded part is used as a second to the first mold-aligned mold insert, here called a matrix. In a second process step a thin film of another polymer with a lower glass-transition temperature than the matrix is positioned between the mold insert and the molded part. The mold insert and substrate are heated up to a temperature in the range of the glass-transition temperature of the thin polymer film, which should be significantly under the glass-transition temperature of the matrix. The third step of this method is simply another embossing cycle structuring the thin film of polymer between the mold insert and the remaining matrix.

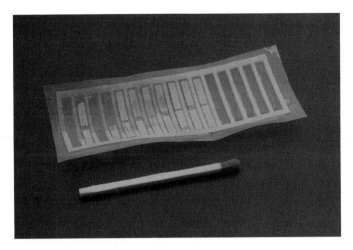

Figure 5.19 Example of molding two polymer foils with different properties with the aim of structuring the top of the composite. The polymer of the remaining layer is cellulose acetate (AC); the structured polymer is polyoxymethylene (POM). The structured layer can be separated from the remaining layer by peeling.

Because of the high accuracy of the alignment of the previous molded matrix compared to the mold insert, the structures of the mold insert will match the inverse structures of the matrix properly. Differences in thermal expansions will be compensated by the relative flexibility of the polymer matrix compared to the mold insert of metal. During the second molding step the mold insert can now cut holes inside the thin polymer film (Fig. 5.21). This principle can be understood like the principle of a shear. The punched areas will remain in the matrix, so the matrix has to be renewed after several punching steps. Depending on the design and the thickness of the polymer film, the demolding of the punched polymer film from the matrix can be difficult and should be accomplished carefully; otherwise damage to the polymer film will occur. Helpful here is the use of a release agent [2].

The advantage of this method is the use of a wide range of mold inserts without any modifications. The material combinations are manifold; besides the differences in softening temperatures, adhesion between the used polymers is not a precondition. Because of the two-step molding, this method is more time-consuming and for series production it has to be improved.

5.5.6 Thermoforming of High-Temperature Polymers by Hot Embossing

In accordance with the molding of through-holes, the concept of two-layer molding can also be used for thermoforming thin films of semicrystalline high-temperature materials like LCP or PEEK. Thermoforming refers to the strain

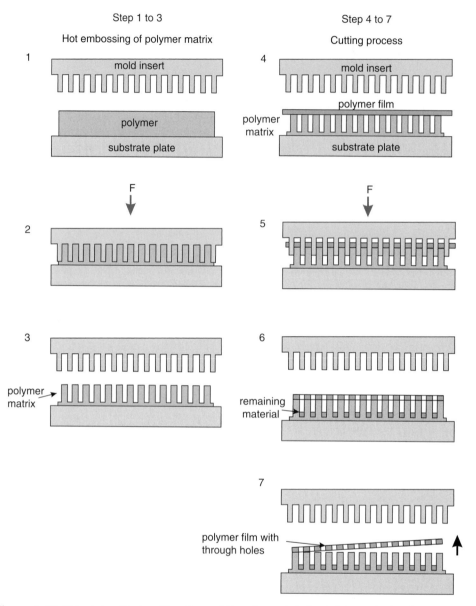

Figure 5.20 Principle of hot cutting for molding through-holes. This method refers to a two-step molding process. In a first step, a high-temperature polymer is structured by a single-sided hot embossing process. In a second step, a thin film of polymer is placed between the mold insert and the molded matrix. By a final molding step, the polymer film is punched, resulting in through-holes in the polymer film. The punched material remains in the matrix.

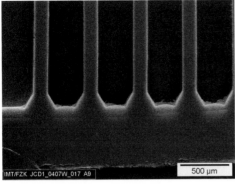

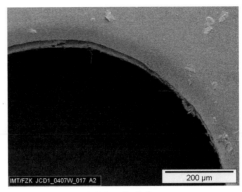

(a) Through holes structured by hot punching

(b) Detailed view

Figure 5.21 Molded through-holes by hot punching. The matrix used for this example was first molded in PSU; the material of the punched polymer film was PMMA. The through-holes are not limited to circular holes, also other geometries can be achieved by this method.

of a polymer, typically by air pressure (Section 2.6). Here the gas pressure is substituted by a polymer melt. The starting point refers to a combination of two polymer films, a thin film of a high-temperature polymer and a second film of a material with a lower softening temperature. A suitable combination is, for example, PMMA as molding material and PEEK as thermoforming material on top. Like the process variation of multilayer molding, the combination of materials has to be selected carefully. The molding temperature has to be set in the range in which the low-temperature polymer is in the melting range and the semicrystalline high-temperature polymer foil is in a temperature range above transition temperature but below melting temperature. In this temperature range the amorphous part is softened, which reduces the elastic modulus of the high-temperature semicrystalline polymer, which supports the strain of the polymer film. Characteristic for this method is the combination of a molding step (PMMA) and a simultaneously thermoforming step of the high-temperature polymer foil (PEEK). The thickness of the foils depends on the process parameters and the necessary molding force. Experiments show that good results of this kind of thermoforming can be achieved up to a film thickness of 100 μm. Here the thickness of the polymer film has to be selected in relation to the size of the cavities. To obtain a three-dimensional shape, the thickness of the polymer foils has to be much lower than the lateral dimensions of the mold cavity; otherwise, the polymer foil fills the cavity completely. Further, the material combination has to be selected such that only a moderate adhesion is effective between both foils after molding because the thin foil has to be separated from the molded low-temperature polymer manually by peeling. With this method a three-dimensional structuring is possible: the back side of the thin polymer foil shows the inverse structure of the front side, which is similar to the mold insert (Figs. 5.22 and 5.23).

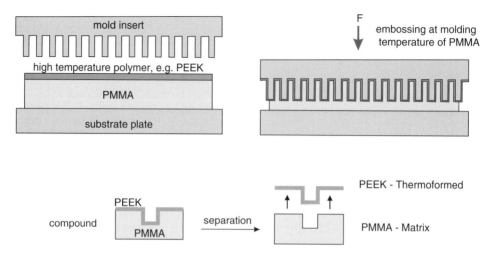

Figure 5.22 Principle of thermoforming of a thin polymer film by hot embossing. Instead of gas, a thin film of high-temperature polymer is formed by a polymer melt of a low-temperature polymer.

Figure 5.23 Thermoformed fluidic structures in a thin film of PEEK. The forming of these structures refers to a molding process with PMMA.

5.6 Micro Embossing Processes

The term *hot embossing* or *embossing* can be used for a large scale of different structuring processes based on an embossing step. The following section will give a short overview of the manifold of these processes. Nevertheless, a clear definition of the terminology cannot be found in literature.

5.6.1 Roller Embossing

Roller embossing, or roll-to-roll embossing, is a well-established modification of the hot embossing process. Using rolls instead of plates, continuous molding can be achieved with advantages regarding the molding times and disadvantages regarding the height and the aspect ratio of the molded structures (Fig. 5.24). Tan et al. [29] used this approach in 1998—called roller nanoimprint lithography—to fabricate sub-100-nm patterns. Two methods were investigated. The cylinder mold method refers to a thin structured metal film bended around a smooth roller. In particular, a compact disc master with a thickness of 100 μm was used. The second method—here called the flat mold method—refers to a structured silicon wafer mold placed on a polymer substrate. A smooth roller mold is rotated over the mold and the deformation of the mold under the pressure of the roll imprints the structures into the polymer. In both methods the roller temperature is set significantly above the glass-transition temperature. For PMMA the roller temperature was set in a range between 170°C and 200°C and the platform temperature in a range of 50–70°C. Roller speeds from 0.5 cm/sec up to 1.5 cm/sec were investigated, and the pressure was set in a range between 300 psi and 4800 psi.

Ng and Wang [25] reported on the hot roller embossing of microfluidic structures with feature sizes down to 50 μm and a corresponding depth up to 30 μm. To fabricate these structures a polymer film was passed between two rollers. The top roller was made of steel wrapped with a structured nickel film with a thickness of 50 μm. The bottom roller, made of rubber, was only a supporting roller. The heat was supplied through the mold and the rollers were applied with force by a clamping unit. The structures were replicated into a PMMA film 1.5 mm thick. The optimal roller temperature was found at 140°C. Embossed depths of more than 30 μm with 100 μm features were achieved at a roller speed of 1.5 mm/sec.

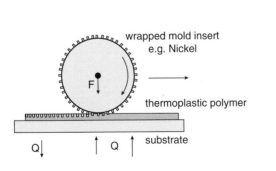

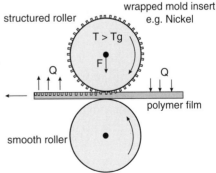

(a) Thermal roller embossing on a flat substrate   (b) Thermal roller embossing with two roles

Figure 5.24 Thermal roller hot embossing. This technology allows a continuous molding. Because of the rotation, the height of the molded structures is limited.

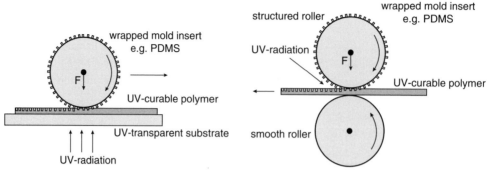

(a) UV roller embossing on a flat substrate   (b) UV roller embossing with two roles

Figure 5.25 UV-roller hot embossing. The technology of roller embossing is also suitable for UV-curing materials. To cure the molded material, the UV radiation has to be applied at the structuring zone.

Roller embossing is not limited to thermal structuring; this technique is also suitable for the structuring of UV-curable materials (Fig. 5.25). Chang et al. [6] used this method for the fabrication of microlens arrays on glass substrates. In a first step a nickel mold with microlens structures is fabricated by electroforming and wrapped in a second step onto a cylinder. During rolling the roller presses and drags the UV-curable polymer on a glass substrate through the rolling zone. At the same time the molded structures are cured by UV radiation and guided into the molding zone. With this technology arrays of polymer microlenses with a diameter of 100 µm, a pitch of 200 µm, and a depth of 21 µm on an array of 100 × 100 were fabricated. A suitable roller speed was found in the range between 0.5 mm/sec and 2 mm/sec. The nickel mold wrapped onto the roll was manufactured by several process steps. First, a silicon wafer was structured by photolithography and deep-reactive ion etching. A polymeric master from the silicon wafer was made by gas-pressurized thermoforming of a polycarbonate film. Depending on the pressure size, the holes of the etched silicon wafer were partially filled, resulting in a convex shape of the polymer film. The formed polymer film was silver-coated and finally copied into a nickel mold insert by electroplating.

Yang et al. [34] fabricated arrays of microlenses by so-called soft roller embossing of a UV-curable resist under the assistance of a gas-pressurized platform. Therefore, a PDMS mold was wrapped onto a cylinder. The fabrication of the mold was similar to the method of Chang described above, with the difference that from the structured polycarbonate film a PDMS casting instead of nickel electroplating was performed. The embossing system was composed of a UV-radiation source, a transparent substrate, the structured roller stamper, a split-coating module, pneumatic cylinders, and a flexible platform with gas-pressurized pad. The soft pad can be deformed by gas pressure and guarantee a conformal contact of a PET substrate between pad and roller. The roller stamping process can be described in three main steps. First, the UV-curable liquid resin is coated onto the soft roller with the microlens array. In a second step, the roller with the resin is brought into contact with the PET film by applying a specific pressure. Finally, the liquid resin is transferred onto the PET film and is cured by UV radiation during rolling. Liu and Chang [17] used the soft roller embossing method without gas-pressurized assistance to fabricate arrays of micro blocks onto a glass substrate. By this design with the structure dimensions of 100 μm × 80 μm and a depth of 40 μm, the process parameters dependent on the structure quality were determined. The maximal depth of 40 μm was achieved only by a decrease in roller velocity down to 0.25 mm/sec.

Theoretical aspects of the roller imprinting process in the case of printing microfluidic devices were investigated by Vijayaraghavan et al. [33]. Especially the effects of process variables like feature spacing, roll radius, roll velocities, molding materials, and design features on the imprint using finite element simulations were analyzed. They found that the feature spacing has the largest impact on the quality of the imprints. The so-called imprint gap—the difference in height between the top of the imprint walls and the undeformed substrate—decreases with increasing feature spacing. For features spaced wider than their width, the imprint gap is negligible. Further, the feature shape does not significantly affect the deformation behavior. The roll radius had a significant impact on the work-piece reaction forces. It was found that larger roll radii result in larger reaction forces, corresponding to the contact area between roll and substrate. Stamping can here be interpreted as an infinite roll radius.

The fabrication of the microstructured roller refers in the cited articles to a fabrication process on flat substrates and additional wrapping around a roller. Jiang et al. [14] presented a method to fabricate a metallic roller mold with microstructures on its surface using a dry film resist. This resist is laminated uniformly onto the surface of an imprinting roller. Using a flexible film photo mask wrapped around the roller, the photo resist can be structured by nonplanar lithography, achievable by rotating of the roller during a UV-lithography step. Finally, the structures in the resist can be transferred into the metallic surface by wet etching. The advantage of this fabrication method is to avoid any sliding problems of wrapped molds during the molding cycle. Further, the metallic surface has a high strength durability and temperature endurance.

Finally, the roller embossing process is not limited to the structuring of polymers. Shan et al. [27,28] demonstrated the structuring of multilayered green ceramic substrates by micro roller embossing using a modified laminating machine. Micro patterns, including channels and electrical passives with a 50 μm line width, were formed over the whole panel area of 150 mm × 150 mm.

5.6.2 Ultrasonic Embossing

Hot embossing is typically characterized by a conductive or convective heat transfer that requires a comparatively long cycle time for the heating of a polymer film over the glass-transition temperature. Lin and Chen [15] showed a hot embossing concept referring to the principle of ultrasonic embossing. An ultrasonic source is located on the top of the mold insert to generate high-frequency vibrations, causing an increase in temperature of a polymer foil up to the melting range. The ultrasonic vibration rapidly increases the temperature at the contact area between mold insert and polymer film, which means that heating times can be reduced significantly. Because of this principle the temperature distribution between the thermal hot embossing principle and the ultrasonic principle is different. In conventional thermal heating concepts for hot embossing, the temperature distribution inside the polymer melt is nearly identical, whereas in ultrasonic embossing the localized heat-effective zones are concentrated on the contact areas of the mold insert and polymer. Therefore, this ultrasonic embossing method is not an isothermal process. The temperature distribution and the polymer flow dependent on the mold geometry and resist thickness during ultrasonic embossing were simulated by Lint and Chen [16]. The lateral and vertical velocity profiles are described through the contribution of Poiseulle flow and Cuette flow. It was found that, because of the higher temperature at the cavity sidewalls, the polymer flow tends to flow along the cavity sidewalls of the mold. Also, the filling mode (single mode and dual mode, Section 6.4.2) of cavities with different diameters was simulated. In ultrasonic vibration both modes are visible. Cavities with wider convexity and thinner resists form a dual-peak mode; in contrast, therefore, a wider cavity and a thicker mold insert form a single peak.

Mekaru et al. [20] implemented an atmospheric hot embossing process on almost the same conditions as vacuum hot embossing combined with ultrasonic vibration (Fig. 5.26). The synergy effect of the conventional load (press force) and ultrasonic vibration can optimize the flow behavior to fill microcavities. In this work, longitudinal waves were generated by an ultrasonic system with a 15 kHz and 900 W output. With this technique, micro pyramids in the size of 100–530 µm and depth up to 260 µm were replicated in polycarbonate. The molding step consists of a conventional heating step of the polymer over the glass-transition temperature and a following molding step in which the mold is pressed into the polymer melt. During the pressing and the following holding

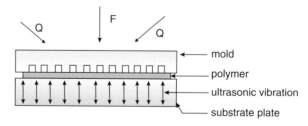

Figure 5.26 Principle of ultrasonic embossing in combination with a conventional thermal embossing process. The substrate plate is exposed by ultrasonic vibration.

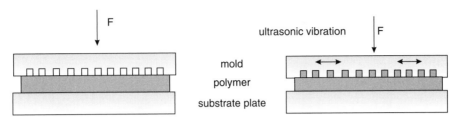

Figure 5.27 Principle of ultrasonic embossing at room temperature. Because of the ultrasonic vibration, the polymer is heated up at the contact area to molding temperature. The vibration can be effected horizontally in the plane of the film or, alternatively, in a vertical direction.

time, the ultrasonic vibration is active. After the holding time, the ultrasonic vibration is stopped and the cooling cycle starts up to a temperature below glass-transition temperature. The demolding can be done with the support of ultrasonic vibration. The result of these experiments shows that contact force and contact times are decreased by applying ultrasonic vibration. With this method it is further possible to mold under atmospheric conditions.

Another approach for ultrasonic molding was presented by Cross et al. [10]. This approach refers to a room-temperature forming of solid films, which are not structured by thermal support (Fig. 5.27).The principle of small amplitude (<10 nm) oscillatory shear forming (SAOSF) is based on two steps. First, a structured mold or a stamper is pressed with a normal contact load (in this case, 50 MPa) into a thin solid film, for example, polystyrene. In a second step, plastic strains are generated by augmenting the normal contact load with a small oscillatory motion in the plane of the film. Finally, after sufficient cycles, mold and film are separated.

5.6.3 Gas-Assisted Embossing

The act of force during the embossing step is mainly characterized by direct contact of rigid plates or roles. The force is typically effected by hydraulic drive units or spindle drives. The contact of rigid plates, for example, requires typically smooth and even surfaces to guarantee a homogeneous contact over the whole molding area, especially if the molding area increases. Further, brittle silicon molds, for example, can be difficult to use in combination with a rigid substrate plate because of the stress concentration during molding caused by uneven substrate plates. The use of gas as an acting load during molding can prevent especially brittle materials from damaging. Pressurized gas is similar to the process of thermoforming. Here the temperature determines the kind of molding process. If the temperature is in the range of the glass-transition temperature of the polymer, the molding is mainly characterized by strain deformation and the molding process is defined as a thermoforming process (see Section 2.6). If the molding temperature increases significantly into the melting state of the polymer, then the forming is mainly characterized by shear, which is a criterion for molding processes. Therefore, the molding or forming temperature is an indication for the classification of the gas-assisted molding process.

Independent from the investigations of micro thermoforming by Truckenmüller [31] in 2003, Chang and Yhang [8] presented a concept for gas-assisted hot embossing with four main molding steps. First, inside a closed chamber with gas inlets and outlets a polymer film is placed onto a microstructured stamper. In a second step, the chamber is closed and the stack of stamper and polymer foil is heated up to molding temperature. After the molding temperature is achieved, the chamber is flooded with nitrogen gas at high pressure. The gas pressure acts uniformly over the whole microstructured stamper and forms the polymer film into the cavity. After cooling, the gas pressure is exhausted and the molded part can be demolded by peeling. By this technique, also double-sided molding was demonstrated using a stack of two molds with a polymer film in between. The gas pressure acts in this case onto a polymer seal film covering the stack (Fig. 5.28). By this technique a 200 μm thick PVC film was structured at 90°C, and a double-sided molding was tested by a 2 mm thick PMMA sheet heated up to a molding temperature of 155°C. The pressure of 40 kg/cm^2 refers to a 4-inch silicon mold with microfluidic channels. Yang et al. [34] used this technology to replicate V-grooves by CO_2-assisted embossing into PMMA at temperatures between 25°C and 130°C on an area of 90 mm × 90 mm. The grooves were characterized by a width of 50°C and a depth of 21 μm. The nickel mold was applied by pressures of 10 bar. Chang et al. [7] fabricated microlens arrays by using gas pressure to form polycarbonate films into etched holes of a silicon mold. An array of 300 × 300 microlenses with a diameter of 150 μm and a pitch of 200 μm were replicated. For polycarbonate, a forming temperature of 150°C and a pressure load of 10–40 kg/cm^2 during 30 seconds up to 90 seconds were implemented. In this example, the temperature range is close to the glass-transition range of polycarbonate, which characterizes the forming mainly by strain deformation. Therefore, this is also an example for the micro thermoforming process (Fig. 5.29).

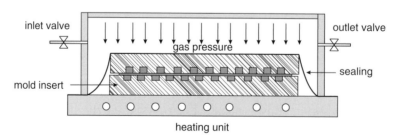

Figure 5.28 Principle of gas-assisted double-sided hot embossing. A seal separates the aligned mold inserts with the polymer between and the gas volume. After heating the polymer over its glass-transition temperature, the gas pressure acts as a press force similar to the conventional hot embossing steps.

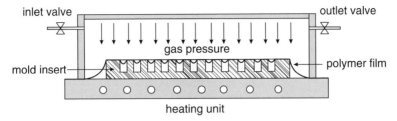

Figure 5.29 Principle of gas-assisted thermoforming of microlenses [7]. A polymer film is placed onto a microstructured stamper with holes, is heated up to forming temperature, and is formed by air pressure into the cavities. The embossing temperature is close to the glass-transition temperature, where the forming is characterized by strain. In this case, the molding principle is defined as thermoforming (Section 2.6) [32].

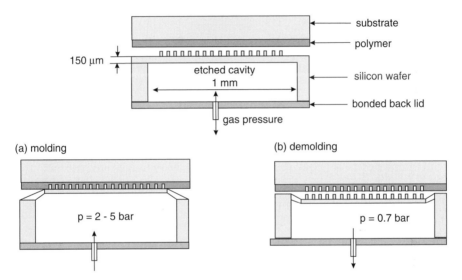

Figure 5.30 Principle of smart stamps for gas-assisted embossing. Small cavities at the back side of a silicon wafer are etched at selected areas to a residual thickness of 150 μm. An bonded back lid forms a so-called intrastamp with an area of around 1 mm × 1 mm. Gas pressure acting on the back side of the stamper can deform the thin residual layer of the silicon wafer. After heating, these stamps can be used for thermal imprinting and also, after cooling, for demolding.

The molding by gas pressure can be miniaturized to single molding elements integrated in a silicon mold (Fig. 5.30). This concept was investigated for thermal nanoimprinting by Pedersen et al. [26]. The concept refers to single air cavities at the back side of a stamp allowing gas pressure to act at the structured areas of a mold. To obtain a minimal bending of the brittle silicon, the back side of the structured areas has be etched to a remaining thickness of around 150 μm. Silicon with this thickness allows a bending under air pressure that is sufficient for the molding of nanostructures. In this case, the depth of the molded structures was in a range below 200 nm. The concept was implemented by bonding a lid containing access holes to the back side of the flexible stamp forming a so-called intrastamp cavity. If after heating of the components gas is pumped into the cavities at the back side, the pressure acts as load for imprinting. Due to the flexibility of the thin remaining layer of the stamper, only moderate pressure is required. If after cooling the back-side cavities are evacuated, the direction of the structured remaining layer changes and the molded structures can be demolded. This concept was implemented for a silicon stamper with more than 1,500 imprint areas of 1 mm × 1 mm separated 0.5 mm from each other. Imprints were performed in PMMA films with a thickness of 305 nm. With this technique, homogeneous imprints across an area of 35 mm diameter were achieved.

5.6.4 UV Embossing

UV embossing is a method similar to the UV nanoimprint process (Section 2.7.2). A UV-curable resin is coated on a substrate, imprinted by a structured mold insert and cured by UV radiation. Molding with UV-curable materials does not require heating systems and, because of the typically low viscosity of the UV-curable materials, only moderate embossing pressures are needed, which results in smaller UV-imprint machines. Chan-Park and Neo [5] used this method for the fabrication of micro cups. A UV-curable resin was embossed at room temperature onto a nickel mold by a rubber-coated laminating roller and in a second step cured by UV radiation. The demolding of the embossed micro cups was achieved by peeling.

5.6.5 Soft Embossing

The process of soft embossing refers to a mold fabricated of elastic materials. A common material is polydimethylsiloxane (PDMS). This elastomeric material can be used because of its thermal stability up to 200°C and its transparency for UV radiation as a mold insert for thermal nanoimprinting, UV nanoimprinting, and hot embossing. Typically, the mold inserts are fabricated by casting from a master, fabricated by lithographic or etching processes. Because of the flexibility of this material, demolding by peeling from a master can be performed without any deformations of structures. Further, PDMS is chemically inert and is characterized by a low surface energy [18]. The properties of PDMS are therefore advantageous for demolding of embossed structures. During molding, an elastic material guarantees a conformal contact between the mold and the polymer independent of unevenness of any substrate plates. Nevertheless, compared to a rigid mold material an elastic mold tends to deform under load, which can result in a loss of precision in lateral dimensions of molded structures, especially if high molding forces are required.

Carvalho et al. [4] replicated microfluidic devices with a feature size of 127 μm and a depth of 50 μm by soft embossing with a silicone mold casted from a micromilled master. Rubber-assisted hot embossing was performed by Nagarajan and Yao [24]. In this work thin polystyrene film with a thickness of 25 μm was patterned with micro grooves in the dimension of 100 μm. In contrast to an elastic mold insert, an elastic rubber was used as the substrate plate (Fig. 5.31). The microstructured rigid mold insert was fabricated by electrical discharge machining. The polymer film is placed between the hard die and the rubber substrate. The metal die is heated up over the softening range of the polymer and pressed into the cold rubber substrate. The rubber substrate deforms under load and acquires the profile of the die, acting as negative counter tool to the positive metal die. By this method a conformal contact between hard mold insert and substrate is achievable, forming the enclosed thin polymer film. The following process steps of force-controlled molding and cooling under load are similar to the process steps of hot embossing. Finally, the ejection of the formed part can be performed by the help of the elastic rubber substrate forming back to the initial shape and ejecting the formed part. The elastic rubber substrate can be used many times. Regarding the temperature range and the forming method, rubber embossing is closely related to the process of micro thermoforming.

Huang et al. [13] used the soft embossing technology for the fabrication of micro-ring optical resonators. First, a master was fabricated by E-beam lithography. From this master the mold was copied by casting of PDMS. This mold was

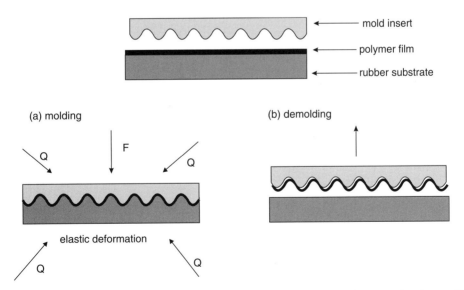

Figure 5.31 Thermoforming of a polymer film on an elastic rubber substrate. A polymer film is placed between a hard mold insert and a rubber substrate. The metal mold insert is heated up over the softening range of the polymer and pressed into the cold rubber substrate. The rubber substrate deforms under load and acquires the profile of the mold insert. The deformation of the rubber substrate is reversible.

finally used for a UV-embossing process to mold a UV-curable resist, which is cured by UV radiation and peeled off from the elastic mold. The ring diameter of the replicated structures was 200 µm, the waveguide width was 2 µm, and the gap between the straight waveguide and the micro ring was 250 nm. The measurement of the optical properties compared to a master device shows that the optical properties of the molded parts are nearly identical, which makes this technology well suited for the replication of optical waveguides and resonator structures.

5.6.6 3D Embossing

The replication by hot embossing mostly refers to so-called 2.5-dimensional structures, structures with the same level of height and vertical sidewalls or structures with demolding angles. Nevertheless, the molding of real three-dimensional (3D) structures is also possible if the structures are characterized by the lack of undercuts that prevent a successful demolding with a conventional technique. Tormen et al. [30] investigated the fabrication of 3D structures by an approach on a combination of lithographic steps and isotropic wet etching. The optical structures were performed on quartz or glass substrates with very accurate shape and nanometer scale surface roughness. The structures were finally replicated in cyclic polyolefin (COC) by hot embossing.

Regarding the combination of rigid mold and thermoplastic polymers, it is obvious that 3D structures with undercuts like dovetails cannot be demolded without damaging the mold insert or the molded structures. A simulation of the demolding cycle (Section 6.4.4) shows that, for structures with a height of 100 µm and a deviation of 0.25 µm from the vertical, the demolding force doubles.

To avoid damaging structures with undercuts, one of the components—the mold insert or the molded part—should have elastic properties so that the demolding can be performed under reversible elastic deformation (Figs. 5.32 and 5.33). In this case the elastic material PDMS has the potential for the replication of 3D structures. Bogdanski et al. [3] investigated the molding of undercut structures over an area of 5×5 cm^2 by structuring PDMS with an etched silicon mold. The undercuts of the structures were in a range between 0.5 µm and 10 µm compared to a vertical trench of the structures of 1 µm up to 15 µm. The replication of these structures in PDMS showed a loss in the sharpness of the edges and a tear off of the structures during demolding if the thickness of the residual layer was in a range of the undercut width. For a thicker residual layer, the connection between structures and residual layer was strong enough to withstand the tensile forces during demolding. Independent of this work, it is also possible to mold and demold structures if a PDMS mold insert is used and the undercut structures are molded in thermoplastic polymers. Finally, the design of undercuts determines the demolding forces and defines the potential of successful replication.

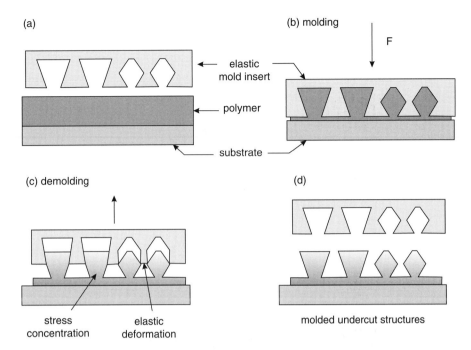

Figure 5.32 Principle of soft embossing of 3D structures using an elastic mold insert. Because of the elasticity of the mold material 3D structures can be demolded. During demolding, the structures of the mold insert are elastically deformed and prevent the molded polymer from deformation.

5.6.7 Hot Embossing of Conducting Paths—MID Hot Embossing

The integration of conduction paths into microsystems is fundamental for the connection of mechanical or fluidic devices to electrical control units or signal analyzers or for the appropriation of a potential difference, for example, in capillary electrophoresis systems (Section 10.3.1).

An example for the integration of conducting paths into molded microstructures by hot embossing was investigated by Heckele and Anna [12]. In a first step, preliminary conducting paths in terms of a gold layer of a few 100 nm in thickness were sputtered through a mask onto a polymer film. In a second step, these thin paths were increased by electroplating of gold up to a thickness between 0.3 µm and 3 µm. In a following step, the polymer film with the paths on top was embossed with a microstructured mold with a structure height of 130 µm. The LIGA mold was characterized by vertical sidewalls and additionally by structures with 45 degree angles. The molding results show that the conduction paths maintain their lateral shape and are not deformed by polymer molding. Further, because of the ductility of gold, the paths do not break when

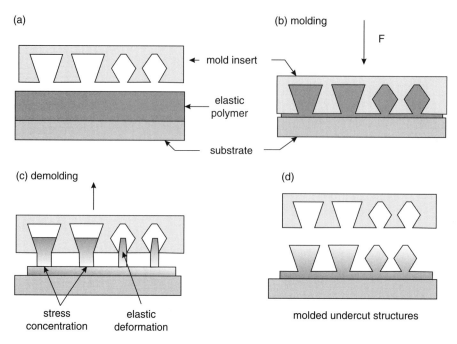

Figure 5.33 Principle of soft embossing of 3D structures using an elastic polymer. Because of the elasticity of the molding material 3D structures can be demolded. Depending on the design and the cross-section, the structures can be torn off during demolding.

passing steps, independent of the angle of the step (45° edges and also 90° edges). In the case of vertical steps, the conducting paths do not follow the sharp edges; the paths are instead submerged into the polymer.

Hot embossing of conduction paths is also an established process for the fabrication of molded interconnected devices (MID) [1]. By this hot embossing technology electronic circuits are embossed on top of already molded polymer components (Fig. 5.34). First, the layout of an electronic circuit is copied into a metal mold insert by mechanical structuring methods, for example, milling. A copper foil is placed between the mold insert and the surface of the molded part. The mold insert is heated up to a molding temperature of 20–40°C over the glass-transition temperature (amorphous polymer) or melting temperature (semicrystalline polymer) of the molded part. If the molding temperature is achieved, the mold insert and the copper foil are pressed by the protruding structures of the mold insert onto the surface of the molded part. The conductor lines are punched out and are connected with the thermoplastic substrate. The connection refers to adhesion because of the roughness of the back side of the copper foil in the range of typically R_a 0.5–2 µm. During the embossing state the polymer melt is displaced, resulting in a bulge beside the conducting paths. The height of this bulge is a function of the polymer material, the temperature of the mold, the embossing time, and the depth of impression.

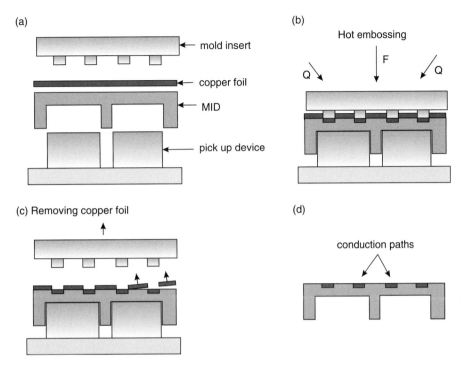

Figure 5.34 Embossing of conduction paths on molded interconnected devices (MID). A copper foil is placed between mold insert and molded part. If after heating the molding temperature is achieved, the mold insert and the copper foil is pressed by structures of the mold insert onto the surface of the molded part. The conductor lines are punched out and are connected with the thermoplastic substrate by adhesion.

References

1. Hahn-Schickard-Gesellschaft für angewandte Forschung e.V., Institut für Mikroaufbautechnik. http://www.hsg-imat.de/, 2008.

2. B. Rapp, M. Worgull, M. Heckele, A. Guber, and D. Herrmann. Mikro-Heissstanzen—Erzeugung von Durchlochstrukturen in ebenen Kunststoffsubstraten. In *Conference in Microsystem Technology*, Freiburg i. Br., Germany, 2005.

3. N. Bogdanski, H. Schulz, M. Wissen, H.-C. Scheer, J. Zajadacz, and K. Zimmer. 3D-hot embossing of undercut structures—an approach to microzippers. *Microelectronic Engineering*, 73–74:190–195, 2004.

4. B. L. Carvalho, E. A. Schilling, N. Schmid, and G. J. Kellog. Soft embossing of microfluidic devices. In *Proceedings of 7th International Conference on Miniaturized Chemical and Biochemical Analysis Systems*, pages 959–962, Squaw Valley, California, USA, 2003.

5. M. B. Chan-Park and W. K. Neo. Ultraviolet embossing for patterning high aspect ratio polymeric microstructures. *Microsystem Technologies*, 9:501–506, 2003.

6. C.-Y. Chang, S.-Y. Yang, L.-S. Huang and J.-H. Chang. Fabrication of plastic microlens array using gas-assisted micro-hot-embossing with a silicon mold. *Infrared Physics & Technology*, 48:163–173, 2006.

7. C. Y. Chang, S. Y. Yang, and J. L. Sheh. A roller embossing process for rapid fabrication of microlens arrays on glass substrates. *Microsystem Technologies*, 12:754–759, 2006.

8. J. H. Chang and S.-Y. Yhang. Gas pressurized hot embossing for transcription of micro features. *Microsystem Technologies*, 10:76–80, 2003.

9. R.-D. Chien. Hot embossing of microfluidic platform. *International Communications in Heat and Mass Transfer*, 33:645–653, 2006.

10. G. L. W. Cross, B. S. O'Connell, H. O. Özer, and J. B. Pethica. Room temperature mechanical thinning and imprinting of solid films. *Nano Letters*, 7(2):357–362, 2007.

11. P. Datta and J. Goettert. Method for polymer hot embossing process development. *Microsystem Technologies*, 13:265–270, 2006.

12. M. Heckele and F. Anna. Hot embossing of microstructures with integrated conduction paths for the production of lab-on-chip systems. In *Proceedings of Spie – DTIP Conference*, volume 4755, pages 670–674, Cannes, France, 2002.

13. Y. Huang, G. T. Paloczi, J. Scheuer, and A. Yariv. Soft lithography replication of polymeric microring optical resonators. *Optics Express*, 11(20):2452–2458, 2003.

14. L.-T. Jiang, T.-C. Huang, C.-Y. Chang, J.-R. Ciou, S.-Y. Yang, and P.-H. Huang. Direct fabrication of rigid microstructures on a metallic roller using a dry film resist. *Journal of Micromechanics and Microengineering*, 18:015004, 2008.

15. C.-H. Lin and R. Chen. Ultrasonic nanoimprint lithography: a new approach to nanopatterning. *Journal of Microlith. Microfab. Microsyst.*, 5:011003, 2006.

16. C.-H. Lin and R. Chen. Effects of mold geometries and imprinted polymer resist thickness on ultrasonic nanoimprint lithography. *Journal of Micromechanics and Microengineering*, 17:1220–1231, 2007.

17. S.-J. Liu and Y.-C. Chang. A novel soft-mold roller embossing method for the rapid fabrication of micro-blocks onto glass substrates. *Journal of Micromechanics and Microengineering*, 17:172–179, 2007.

18. C. Khan Malek, J.-R. Coudevylle, J.-C. Jeannot, and R. Duffait. Revisiting micro hot embossing with moulds in non-conventional materials. *Microsystem Technologies*, 13:475–481, 2007.

19. Ch. Mehne. *Grossformatige Abformung mikrostrukturierter Formeinsätze durch Heissprägen*. PhD thesis, University of Karlsruhe, Institute for Microstructure Technology, 2007.

20. H. Mekaru, O. Nakamura, O. Maruyama, R. Maeda, and T. Hattori. Development of precision transfer technology of atmospheric hot embossing by ultrasonic vibration. *Microsystem Technologies*, 13:385–391, 2006.

21. A. Michel, R. Ruprecht, M. Harmening, and W. Bacher. *Abformung von Mikrostrukturen auf prozessierten Wafern*. Scientific report FZKA 5171. Forschungszentrum Karlsruhe, 1993.

22. K.-D. Mueller. *Herstellung von beweglichen metallischen Mikrostrukturen auf Siliziumwafern*. Scientific report FZKA 6254. Forschungszentrum Karlsruhe, 1999.

23. T. Veres, S. C. Jakeway, H. J. Crabtree, N. S. Cameron, and H. Roberge. High fidelity, high yield production of microfluidic devices by hot embossing lithography: rheology and stiction. *Lab Chip*, 6:936–941, 2006.

24. P. Nagarajan and D. Yao. Rubber-assisted hot embossing for structuring thin film polymeric films. In *IMECE2006-15297, Mechanical Engineering Congress and Exposition, Chicago, Illinois*. ASME International, 2006.

25. S. H. Ng and Z. F. Wang. Hot roller embossing for the creation of microfluidic devices. In *Proceedings of DTIP Conference 2008*, Nice, France, EDA Publishing. 2008.

26. R. H. Pedersen, O. Hansen, and A. Kristensen. A compact system for large-area thermal nanoimprint lithography using smart stamps. *Journal of Micromechanics and Microengineering*, 18:055018, 2008.

27. X. Shan, Y. C. Soh, C. W. P. Shi, C. K. Tay, K. M. Chua, and C. W. Lu. Large-area patterning of multilayered green ceramic substrates using micro roller embossing. *Journal of Micromechanics and Microengineering*, 18:065007, 2008.

28. X. Shan, Y. C. Soh, C. W. P. Shi, C. K. Tay, and C. W. Lu. Large area roller embossing of multilayered ceramic green composites. In *Proc. of DTIP-Conference*, Nice/France, 2008.

29. H. Tan, A. Gilbertson, and S. Y. Chou. Roller nanoimprint lithography. *Journal Vac. Sci. Technol.* B, 16(6):3926–3928, 1998.

30. M. Tormen, A. Carpentiero, E. Ferrari, D. Cojoc, and E. Di Fabrizio. Novel fabrication method for three-dimensional nanostructuring: an application to micro-optics. *Nanotechnology*, 18:385301, 2007.

31. R. Truckenmüller, Z. Rummler, Th. Schaller, and W. K. Schomburg. Low-cost production of single-use polymer capillary electrophoresis structures by microthermoforming. In *12th Micromechanics Europe Workshop (MME)*, pages 39–42, Cork, Ireland, 2001.

32. R. Truckenmüller, Z. Rummler, Th. Schaller, and W. K. Schomburg. Low-cost thermoforming of micro fluidic analysis chips. *Journal of Micromechanics and Microengineering*, 12:375–379, 2002.

33. A. Vijayaraghavan, S. Hayse-Gregson, R. Valdez, and D. Dornfeld. Analysis of process variable effects on the roller imprinting process. Technical report, Laboratory for Manufacturing and Sustainability, Green Manufacturing Group, University of Califonia, Berkeley, 2008.

34. S.-Y. Yang, F.-S. Cheng, S.-W. Xu, P.-H. Huang, and T.-C. Huang. Fabrication of microlens arrays using UV micro-stamping with soft roller and gas pressurized platform. *Microelectronic Engineering*, 85:603–609, 2008.

35. D. Yao and R. Kuduva-Raman-Thanumoorthy. An enlarged process window for hot embossing. *Journal of Micromechanics and Microengineering*, 18:045023, 2008.

6 Modeling and Process Simulation

The aim of this chapter is to enhance the in-depth understanding of the hot embossing process. Theoretical aspects will be covered as well as practical experience. In addition, the knowledge of polymer material behavior will be conveyed, which will be helpful in finding the adequate process parameter for a selected design. Furthermore, the theoretical knowledge provided will be helpful in designing microstructured molds and will demonstrate both the potentials and the limits of hot embossing processes.

In a first step, the process will be described with a simple analytical model. With this model, basic correlations can already be shown. In a second step, a detailed process simulation referring to the material behavior of thermoplastic polymers will reveal more complex correlations. The results obtained will then be used for further developments in the field of hot embossing. This chapter will present current developments in modeling and simulation and demonstrate the potentials of process simulation, but also the limits, especially as far as the molding of microstructures is concerned.

6.1 Analytical Model—Squeeze Flow

To give the reader a first idea of the influence of force, viscosity, and molding velocity on the thickness of the residual layer, a simple analytical model can be used to describe the molding process. This model is known as the squeezing flow between two parallel disks. In literature, several approaches can be found to describe the squeezing flow of viscoelastic materials [1,23,24,37]. The fundamental relationships can be described with a Newtonian fluid between two discs.

6.1.1 Squeezing Flow of a Newtonian Fluid

The approach is based on two flat, circular discs with the radius R and a Newtonian fluid with the viscosity η in between. The gap between the two discs is $2h_0$. To the circular disks, a force of F is applied (Fig. 6.1).

The force F exerts a pressure on the fluid, which induces compressive stresses in the fluid. These stresses cause a radial flow of the fluid from the disc center outwards to its circumference. The flow is based on a pressure difference dp. Under the influence of the force dF, the gap between the two discs is decreased by dh. The volume flow $\dot{V}$ leaving the gap between the discs has to be equal to the displaced volume per time unit caused by the motion of the two discs.

$$\dot{V} = -\pi R^2 \dot{h} = 2\pi r \int_{-h}^{h} \nu_r dz \qquad (6.1)$$

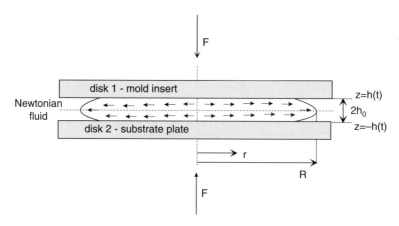

Figure 6.1 A Newtonian fluid with the viscosity η is compressed by the force F acting on circular discs.

Because of a decreasing gap and an increasing displaced volume, the required force to maintain a constant motion of the disc will increase significantly. The equation to describe this behavior can be deduced under the following boundary conditions.

- The problem will treated as a steady-state hydrodynamic problem.
- The radial velocity of the fluid is a function of the radius r and the vertical distance of both plates z: $\nu_r = \nu_r(r, z)$.
- The vertical velocity of the fluid depends only on the distance: $\nu_z = \nu_z(z)$.
- The pressure p is a function of the radius: $p = p(r)$.
- The ambient pressure is p_a.

First the equations of change have to be defined:

$$\text{Continuity} : \frac{1}{r}\frac{\delta}{\delta r}(r\nu_r) + \frac{d\nu_z}{dz} = 0 \tag{6.2}$$

$$\text{Motion} : -\frac{dp}{dr} + \eta\left[\frac{\delta}{\delta r}\left(\frac{1}{r}\frac{\delta}{\delta r}(r\nu_r)\right) + \frac{\delta^2\nu_r}{\delta z^2}\right] = 0 \tag{6.3}$$

Because of the postulate that v_z does not depend on r and has a form like $v_r = rf(z)$, the equation for the motion can be described as

$$-\frac{dp}{dr} + \eta\left[\frac{\delta^2\nu_r}{\delta z^2}\right] = 0 \tag{6.4}$$

The integration of the equation of continuity over $z = 0$ to $z = h$, and over the radius from $r = 0$ to $r = r$, results in

$$2\pi r \int_0^h \nu_r dz + \pi r^2 \dot{h} = 0 \tag{6.5}$$

with

$$\dot{h} = \frac{dh}{dt} = \nu_z(z = h) \tag{6.6}$$

The integration of the equation of the motion with respect to z under the following boundary conditions $f = 0$ at $z = h$ and $f'(z) = 0$ at $z = 0$ results in

$$\nu_r = rf = \frac{h^2 r}{2}\left(-\frac{1}{\eta r}\frac{dp}{dr}\right)\left[1 - \left(\frac{z}{h}\right)^2\right] \tag{6.7}$$

Substitution of Equation 6.7 into Equation 6.5 results in

$$-\frac{1}{\eta r}\frac{dp}{dr} = -\frac{3\dot{h}}{2h^3} \tag{6.8}$$

The combination of Equations 6.7 and 6.5 results in an equation describing the radial velocity of the fluid in terms of the motion of the plates, $\dot{h} = dh/dt$.

$$\nu_r = \frac{3}{4}\frac{-\dot{h}}{h}r\left[1 - \left(\frac{z}{h}\right)^2\right] \tag{6.9}$$

The integration of Equation 6.8 results in an equation for the radial pressure distribution:

$$p - p_a = -\frac{3\dot{h}\eta R^2}{4h^3}\left[1 - \left(\frac{r}{R}\right)^2\right] \tag{6.10}$$

The calculation of the required force $F(t)$ to be applied in order to maintain the disc motion $h(t)$ requires first the calculation of the shear stress distribution τ_{zz} over the height h.

$$\tau_{zz} = -2\eta\frac{\delta\nu_z}{\delta z} = 2\eta\left(\frac{1}{r}\frac{\delta}{\delta r}(r\nu_r)\right) = 3\eta\left(\frac{-\dot{h}}{h}\right)\left[1 - \left(\frac{z}{h}\right)^2\right] \tag{6.11}$$

With this equation for the shear stress, the required force for a constant motion of the plates is calculated to:

$$\begin{aligned}
F &= \int_0^{2\pi}\int_0^R (p - p_a + \tau_{zz})|_{z=h}\, r\, dr\, d\theta = 2\pi R^2 \int_0^1 (p - p_a)\left(\frac{r}{R}\right)d\left(\frac{r}{R}\right) \\
&= -\frac{3\pi R^4 \eta \dot{h}}{8h^3}
\end{aligned} \tag{6.12}$$

Equation 6.12 is called the Stefan equation [1] and describes the required force for a constant molding velocity. In the case of a constant force, Equation 6.12 has to be solved for $h(t)$ and results in:

$$\frac{1}{h^2} - \frac{1}{h_0^2} = \frac{16Ft}{3\pi R^4 \eta} \qquad (6.13)$$

Equation 6.13 describes the distance of both plates under a constant force of the elapsed time t.

Referring to an initial thickness of $h_0 = 1$, Equation 6.13 can describe the achievable thickness of a residual layer under a constant molding force:

$$h(t) = \sqrt{\frac{3\pi R^4 \eta}{16F \cdot t}} \qquad (6.14)$$

This equation can be transferred to the hot embossing process with regard to molding force, molding velocity, and thickness of the residual layer. Therefore, fundamental relationships of molding can be derived. Otto et al. [30] developed, on the basis of squeeze flow, an equation for the residual thickness under the consideration of elevated and recessed structures. Bogdanski et al. [5] modified the equation of squeeze flow for a number of periodically arranged cavities with defined width and height. Independent of the modifications, the fundamental relationships can be described.

6.1.1.1 Velocity-Controlled Molding

- For the molding step of velocity-controlled molding, the molding force is proportional to the fourth power of the radius.

$$F \propto R^4 \qquad (6.15)$$

- The molding force is proportional to the reciprocal value of the third power of the thickness of the achievable residual layer.

$$F \propto \frac{1}{h^3} \qquad (6.16)$$

- The molding force is only linearly proportional to the molding velocity and the viscosity of the (Newtonian) fluid.

$$F \propto \dot{h} \qquad (6.17)$$

$$F \propto \eta \qquad (6.18)$$

These fundamentals have to be taken into account, for example, when the molding area increases. On the one hand, to obtain the same thickness of the residual layer for an area 2 times larger, a molding force 16 times higher is needed. On the other hand, if the thickness of the residual layer has to be halved, the necessary force is 8 times higher. This fundamental relationship shows the significant dependence of the force on the area and the thickness of the molded part. It is one of the reasons why large and stiff embossing machines are needed for molding microstructured parts on large areas with a thin residual layer.

6.1.1.2 Force-Controlled Molding

Under constant force during the force-controlled molding step, the relationships can be described as follows.

- The force and the elapsed time are responsible for a thin residual layer. The product of force F and elapsed time t is a characteristic value.

$$h(t) \propto \sqrt{\frac{1}{F \cdot t}} \tag{6.19}$$

- The achievable thickness is proportional to the second power of the radius.

$$h(t) \propto R^2 \tag{6.20}$$

The equations above show that the geometric relationships especially are determining factors regarding the achievable height of the residual layer and the required force for velocity-controlled molding. Another fundamental aspect is the pressure distribution over the molding area. It is important for the filling of microcavities and the influence on shrinkage. For the model described here, the pressure distribution can be deduced as a function of the radius. The pressure distribution is similar to a parabolic shape, with the maximum pressure being encountered in the center of the discs and ambient pressure at the flow fronts.

Finally, this model refers to a Newtonian fluid and does not consider the shear thinning behavior of polymers. In the range of small shear velocities, the behavior of a shear thinning fluid is nearly similar to that of a Newtonian fluid, and this model can be used to approximate the molding behavior.

6.2 Process Simulation in Polymer Processing

Process simulation in polymer processing is an established procedure nowadays in science and industry. In science, a number of simulation tools specified and optimized for selected processes exist. For injection molding in particular, commercial FEM-simulation tools with comfortable graphic interfaces are available [3]. Process simulation focuses on mold filling and the reduction of shrinkage and warpage. The simulation results are fundamental for the determination of the molding window and the design of molding tools. Moreover, prior process simulation can save time and costs and allows the adaptation of the process parameters to sophisticated designs. Nevertheless, it has to be taken into account that every process simulation is also time-consuming, and modeling of the whole molding process should result in reasonable idealizations. The influence of idealizations on the results will vary for every process, simulation tool, and molding design. Therefore, the results should be verified by experimental investigations, if possible. Still, simulations are an excellent tool for the prediction of molding behavior.

6.2.1 Simulation of Macroscopic Processes

Representative features of simulation tools shall now be highlighted using the common injection molding process as an example. Essential points of interest are the filling of cavities and the related physical properties like shrinkage, warpage, stress, and molecule orientation. To simulate the flow behavior correctly, the boundary conditions have to be considered as much as possible. The cooling behavior of a polymer melt influences the quality of the molded part and can be considered by modeling the thermal behavior of the polymer, the mold, the cooling liquid, and the arrangement and geometry of the cooling system in the mold. Apart from modeling the boundary conditions, the process parameters like injection pressure, injection speed or injection profile, melt, and mold temperature can be set by the user. The quality of the simulation is also determined by the quality of modeling of the polymer used, especially its flow behavior. A detailed measurement of the temperature and shear-dependent viscosity, the determination of the time-temperature shift function, or the measurement of the p-v-T behavior is essential. Thermal material properties of the polymer and the mold material are fundamental in characterizing heat transfer and heat conduction effects during cooling.

Which steps are generally necessary for a process simulation? The details depend on the process, but in general the following aspects should be taken into account by a simulation.

- First the part should be modeled and, for analysis in an FEM tool, be meshed. The geometry and its complexity will define the kind of modeling. Typically the detailed results will be achieved by three-dimensional (3D) modeling but the expenses increase. In some cases sufficient results will be obtainable with a two-dimensional model or a mid-plane model.

- The boundary conditions of the molding process have to be defined. In detail, theses condition depend on the process, but in some kinds of polymer replication the characteristics of the individual molding machines, like the cooling system, heat conduction of the mold, and mold material, have to be characterized. Especially the modeling of the cooling system is required to calculate the heat transfer during cooling, which will influence the shrinkage, warpage, and crystallinity of the molded part.
- The material properties of molding materials have to be characterized: the shear-dependent viscosity, the time-temperature behavior of the material, the p-v-T behavior as an important behavior to determine shrinkage, the heat capacity of the material, and, if required, the viscoelastic behavior. Measurements of the shrinkage behavior of the material under typical process conditions will help to improve the shrinkage models in the simulation tool. Finally, the material properties should be determined as a function of temperature, especially in the temperature range of the replication process. In this case, the quality of the measured material properties will further influence the results of the simulation. In commercial simulation tools, typically properties for different polymers from an integrated database can be used.
- The process parameters defining the molding step have to be defined: molding temperature, demolding temperature, molding pressure, molding time, and packing profile. Depending on the process, these parameters will vary. For injection molding, e.g., the injection time or the volume flow, or an injection profile can be defined; for hot embossing, e.g., the velocity profile during embossing or the time of dwell pressure and the demolding velocity are important parameters.
- The results of any simulation should be validated by the molding process. In general, a number of results will be obtained during a simulation run. These include the volumetric shrinkage of the part, molecule orientation, warpage, distribution of shear, stress, and pressure. The visualization of the filling of cavities is essential because, e.g., the runner system or other molding process–specific conditions can be optimized. Further, with systematic variations of process parameters in simulation tools, specific molding windows can be determined.

6.2.2 Simulation of Microscopic Polymer Processes

A commercial simulation tool for the hot embossing process or an adapted tool for the micro injection molding process is not available. Commercial tools are optimized and validated for molding macroscopic parts with established processes. If they are adapted to microscopic boundary conditions, the results, especially quantitative results, should be validated. Process simulation of microscopic replication processes therefore is still a subject of scientific research. Nevertheless, qualitative results of such simulation tools can give precious information about the macroscopic parts and in particular about the residual layer as a carrier of the microstructures.

Compared to the simulation of polymer processing of macroscopic parts, process simulation of molding microstructures is a subject of constant development. This development focuses on the complexity of microstructured parts and the validity of established material models and material properties in the micron range. In detail, the following aspects are considered.

- The mesh size increases for complex microstructured molded parts. Compared to macroscopic molding, the ratio between volume and surface increases. This effect results in a significant increase of nodes, especially when a mold is meshed in 3D. To mesh an area of a typical LIGA mold of 26×66 mm^2 with a mesh size of 5 microns, e.g., approximately 63×10^6 nodes are necessary. The size increases dramatically when an 8-inch microstructured molded part is to be meshed in 3D. The consideration of symmetry can reduce the size, but to mesh a complete microstructured mold insert with an adaptive resolution to obtain meaningful results, the number of nodes must be set in the range of 10^6. Simulation of a molding process with such high numbers of nodes makes it necessary to use high CPU capacity. Today, typical personal computers are not suitable for such a complex simulation. However, when the model can be reduced to a two-dimensional model with significant boundary conditions, the number of nodes can be reduced dramatically and the computing time decreases, such that a personal computer can be used to obtain first results.

- Every simulation refers to a material model based on the measurement of material properties. Models to describe the behavior of a polymer melt and the polymer in the solid state are state of the art and are established in many simulation tools. Viscoelastic behavior, for instance, can be described by arrangements of springs and dashpots, and time dependency can be described by a time-temperature shift function like WLF and Arrhenius. These models refer to macroscopic dimensions, and they are also validated for the macroscopic world. To adapt this behavior to the micron range, the models should be updated, but the lack of miniaturized sensors and measurement systems for the systematic determination of the flow of polymer melts in microscopic dimensions makes it difficult to adapt the existing models [4]. Hence, existing models are used to describe the material behavior also in microscopic dimensions.

- Thermal material properties, especially the thermal behavior of a polymer melt, are also fundamental for process simulation. Properties like thermal conductivity, heat capacity, heat transfer, or thermal expansion of solid parts can also be used in the microscopic range.

- However, the flow of a polymer melt in microscopic dimensions may differ from macroscopic flow. For example, the boundary layer in a flow channel has a higher influence when the cross-section of the channel decreases. Flow behavior in micro channels is additionally characterized by capillary flow. This effect is also encountered when molding a polymer melt with low viscosity.

- The physical strength of the molded material is also an important aspect when simulating the demolding cycle. During demolding of microstructures, especially free-standing microstructures with high aspect ratios, the microstructures are subject to tensile forces, which result in strain and in the worst case in the damage of structures. The stress-strain behavior is well-known in the macroscopic range, even at higher temperatures up to the glass-transition range. To measure the stress-strain behavior in the microscopic range, an increasing effort is required. Apart from the microscopic tensile specimen in the dimensions of the microstructures, a measurement system of highest resolution is necessary to measure the tensile force and the related strain.

All aspects mentioned above result in a high complexity of modeling microstructured parts and a certain uncertainty of the simulation results. To improve the results, the material properties may be measured in the micron range, which results in a high measurement expenditure. Not every property of a polymer can be determined in the micron range because of the lack of microscopic sensors of high spatial resolution. The lack of miniaturized sensors also is a problem when validating the simulation results. For example, there is no sensor to measure the pressure distribution in a microcavity of a few microns. Consequently, some of the simulation results cannot be validated quantitatively; they can only be checked for plausibility. Nevertheless, the use of commercial simulation tools allows one to obtain initial information about the flow and the thickness of the residual layer as a carrier of microstructures, for example.

6.2.3 Commercial Simulation Programs

Commercially available simulation software covers macroscopic molding processes, mostly injection molding and the related processes like injection compression molding or chip encapsulation. However, these tools can be used in a first step to simulate the characteristic molding steps. The process steps of micro injection molding are, in an initial view, partially similar to conventional injection molding steps. Hence, under selected aspects commercial software can be used. Nevertheless, the simulation results should be validated. Use of a commercial simulation tool for hot embossing is much more difficult. A special tool that refers to every process step and all boundary conditions of hot embossing and its process variations is not commercially available. Some commercial tools, though mostly optimized for molding in macroscopic dimensions, are flexible enough to set the boundary conditions such that hot embossing conditions are approximated. With the help of such an adapted tool, the different steps can be simulated, for example, the filling of cavities or the thickness of a residual layer. In particular, the flow behavior can be predicted. The injection compression feature of the commercial software MOLDFLOW [3], for instance, can be used to describe the two-step embossing process. A detailed analysis of the available software reveals other combinations. Nevertheless, developments from scientific institutes are also partially commercially available, for example, the software EXPRESS [2], a simulation tool describing the filling behavior during embossing of fiber-filled polymers. This simulation tool considers the position of the semifinished product in relation to the volume of the molded part. Another tool will be commercially available in the near future. Based on a cooperative effort between Forschungszentrum Karlsruhe, Germany, and the Industrial Materials Institute NRC, Canada, a basic 3D simulation tool for hot embossing was developed [39]. This tool allows the simulation of the characteristic molding steps, especially the cooling and demolding step.

6.3 Process Simulation of Micro Hot Embossing

After general remarks, the aim of this section is to analyze every process step of hot embossing: the heating step, the filling of microcavities, the velocity- and force-controlled molding, the cooling step, and finally, the demolding of the structures to complete the process chain. Analysis will cover different molding tools, commercial tools, and special developments, and also different geometric models. The lack of a special simulation tool makes it necessary to split the simulation into several parts [38]. Commercial simulation software is not suited to simulate the demolding behavior. However, the demolding step is of importance because of the high risk of damaging structures during this process step. Due to the lack of an adapted simulation tool, this step has to be simulated by a commercial FE tool in terms of the characteristic properties of demolding in hot embossing, demolding supported by the adhesion of a substrate plate, and differences in shrinkage between the polymer and mold insert [10,11,13,16,26,41]. Besides the characterization of the viscoelastic material properties, the thermal behavior, and the determining of friction coefficients, the geometric modeling of mold insert and substrate plate is also required.

6.3.1 Modeling of Typical Microstructured Parts

The challenge of geometrical modeling is the large number of nodes required for modeling a complete microstructured molded part. A typical molded part consists of microstructures and the remaining layer as a carrier of the microstructures. Due to the different lateral dimensions of the residual layer and the microstructures, aspects concerning the whole part can be analyzed in a first step by concerning only the component with the largest dimension, here the residual layer without microstructures. In detail, the following differentiation can be made.

- An elementary design is only a flat circular molded plate with typical properties of a molded part like diameter and thickness. This model represents the residual layer without microstructures and can be used to simulate elementary correlations like pressure distribution, elementary flow behavior, shrinkage, and warpage.
- The influence of microstructures on this elementary behavior can be analyzed by modeling an elementary layer with typical dimensions enlarged by a systematic arrangement of single microstructures. These microstructures may have different geometries and arrangements. In a first analysis, however, freestanding structures that are systematically arranged over the whole area allow the analysis of the influence of the position of the structures on the center of the molded part (Fig. 6.2).
- Modeling of a real geometry eventually allows the analysis of a design in detail and to validate the simulation results quantitatively. This analysis requires high computing power and measurement systems of high resolution. Detailed analysis will probably show more details and effects that cannot be predicted by simpler models (Fig. 6.2).

A common feature of all models is the consideration of the mold insert and substrate plate as parts of the demolding simulation. As an approximation, these components can be modeled as rigid elements without any deformations. Friction between the mold insert and polymer, and especially adhesion and friction between the polymer and substrate plate, have to be considered.

6.3.2 Modeling of Process Steps

Apart from the geometry, characteristic process steps of a hot embossing cycle have to be modeled by defining the related load steps.

- The positioning of the polymer foil and the heating of the polymer sheet will cause undefined shapes because of the retardation effects of the typically oriented polymer foils. This effect can be countered

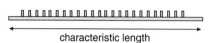

characteristic length

(a) 2-dimensional geometry with arranged microstructures

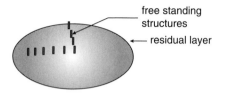

(b) Circular disk with arranged microstructures

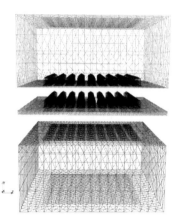

(c) 3-d modeling with mold insert and substrate plate

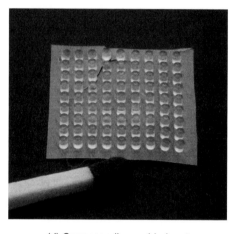

(d) Corresponding molded part

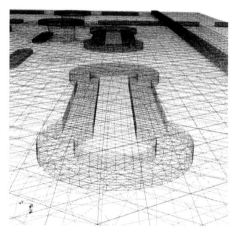

(e) Large area modeling

(f) Corresponding 8 inch microstructured mold insert

Figure 6.2 The geometric models are split into several groups, ranging from a simple 2D up to a complex 3D model of a real mold insert. Apart from the molded part, the mold and the substrate plate have to be modeled.

by the use of touch force during heating. To simulate this effect theoretically, the history of the polymer foil has to be considered, which is nearly impossible in practice. Therefore, the process simulation typically starts with a defined layer of polymer melt at defined temperature. This simplification is already a first idealization.

- The molding has to consider both load steps—the velocity-controlled molding, with the molding velocity being constant, and the force-controlled molding, with a constant dwell pressure. These two steps are similar to injection compression molding. Hence, commercial simulation programs can be used and the process parameters like embossing velocities and the isothermal embossing cycle have to be adapted.

- The cooling of the molded part starts after an isothermal holding time under dwell pressure. The cooling can be modeled by the consideration of a defined cooling rate or, in a simple case, by a temperature profile of the molded part. This assumption is suitable if the mold insert and the substrate plate will be cooled uniformly without any gradient over the area even in the non-stationary cooling state. An ideal simulation will consider the cooling system, the flow channels, the cooling fluid, the convective heat transfer, the heat conduction of the material, and finally the heat transfer between the interface of mold insert and cooling unit dependent on the pressure. Especially the contact pressure will influence the heat transfer between the plates. Nevertheless, all these factors can be theoretically determined and integrated in a simulation tool, but this will increase the complexity of a simulation tool and can decrease the flexibility of its area of application. An idealized approach is therefore to assume a cooling profile acting on the molded part that can be suitable for thin residual layers and a homogeneous temperature distribution in the mold insert and substrate plate during cooling. This approach is underlined by the comparatively small heat capacity of the molded part compared to the heat capacity of the mold insert and tool.

- The demolding step is characterized first by a switch from force-controlled movement during cooling to a velocity-controlled relative movement of the substrate plate and mold insert. This movement is defined by the demolding velocity. The modeling of demolding differs from the filling and cooling because of the switch between a material model suitable for the simulation of the flow behavior of a polymer melt to a model that is suitable to describe the behavior of the solid material during demolding near the glass-transition range. An important aspect is the transfer of stress induced by the molding from one model to another model. Nevertheless, if the molding temperature is in a range in which the relaxation times of a polymer fit the typical molding times, the stress will be reduced and an assumption can be made that the stress during demolding is caused only by the differences in thermal behavior during cooling from molding temperature to demolding temperature.

- Finally, all the aspects described here consider only the pure embossing process. The technical implementation of the hot embossing process will cause a lot of factors that can influence the process significantly. In a process simulation these factors—like uniform heating, uniform surface roughness of mold inserts, undercuts of microcavities, uniform distribution of polymer melt, bending of mold insert and substrate plate under load, or unevenness of mold inserts—can result in a falsification of the simulation results. Therefore, the validation of idealized simulation results should be considered under these boundary conditions.

6.3.3 Modeling of Material Behavior

The modeling of the material behavior for a process simulation requires the modeling of the behavior of the polymer during melt, cooling, and demolding. Further, the shear thinning behavior and the viscoelastic properties and the p-v-T-behavior of the polymer have to be taken into account. Here, an amorphous PMMA, BASF Lucryl G77Q11, shall be described in detail. This material may be considered a representative thermoplastic amorphous polymer material. Hence, the qualitative results may be transferred to other polymer materials. These measurements refer to established measurement techniques for determining the viscosity behavior.

6.3.3.1 Shear Thinning Material Model for the Flow Behavior

The thermal molding window of hot embossing contains the temperature range beginning at the softening range of the polymer up to the range of a polymer melt with viscous flow properties. The flow behavior can be characterized by the shear and temperature-dependent viscosity of the melt modeled with a viscosity model of Cross or Carreau, considering the shear thinning behavior (Section 3.3.1). Under consideration of the typical shear velocities during hot embossing in the range between 0.1 and 100 per second, the temperature-dependent behavior of the viscosity is visualized in Fig. 6.3.

The flow of polymer melts in the sub-micro and nano scale could change the material properties. Leveder et al. [25] used the nanoimprint process to characterize the flow properties above glass-transition temperature of thin polymer resist in a range below 100 nm. They found that the order of magnitude of viscosity measured at nano scale was higher than the macroscopic viscosity of polymers with a similar molecular weight.

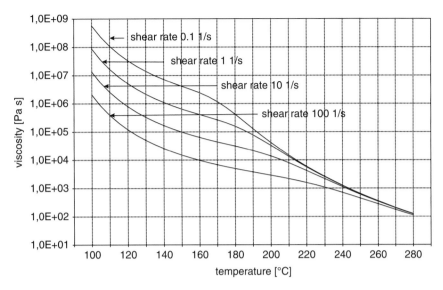

Figure 6.3 Viscosity model of PMMA. The model describes the behavior of PMMA Lucryl G77Q11 dependent on temperature and shear velocity. Typical shear velocities during hot embossing are in the range between 0.1 and 100 per second.

6.3.3.2 Material Model for the Demolding Behavior

After filling the microcavities the material is cooled down to demolding temperature, a temperature range near or below the glass-transition temperature of the polymer. This behavior is determined by relaxation and creeping of the material under load and under different temperatures. In this case, the relaxation modulus of the material has to be determined under consideration of the time-temperature shift using the WLF equation. Especially the time-temperature shift is required to characterize the relaxation behavior over the temperature range of cooling and demolding. The determining of the relaxation behavior can be done by systematic DMTA measurements (Section 3.4.3). The measurement results in a time- and temperature-dependent Young's modulus, here the relaxation modulus (Fig. 6.4).

A significant relaxation during typical process times of a few minutes can be found in the temperature range of about 170°C, a temperature of 65°C above the glass-transition temperature of this polymer. At 170°C, a relaxation of 10^4 Pa can be observed in approximately 300 seconds. At 160°C, the same value of relaxation is reached after 2.8 hours; at 180°C, only 100 seconds are necessary. To reduce the residual stress during molding, the temperature has to be set such that the relaxation times are shorter than the process times. In this concrete case, the molding temperature should be set up to 180°C. The relaxation behavior of molded parts was also investigated by Ding et al. [7]. They studied the evolution of polystyrene grating patterns using in situ light diffraction during

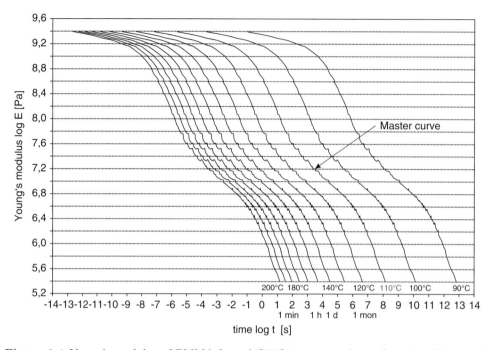

Figure 6.4 Young's modulus of PMMA Lucryl G77Q11 measured as a function of time and temperature. The relaxation time depends largely on temperature and is characterized by the WLF time-temperature shift. At temperatures of 170°C and higher, the relaxation times are in the range of or lower than the process times.

thermal annealing. It was found that the decay rate shifts to a shorter time with increasing annealing temperature.

From the Young's modulus measured, the shear modulus G and compression modulus K can be calculated taking into account the temperature-dependent Poisson value $\mu(T)$.

$$G(t, T) = \frac{E(t, T)}{2(1 + \mu(T))} \tag{6.21}$$

$$K(t, T) = \frac{E(t, T)}{3(1 - 2\mu(T))} \tag{6.22}$$

To describe the viscoelastic material behavior mathematically, a general Maxwell model—a parallel connection of a spring and a dashpot arranged in series—can be used (Section 3.3.3). To obtain a model, the measured relaxation curves can be fitted to the theoretical curves of a generalized Maxwell model. The parameter obtained by the curve fitting determines the spring constant and the viscosity of each Maxwell element. A model with 10 elements describes the viscoelastic material behavior in a time range over 12 dimensions. This time

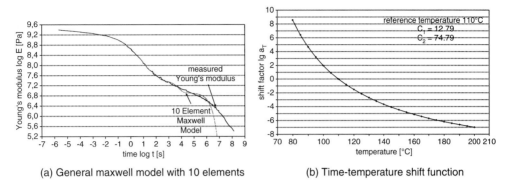

(a) General maxwell model with 10 elements

(b) Time-temperature shift function

Figure 6.5 The viscoelastic behavior can be described by a general Maxwell model with, e.g., 10 single Maxwell elements. The behavior can be characterized over a period of 12 dimensions.

range is sufficient to characterize the behavior during typical process times of several minutes (Fig. 6.5)

6.3.3.3 Material Strength during Demolding

The simulation results, especially the calculated equivalent stress of the structures during demolding, correspond to the strength of the polymer under demolding conditions. To characterize the strength of the polymer material during demolding, systematic measurements of the tensile behavior are required. The demolding process is characterized by tensile forces, applied on the structures at a temperature range near the glass-transition temperature of the polymer. Databases results of tensile measurements in this temperature range and under typical tensile velocities compared to demolding velocities are still lacking. The material behavior under these boundary conditions has to be determined by tensile test under constant temperatures and velocities with a specimen molded by hot embossing. Results of the measurements have already been described in Section 3.6.3.

6.3.3.4 Friction between Mold and Polymer

The simulation of the demolding is characterized by the determination of the friction between mold and polymer and also between polymer and the rough substrate plate. Especially the adhesion and the friction on the rough substrate plate in the case of single-sided molding influences the demolding behavior significantly. In this case, the normal forces acting on the sidewalls of the micro-structured mold are also determined by adhesion and static friction of a thin residual layer on the substrate plate. High adhesion and high friction on the substrate partially compensates the thermal contraction during cooling and reduces the contact stress on the sidewalls.

The determination of friction between mold insert and polymer and substrate plate and polymer is determined by a previous molding step in which the liquid polymer is pressed into the roughness of the surfaces of mold and polymer. Therefore, typical measurements regarding the DIN/ISO instructions are not suitable for the determination of friction coefficients. In this case, the boundary condition of the polymer replication process has to be considered. Nevertheless, the difficulties in a systematic measurement of friction by the measurement of demolding forces under real hot embossing conditions can be seen by the unknown normal force acting on the sidewalls of a real mold insert during demolding. Therefore, the measurement has to be transferred from the hot embossing process to a measurement system that allows the simulation of the previous hot embossing cycle under typical hot embossing conditions and under defined temperatures, forces, and surface roughnesses. The surface roughnesses correspond to the typical roughness of the different kinds of mold inserts. A detailed description of the measurement system can be found in Section 3.7. Ferreira et al. [9] have already investigated the friction for injection molding conditions by the development of a measuring technique for tensile testing machines. They found that the surface roughness and the presence of release agent are the most important parameters. Melt and mold temperature were identified as second-order parameters.

Independent of this work with the systematic experiments done by the friction measurement technique, the friction coefficients in relation to the process parameters and surface roughness can be determined. Molding temperature and molding force especially show a significant influence on the static-friction coefficient [40] (Fig. 6.6). The temperature and also the molding pressure are responsible for the form fitting between polymer melt and surface roughness. This form fitting increases if the viscosity decreases and the pressure acting

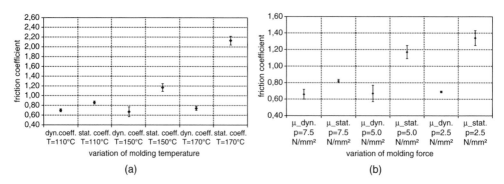

Figure 6.6 Measurement of friction coefficients dependent on (a) molding temperatures and (b) molding pressure. An increase in molding temperature and molding force increases also the static-friction coefficient.

on the melt increases. Molding temperature and molding force influence mainly the static-friction coefficient; the dynamic-friction coefficient is influenced moderately. Nevertheless, the static-friction coefficient is responsible for the breakaway torque during demolding and corresponds to the damage of structures during demolding.

6.4 Process Analysis

The aim of this section is to analyze every process step of hot embossing and to present the characteristic correlations based on simulation results.

6.4.1 The Process Step of Heating

The process step of heating is characterized by heat transfer from the heating system to the polymer foil. Besides the heat transfer over different contact areas, the heat flow inside the material also determines the heating rate.

Regarding the basic correlations of heat transition and heat conduction, the main influence factors can be deduced.

The stationary heat flow determined by heat conduction is characterized by Fourier's law:

$$j_q = -\lambda \frac{\delta \vartheta}{\delta n} \tag{6.23}$$

where j_q is the heat flux density, λ is the coefficient of thermal conduction, n is the direction in space, and $\delta \vartheta / \delta n$ is the temperature gradient in the direction in space.

The heat flux density can be determined by

$$c\rho \frac{\delta \vartheta}{\delta t} = \dot{f} - \left(\frac{\delta j_{qx}}{\delta x} + \frac{\delta j_{qy}}{\delta y} + \frac{\delta j_{qz}}{\delta z} \right) \tag{6.24}$$

and the temperature field by

$$c\rho \frac{\delta \vartheta}{\delta t} = \dot{f} - \lambda \left(\frac{\delta^2 \vartheta}{\delta x^2} + \frac{\delta^2 \vartheta}{\delta y^2} + \frac{\delta^2 \vartheta}{\delta z^2} \right) \tag{6.25}$$

where λ is the coefficient of thermal conduction, f is the density of energy in the source of heat or heat sinks, c is the heat capacity, and ρ is the density.

Because of the difficulties in solving the equation for real problems, parameters are defined characterizing the heat conduction of a material. The parameter characterizing the material behavior for heating and cooling is defined by the "heat penetration coefficient" b

$$b = \sqrt{\lambda c_p \rho} \tag{6.26}$$

and the thermal diffusivity characterizing the material behavior for temperature cycles:

$$a = \frac{\lambda}{c_p \rho} \tag{6.27}$$

Besides the heat conduction, the heat transition between different contact areas also influences the heat flow. The heat flow between n plates is calculated by:

$$\dot{Q} = k \cdot A \cdot \Delta \vartheta \tag{6.28}$$

with the heat transition coefficient k

$$k = \frac{1}{\frac{1}{\alpha_1} + \sum_1^n \left(\frac{s_i}{\lambda_i}\right) + \frac{1}{\alpha_2}} \tag{6.29}$$

where λ is the coefficient of thermal conduction, s is the thickness of the layer, α is the heat transmission resistance at start (1) and end (2) of heat flow, and A is the contact area.

The heat transfer coefficient k determines the heat transmission and depends on the relation between the thickness of the transmission layer and the coefficient of the thermal conduction of this layer. Here, only a small air gap will reduce the heat transmission significantly because of the small heat conduction coefficient of air. This requires a high grade of evenness of the contact area and a low surface roughness of the contact partners. If the molding force acts on the contact partner, the micro gaps and the unevennesses will be closed under pressure and an increase of the heat conduction will be observed. The increase in heat conduction can be observed regarding the temperature curves of a typical hot embossing cycle.

The selection of the material used for the fabrication of heating and cooling plates refers to the coefficient of thermal conduction and the stiffness characterized by the compression modulus of the material. Table 6.1 shows characteristic material data. For short heating times, the material should be characterized by a material with a high coefficient of thermal conduction and also a high stiffness against the cyclic thermal and mechanical load during embossing. To achieve a good compromise between stiffness and heat conduction, specific alloys are typically used.

For a process simulation the heat distribution over time in a mold insert and substrate plate can be calculated by finite element analysis. Nevertheless, these

Table 6.1 Thermal and mechanical properties of tool materials

Material	λ (W/m K)	a ($\times 10^6$ m^2/sec)
Aluminum	221	88.9
Copper	393	113.3
Steel 0.6 C	46	12.8
Polystyrole	0.17	0.125
Air	0.026	21.8

Copper is, for example, a material with excellent heat transfer properties, but it deforms under high molding load. Steel has a high stiffness but bad heat conductivity. Specific alloys are typically used as a compromise between stiffness and heat conduction.

calculations are specific for a certain heating system and cannot be transferred to other systems. Regarding the coefficients of thermal conduction, it can be assumed that if the stationary state is achieved, a nearly homogeneous temperature distribution in the surface of mold and substrate plate is attained. These assumptions can be confirmed by temperature measurements under the surface of a substrate plate and a mold insert (Section 8.2.1). For a process simulation, therefore, the molding process can start at a constant temperature.

6.4.2 The Process Step of Molding

The flow behavior during molding is one the most investigated parts of hot embossing or nanoimprinting. Cross [6] surveyed the developments in nanoimprinting with the focus on the flow behavior and the filling of cavities. The filling of microcavities and the theory for flow in the micro and nanoscopic range is a basic research topic because of the differences during macroscopic and microscopic flow behavior. Systematic experiments regarding the flow behavior were investigated by Scheer and Schulz [33] and Heyderman et al. [14]. Juang et al. [17,18] investigate theoretically and by experimentation the flow behavior during filling microcavities under isothermal and non-isothermal conditions and the effects the processing conditions have on the part quality. They found that the flow profile differs depending on the temperature conditions. Under isothermal molding conditions, the quality of the part depends strongly on the processing conditions and parameters. Non-isothermal molding results in an inhomogeneous temperature inside the polymer and a different flow profile compared to the flow profile of isothermal embossing [17]. Schift et al. [35] investigated micro- and nano-rheological phenomena occurring during hot embossing of thin polymer films. These flow phenomena can be attributed to electrostatic interaction, capillary action, and surface energy minimization effects. Lei et al. [22] investigated the contact stress between mold and polymer and wall profiles during molding micro channels in PMMA in a temperature range of 120°C, near the glass-transition temperature of the polymer.

Nevertheless, this process step is characterized by a complex flow behavior of the polymer melt with two different, partly overlapping, drifts of the polymer melt. On the one hand, the filling of microcavities is generally associated with a flow in the direction of the acting press force. On the other hand, a radial squeeze flow occurs, as a polymer melt is squeezed between two plates. The squeeze flow, the behavior of which is described above in an analytical way, can be used to characterize the flow in the residual layer after the microcavities are filled. To understand the complex flow, the simulation of molding behavior is split into a section modeling the general flow behavior of the residual layer without any cavities and a second section describing the mold filling. Finally, the overlapping of both flows characterizes the flow behavior during molding.

Because commercial hot embossing software is lacking, general aspects can be shown by the commercial simulation of the injection compression molding process. Its compression state is similar to the compression state of hot embossing. The previous injection process can be set, such that the distribution of the polymer melt at the end of the injection step is similar to the molten polymer of the foil.

6.4.2.1 Squeeze Flow Behavior

As mentioned above, hot embossing is characterized by a squeeze flow. Nevertheless, independent of the analytical approach, a detailed simulation allows one to determine and visualize further correlations like the direction of flow, radial flow velocities, the shear velocity, and the pressure distribution during the velocity-controlled molding and the force-controlled molding step. Further, the thickness of the residual layer under the influence of molding force, molding temperature, and molding time can be determined. Finally, all these results correspond to a viscoelastic material model regarding shear thinning flow behavior.

The model used for the analysis is, first, a simple disc with a diameter of 80 mm corresponding to the size of a 4-inch wafer. Further, a rectangular geometry in the size of 26×66 mm^2 was selected to correspond to the size of a typical LIGA mold insert.

In addition to the simulation results, the characteristic flow can be visualized in an experimental way. For this purpose, polymer foils are marked with selected colored lines and are put in a stack, such that different-colored lines define different layers of the polymer stack. After heating up under touch force and after the molding with two flat plates, the flow of polymer can be retraced in different layers (Fig. 6.7). Retracing of polymer flow is a helpful method to analyze flow behavior. These colored markers allow the visualization of comparatively large

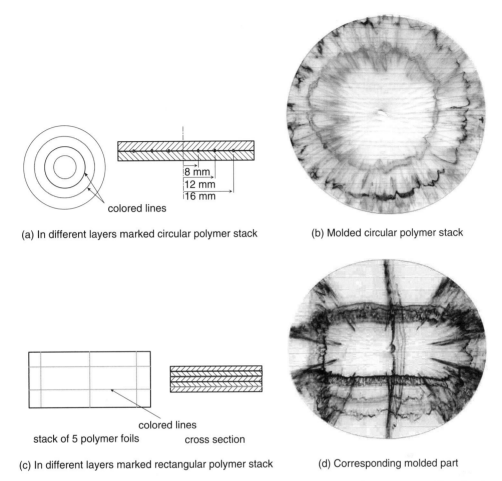

(a) In different layers marked circular polymer stack

(b) Molded circular polymer stack

(c) In different layers marked rectangular polymer stack

(d) Corresponding molded part

Figure 6.7 Characteristic flow behavior of a circular and rectangular geometry. The circular or rectangular foils are marked by colored lines in different layers. After molding, the flow behavior of the melt can be retraced by these lines.

flow distances. The visualization of the smallest deformations during cavity filling can be difficult. Here, the method developed by Macintyre and Thoms [27] can help to visualize the flow of polymer also in the sub-micron scale. They embedded a fiducial grid with high resolution in a second layer under the molding resist. Molding of this stack deforms the resist and also the grid. By an analysis of the grid deformation after molding, conclusions to the flow behavior can be drawn. Nevertheless, the ideal method is on-line viewing during filling, already achieved for injection molding [42]. Because of the flow of force in hot embossing, the visualization of the polymer flow in a mold is challenging, especially the integration of optical systems.

The molded parts portray the expected radial flow behavior. The analysis of the color distribution after molding shows that the flow in different layers is not

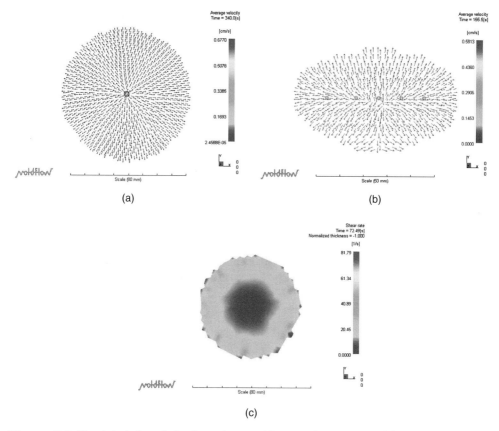

(a)   (b)

(c)

Figure 6.8 Simulated flow behavior when molding a flat layer of (a) circular and (b) rectangular geometry. The typical squeeze flow during embossing can be visualized by the flow velocities of the polymer melt during molding. The velocity vectors are oriented toward the peripheral regions. (c) This figure illustrates flow at the end of the velocity-controlled molding step with a typical molding velocity of 1 mm/min. Under the assumption of a constant squeeze velocity, the velocity of the polymer melt in the gap between the two plates increases during molding. A maximum shear velocity in a range of 40 per second was calculated.

uniform, which can result from a radial flow profile. The flow length varies over the height of the molded disc. Further, these experiments show the flow behavior caused by an angular deviation of both plates. This angular deviation results in a wedge-shaped gap that increases the flow resistance. Because of this, the disc does not show a perfect circular form. The corresponding simulation shows the flow behavior while embossing a flat residual layer. As visualized by the experiments, the flow is characterized by a radial flow from the center of the molten polymer toward the peripheral regions. By a typical molding velocity of 1 mm/min, the shear velocity during molding is in a range of 40 per second (Fig. 6.8).

6.4.2.2 Characteristic Pressure Distribution

The act of compression force results in a pressure distribution in the melt with the effect of the melt squeezing out during the velocity- and force-controlled molding state. The squeezing of the melt is responsible for the final thickness of the residual layer of a molded part. The final thickness is based on the equivalence between the compression force and the flow resistance in the gap under consideration of the viscoelastic behavior of the polymer melt, which also results in an asymptotic decrease in thickness.

Knowledge of the pressure distribution during molding allows one to estimate the filling behavior of microcavities arranged on the remaining layer. Local pressure at a defined position may be considered the pressure available to fill a microcavity at this position. For a typical molding geometry of a circular plate, this pressure distribution can be simulated over a molding cycle. Pressure in the residual layer depends on geometry and time, which relates to the viscoelastic behavior of the polymer melt. In Fig. 6.9, the pressure inside a molded circular plate is calculated considering the two steps of isothermal molding, constant velocity, and constant force.

During the velocity-controlled molding step, the pressure gradient increases and reaches the maximum at the end of this molding step. After switching to force-controlled molding, the pressure gradient decreases with time, as the molded area increases due to the creeping of the melt under a constant load

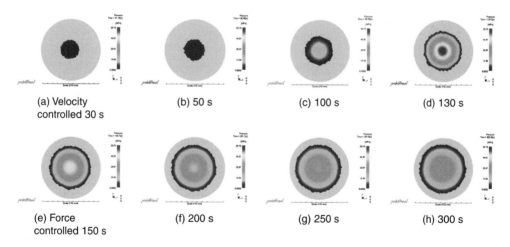

(a) Velocity   (b) 50 s   (c) 100 s   (d) 130 s
controlled 30 s

(e) Force   (f) 200 s   (g) 250 s   (h) 300 s
controlled 150 s

Figure 6.9 Characteristic pressure distribution inside a circular disc during the two steps of molding (molding velocity, 0.5 mm/min; molding force, 100 kN; molding temperature, 180°C; press holding time, 300 s). The circular disc has a maximum diameter of 80 mm. Characteristic is the time dependence that results in a maximum pressure gradient at the end of the velocity-controlled molding state and a decrease in the pressure gradient in the force-controlled molding state.

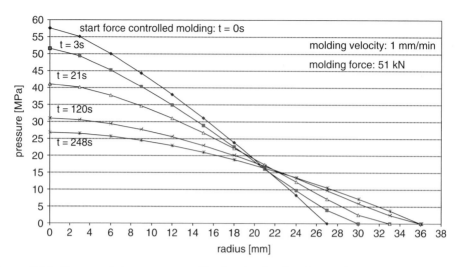

Figure 6.10 Characteristic pressure profile of a circular disc during force-controlled molding. The pressure gradient decreases with time. Similar to this effect, the diameter of the molded disc also increases by the effect of creep under load.

(Fig. 6.10). This effect is illustrated more clearly by Fig. 6.11, in which the measured pressure profile versus the radius of the disc is presented.

For the geometry of a circular disk, a parabolic pressure profile can be observed, with the maximum pressure being encountered in the center of the disc. At the circular flow fronts, the pressure corresponds to ambient pressure. The decrease of the pressure gradient during the force-controlled molding step is not linear with time. At the beginning of this step, a significant decrease in the gradient within the first seconds can be calculated. At the end of the holding time of 280 seconds, the maximum pressure decreases by about 57%.

The decrease of the pressure after switching to the force-controlled molding can be illustrated by Fig. 6.12. The pressure in the center of the disc is plotted over the molding time for different molding velocities. The decrease of pressure shows an asymptotic shape of the curve. Independent of this simulation Mehne [46] has measured the pressure distribution between two flat plates during embossing. Because of the different press force the values cannot be compared directly, but the shape of the pressure curve is similar.

Simulation shows that a higher molding velocity results in a higher pressure in the center of the disc at the end of the velocity-controlled molding step. In Fig. 6.13, the pressure profiles relating to the molding velocities of 1 mm/min and 10 mm/min are compared. At the beginning of the force-controlled step, both profiles differ. At the end of the holding time of 280 seconds, however, the pressure differences between both profiles are balanced. The pressure profile at the end of the holding time is independent of the molding velocity. However, an appropriate holding time is a prerequisite.

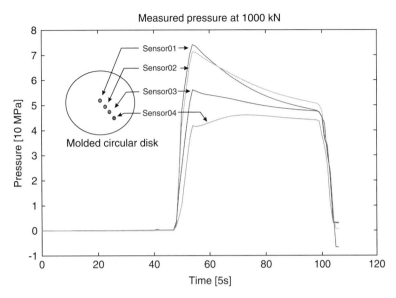

Figure 6.11 Measured pressure profile in a circular disc. The measurements of the radially arranged pressure sensors show a similar time-dependent pressure profile at the center of the disc.

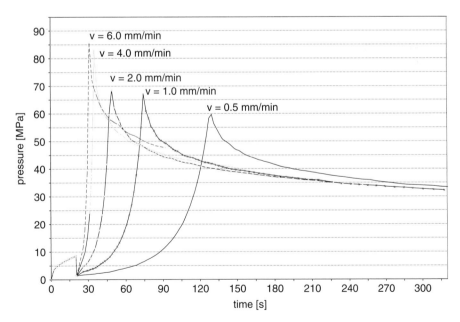

Figure 6.12 Characteristic pressure in the center of the disc over the molding process. After the switch to force-controlled molding, the pressure decreases in an asymptotic form, independent of the molding velocity.

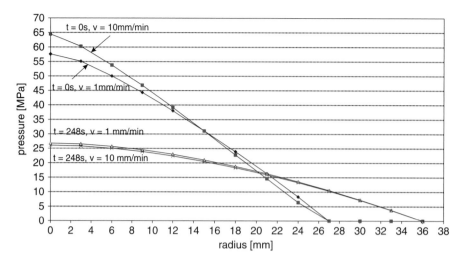

Figure 6.13 Pressure profiles of a circular disc during force-controlled molding as a function of the velocity of the velocity-controlled molding step. A higher molding velocity increases the pressure gradient at the beginning of the force-controlled molding step. With progressing time, however, the gradients are balanced.

6.4.2.3 Thickness of the Residual Layer

For many applications, the residual layer has to be as thin as possible. As already seen in Fig. 6.11, the diameter of the residual layer increases during force-controlled molding, which results in a decrease in thickness. Apart from the volume of the polymer material used, the process parameters of molding temperature, press force, molding velocity, and holding time influence the thickness of the remaining layer. An effective way to reduce the thickness of the residual layer is to use a high molding temperature. As the viscosity of most thermoplastic polymer melts depends on temperature, it will decrease with an increasing temperature, which will result in a reduction of the pressure needed to obtain a flow in the gap between two plates. Besides the process parameters, the initial thickness is, in praxis, an influence factor. Lee et al. [21] showed the dependency of the initial thickness and the final residual layer by imprint experiments and measured the thickness with specular X-ray reflectivity methods.

The influence of the process parameters of press force, molding velocity, and holding time is illustrated in Fig. 6.14. The thickness of the remaining layer may be represented by the calculated time-dependent position of the crossbars of the hot embossing machine during the force-controlled molding step. Irrespective of the molding velocity and press force, a significant reduction of the residual layer can be obtained in the first seconds of the force-controlled molding state. By way of example, a time period of approximately 120 seconds is calculated. With further time, the reduction of the residual layer decreases because of the decreasing gap and the related increase in pressure drop. This behavior results in the

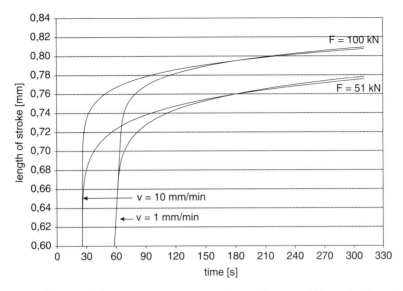

Figure 6.14 Influence of the process parameters of press force, molding velocity, and holding time on the thickness of the residual layer. To make this influence obvious, the press force is set at 51–100 kN and the molding velocity at 1–10 mm/min. The values refer to a disc of about 80 mm in diameter and a molding temperature of 180°C for PMMA.

asymptotic shape of the curve. For the final thickness of the residual layer, only the parameters of temperature, press force, and holding time are responsible. The influence of the molding velocity is lost with further time periods. Only for short periods after the switch from velocity-controlled molding to force-controlled molding does the influence of the molding velocity have to be taken into account.

When transferring the behavior described above to a real microstructured mold, it is evident that the filling of the microcavities is part of the velocity-controlled molding step, because the maximum pressure to fill microcavities is reached at the end of this step. The maximum pressure is determined by the selected molding force and the dynamic part represented by the molding velocity. After switching to the force-controlled molding step, the pressure decreases. During holding time, only the thickness of the residual layer is reduced.

6.4.2.4 Filling of Microcavities—Pressure Drop Estimation

The filling of microcavities by the hot embossing process is characterized by short flow distances, that is, the height of the cavities. Typical flow distances are in a wide range from as few as ten nanometers over typically a few microns up to a height of approximately 1 mm. Cross-sections vary also from the nano range up to a cavity width of typically several hundred microns. Interest focuses on the description of the filling behavior [34]. Heyderman et al. [14] visualized the different filing states of a 100 nm-high structure by optical images and AFM measurements. Hirai et al. [15] investigated by numerical simulation and experiments the filling of cavities with a 100 nm width and up to 860 nm height. They found that, for filling cavities, the required pressure increases when the initial thickness of the polymer decreases to less than about two times the cavity depth of the mold insert. Mohamed et al. [29] investigated the molding of structures around 60 nm in height in a 100 nm PMMA film below the glass-transition temperature.

Independent of different approaches to describe the filling of microcavities [36,44,45], the general particularities of the polymer flow in micro channels shall be discussed. In addition to the macroscopic flow, other aspects have to be taken into account [19,43]:

- The change of the viscosity in micro channels: Eringen and Okada [8] indicated that in micro channels in the dimension of tens of micrometers the viscosity at the channel wall is 50–80% higher compared to the bulk viscosity of the fluid. This increase can be explained by collective molecular motion effects or by the immobility of the layer of molecules in contact with the solid channel wall. Eringen and Okada further developed a theory allowing, for viscous, fluids the determination of the viscosity dependent on molecule orientaton effects and the gyration radius of the fluids.
- Wall slip: In contrast to non-slip boundary conditions, polymer melts can slip over solid surfaces if the shear stress exceeds the critical stress (usually 0.1 MPa for injection molding) [43]. Also, at low shear stress wall slip can be observed. A linear slip velocity (linearly proportional to the wall stress) can be determined in a range as a fraction of 1 μm/sec.
- Surface tension effect: Surface tension is a driving effect for the capillary effect. Its influence on cavity filling can be estimated by the equivalent pressure change Δp by the following equation:

$$\Delta p = \frac{\lambda \cos \theta}{r} \tag{6.30}$$

  where λ is the surface tension of the liquid, r is the radius of capillary, and θ is the contact angle of the liquid with the capillary tube. Regarding the surface tension of polystyrene, it can be shown that the equivalent pressure can be neglected for capillary sizes over 1 μm compared to typical filling pressures in the range of a few megapascals. For a 100 nm channel, an equivalent pressure of around 0.5 MPa can be calculated.

Finally, the effect of micro-scale viscosity and wall slip are considered by a flow simulation with polystyrene [43]. It can be shown that the predicted filling pressure is 20–30% higher than in standard filling. Independent of these results, a coarse estimation of the required filling pressure can be given by a simple model, extended with a model for the shear thinning behavior of the melt. Nevertheless, the calculated pressure drops should be seen under the aspects named above.

For structures with higher aspect ratios and therefore longer flow paths, basic relationships and a first estimation of the pressure necessary to fill a cavity can be given by an analytical approach. Detailed information requires a FEM analysis. Here, both the cases of the analytical approach and a FEM analysis will be presented.

In case the cavities are considered to be flow channels with a flow distance much larger than the diameter, the well-known equations to describe the pressure drop in channels can be used. This model is only valid for Newtonian fluids, such that the equations have to be adapted to describe the shear thinning and time-dependent behavior of a polymer melt. Under the preconditions of laminar flow behavior and homogeneous polymer melts, the model of the representative viscosity [28] (Section 3.3.2) can be used. The calculated pressure drop can be used as the minimum pressure needed to fill the cavity. Here, the pressure drop is estimated as a function of molding velocity and temperature-dependent viscosity, with circular cavities being used as an example. The model to approximate the pressure drop is illustrated in Fig. 6.15. Assuming that the thickness of the polymer melt prior to molding is greater than the height of the cavity, the

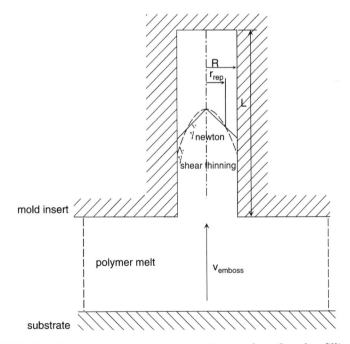

Figure 6.15 Model of the representative viscosity to describe the filling of circular microcavities. This model refers to a representative point at which the shear velocities of a Newtonian fluid and a shear thinning polymer are equal. The flow velocity in the capillary can be set to the molding velocity in the first approximation.

velocity of the polymer melt in the circular capillary can be set to the molding velocity in the first approximation.

The material used for the description is PMMA. The shear-dependent behavior can be described by the Cross-WLF model (Section 3.3.1).

$$\eta(\dot{\gamma}) = \frac{\eta_0}{1 + \left(\frac{\eta_0 \dot{\gamma}}{\tau^*}\right)^{(1-n)}} \tag{6.31}$$

with

$$n = \frac{1}{m} = \frac{\Delta(\log \tau)}{\Delta(\log \dot{\gamma})} \tag{6.32}$$

where η_0 is the viscosity at zero rate of shear, $\dot{\gamma}$ is the shear velocity, τ^* is the shear stress at the intersection of the two lines defining the change between Newtonian fluid and shear thinning behavior, and n is the reciprocal flow exponent.

To approximate the pressure drop, the shear velocity in a cavity has to be estimated. The shear velocity $\dot{\gamma}$ at the representative point is defined by

$$\dot{\gamma}_{rep} = \dot{\gamma}_{newtonian}(r_{rep}) = \frac{4\dot{V}}{\pi R^3} \cdot e_0 \tag{6.33}$$

with the representative point

$$e_0 = \frac{r_{rep}}{R} \approx \frac{\pi}{4} \tag{6.34}$$

where $\dot{V}$ is the volume flow, R is the radius of the capillary, and $\dot{\gamma}$ is the shear velocity.

Under the assumption of a constant volume flow in the microcavities, typical shear velocities are in the range of 1 up to 100 per second, depending on the molding velocity (here, 1 mm/sec up to 20 mm/sec) and the diameter of the capillary. These values are significant when considering the limiting value for PMMA of around 40,000 per second. With the shear velocity at the representative point and the material model of PMMA, the resulting pressure drop in the cavity can be estimated by

$$\Delta p = \frac{8\dot{V}L\eta(\dot{\gamma})}{\pi R^4} = \frac{8v_{mold}L\eta(\dot{\gamma})}{R^2} \tag{6.35}$$

Based on this model already, the most influencing parameters are

- the radius of the cavity: $\Delta p \propto 1/R^2$
- the molding velocity: $\Delta p \propto v$
- the shear- and temperature-dependent viscosity
- the temperature of the polymer (WLF time-temperature shift)

Figures 6.16 and 6.17 visualize the pressure drop for 100 μm–high circular cavities with different diameters as a function of the molding velocity and the temperature. Due to the viscoelastic behavior of the polymer melt, the pressure drop does not increase linearly with the increase in molding velocity. A significant influence is exerted by the molding temperature. In this example, an increase in temperature of about 20°C reduces the pressure needed for filling a cavity by about 50%. Hence, it is recommended to fill cavities, especially cavities with high aspect ratios, at a higher temperature, supported by low molding velocities. Nevertheless, a higher molding velocity increases the pressure necessary to fill the cavities, irrespective of the reduction of viscosity by the shear thinning behavior.

This calculated pressure drop may be considered an estimation of the minimum pressure needed to fill a cavity. However, the pressure is not high enough to avoid sink marks and reduce shrinkage during cooling. Consequently, a higher pressure in the cavities is recommended during the hot embossing cycle.

6.4.2.5 Filling of Microcavities—Flow Analysis

A real mold insert is characterized by an arrangement of different cavities with different geometries. The simulation of filling a single independent cavity

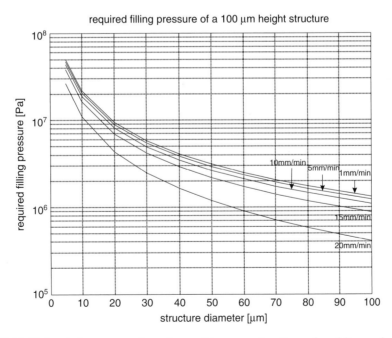

Figure 6.16 Estimated pressure drop in cavities as a function of molding velocity. The influence of molding velocity is not linear because of the shear thinning behavior of the polymer melt, which reduces the viscosity with an increase in shear velocity.

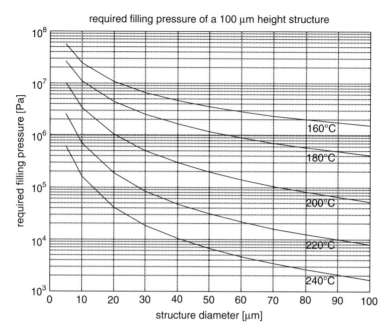

Figure 6.17 Estimated pressure drop in cavities as a function of molding temperature. An increase in temperature will reduce the pressure necessary to fill cavities in an effective way. Here, the pressure drop can be halved by an increase in temperature of around 20°C.

as described above is only a part of the simulation to characterize the filling behavior of a typical microstructured mold insert, because the filling behavior is influenced by the pressure distribution in the remaining layer and the arrangement of cavities with different geometries. A detailed filling simulation of microstructured molds requires much CPU power, is time-consuming, and varies for every design. To obtain general information on the filling behavior during the compression state, a characteristically geometric representation of a typical mold insert has to be chosen. Here, freestanding molded microstructures systematically arranged on a circular area of 4 inches or, alternatively, on a rectangular area of 26×66 mm^2, historically defined as the LIGA format, are selected for simulation (Figs. 6.18 and 6.19).

The filling behavior is characterized by the time-dependent flow and the related pressure in the melt. As described above, the pressure distribution over the remaining layer is parabolic, which characterizes the pressure available for filling a cavity at a defined distance from the center of the mold, where the maximum pressure can be observed. Considering this characteristic pressure distribution, the filling of an arrangement of microcavities of equal geometry starts in the center of the mold and ends in the peripheral areas. Hence, structures or cavities with the highest pressure needed for filling should be arranged in the center of the mold.

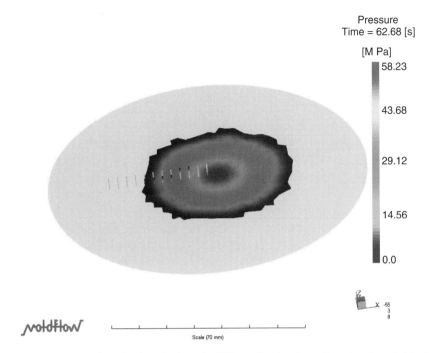

Pressure
Time = 62.68 [s]

[M Pa]
58.23

43.68

29.12

14.56

0.0

Scale (70 mm)

Figure 6.18 Pressure distribution during the filling of a circular microstructured 4-inch mold insert. The pressure distribution is parabolic, with the maximum in the center of the mold. This distribution is fundamental for filling the cavities at a defined distance from the center of the mold. According to the pressure distribution, the filling of equal cavities starts in the center of the mold and ends in the peripheral areas.

Besides the pressure distribution over the whole molding area, the filling profile determines the flow into the cavity. Rowland et al. [20,31,32] analyzed and simulated polymer flow during nanoimprinting with the focus on filling micro- and nanocavities. The filling profile during cavity filling is characterized by two different filling modes: single peak and dual peak (Fig. 6.20). Rowland et al. investigated the filling profiles dependent on the geometry, the initial film thickness, the height and the width of the cavities, the arrangement of different cavities and their influence to each other, and finally the viscosity of the polymer melt. Basic correlations were investigated by a temperature range around the glass-transition temperature of an amorphous polymer. The filling behavior was described by dimensionless numbers. Three parameters can predict the polymer deformation mode: directional flow ratio, polymer supply ratio, and capillary number. The directional flow ratio is defined as the cavity width to the initial thickness of the polymer film and determines the stress distribution inside the polymer during filling (single or dual peak) and further determines vertical or lateral flow. The polymer supply ratio is defined as the width indented to the height of the cavity and predicts the influence of squeeze flow and relative filling times. The capillary number characterizes the molding and

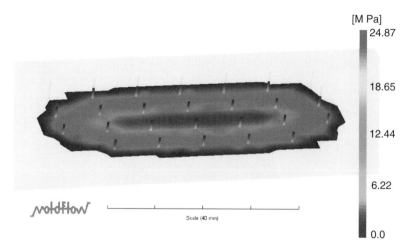

Figure 6.19 Pressure distribution during the filling of a rectangular microstructured mold insert. As in the case of circular geometry, pressure distribution is characterized by the maximum pressure occurring in the center of the mold and decreasing to ambient pressure at the edges. The filling of cavities is determined by this pressure distribution.

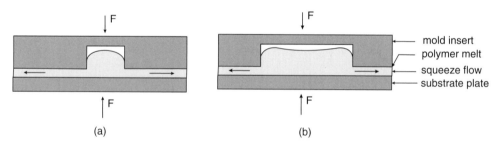

Figure 6.20 Different kinds of filling modes: (a) single-peak profiles were observed at moderate loads; (b) dual-peak profiles were observed in high loads characterized by high flow velocities. The double peak appears under the rapidly applied load due to the viscoelastic and time-dependent response of the polymer.

material properties. Scheer et al. [12] investigated the profile of flow during nanoimprint. Results indicate that, in the first moment, the stamp intrusion into the polymer melt is fast due to shear thinning at the periphery of the feature.

Finally, the flow and the cavity filling can be imagined by a simple model. The first level of pressure distribution is characterized by the squeeze flow of the macroscopic area. The arranged microcavities each require an individual filling pressure. A small cavity requires a higher pressure than a large cavity. The required pressure is delivered by the squeeze flow. Or in a simple rule: The polymer flows first in the direction with the lowest flow resistance.

6.4.3 The Process Step of Cooling

The cooling step of a polymer molding process is of importance, as cooling influences the properties of the molded part. Cooling velocity influences the morphological structure of semicrystalline polymers, the percentage of crystallinity, and, thus, the physical behavior of a molded part. In addition, the dwell pressure and holding time influence the shrinkage of the molded part during cooling and determine the demolding forces acting on the microstructures during demolding. Apart from its influence on demolding forces, an adequate dwell pressure is needed to avoid sink marks during cooling.

Like the heating step, the cooling of microstructures represents an individual solution for every machine in terms of the tools used, the mold inserts, and the kind of polymer. Hence, detailed analysis should be based on concrete designs. Here, it is focused on ideal uniform cooling of the mold, substrate plate, and molded part. Measurements show that uniform cooling can be assumed when using standard molding tools from hot embossing machines, but on a smaller molding area than the maximum area available. In Section 8.2.1, heat conduction in a molding tool is discussed briefly.

During the cooling step, the difference in shrinkage between polymer, mold insert, and substrate plate becomes important. To analyze this influence, the model has to consider the characteristic dimensions of typical molded parts. To represent the wide range of designs, a circular microstructured area with a diameter of 66 mm is selected as a suitable model. On the residual layer, freestanding structures are arranged systematically. To study the difference in thermal expansion, mold and substrate plate have to be modeled as rigid bodies. Between these rigid bodies and the molded part, contact elements are defined that describe the behavior of the interacting bodies (Fig. 6.21).

During the cooling stage of hot embossing, the shrinkage of the polymer and, hence, the demolding force are influenced by the dwell pressure, the thickness of the residual layer, and the adhesion of the residual layer to the substrate plate. Adhesion can be simulated in the range between an ideal adhesion (fixed) and sliding on a rough surface, assuming Coulomb's friction. The influence of the contact with the substrate plate and the dwell pressure, here force, is illustrated by Fig. 6.22. For a thickness of the residual layer of 100 μm and a defined friction, thermal expansion of a circular residual layer is presented. With an increase of the press force during cooling, thermal expansion is reduced and is set to a minimum by an ideal adhesion between residual layer and substrate plate. Besides the dwell pressure, the thickness of the residual layer affects the shrinkage. An increase in thickness increases the shrinkage, which can only be compensated by an increase in dwell pressure (Fig. 6.23).

These effects reflect the importance of the use of an adequate dwell pressure for every design to reduce shrinkage to a minimum value. Relevance of the dwell

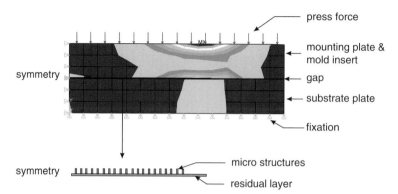

Figure 6.21 Representative model to analyze the cooling and demolding step of hot embossing. The model is based on a circular area with a diameter of 66 mm and a thickness of the residual layer of 100 μm. On this residual layer, freestanding microstructures are arranged systematically (height, 200 μm; width, 20 μm). Mold and substrate plate are modeled as rigid bodies, such that the thermal expansion coefficient of the mold has to be set to the difference between the mold and the polymer. Due to symmetry conditions, only half of the geometry is printed.

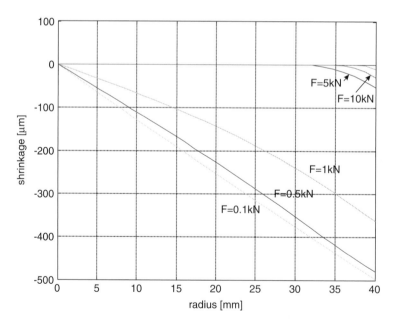

Figure 6.22 Lateral shrinkage of a circular residual layer (thickness, 100 μm; diameter, 80 mm) during cooling from 200°C down to 80°C. The press force varies from 0.1 kN up to 10 kN. With increasing press force, lateral shrinkage is reduced and reaches a minimum when the residual layer is completely fixed to the substrate plate. A higher press force restricts the shrinkage during cooling to the peripheral areas of the molded part.

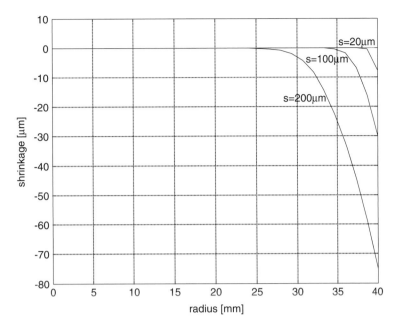

Figure 6.23 Lateral shrinkage of a circular residual layer during cooling from 200°C down to 80°C using a press force of 10 kN. At a constant press force, shrinkage increases with an increase in the thickness of the residual layer.

pressure and the adhesion to the substrate plate is underlined by the fact that shrinkage is one of the major factors determining the demolding forces.

6.4.4 The Process Step of Demolding

Demolding of microstructures is gaining importance because of the risk of damaging structures, especially filigree freestanding structures with high aspect ratios. Apart from the design of the microstructures and the quality of the mold insert, the contact stress between structures and mold insert during cooling and directly at the beginning of the demolding step after changing from dwell pressure to velocity-controlled demolding is responsible for successful demolding.

6.4.4.1 Contact Stress

To analyze the demolding of microstructures, the distribution of contact stress should be studied first.

Contact Stress under the Impact of Dwell Pressure   As already shown in Section 4.4.3, for a simple circular plate, shrinkage can be reduced by dwell pressure and adequate adhesion to the substrate plate. This behavior may also be transferred to the model with systematically arranged structures, which results in moderate contact stress between mold and polymer under the impact of dwell pressure. As shrinkage is reduced under dwell pressure, the distribution of contact stress is nearly homogeneous for the systematically arranged free-standing microstructures.

Contact Stress at the Beginning of Demolding   The behavior described above changes when the process switches from dwell pressure to the velocity-controlled demolding step. If the dwell pressure ceases, shrinkage is caused in the molded part, with the resulting contact stress now being a function of the thickness of the residual layer and the adhesion to and friction of this layer on the substrate plate. This behavior, the loss of dwell pressure, and the reduction of shrinkage result in a high contact stress that acts in particular on the structures arranged in the peripheral regions of the molded part. Figure 6.24 shows the contact stress of the last seven structures of the model selected at the beginning of the demolding step. The high contact stress caused in the last structures arranged in the edge region must be noted. A high compressive stress in the direction toward the center of the molded part is found at the intersection between structure and residual layer. At the same time, all structures are exposed to a tensile contact stress caused by the demolding motion. Due to the higher compressive contact stress in the last structure, the tensile contact

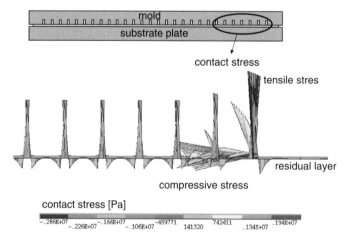

Figure 6.24 Contact stress between molded part and mold insert under dwell pressure at the beginning of demolding. This figure shows the contact stress of the last seven structures arranged on a circular residual layer. Positive values characterize the compression stress; negative values denote tensile stress. The structure arranged on the residual layer with the largest distance from the center of the molded part is exposed to maximum contact stress. At the intersection between the structure and the residual layer, high compressive stress occurs. Parallel to the maximum compression stress, this structure is exposed to the maximum tensile stress during demolding.

stress increases in the structures in the peripheral area and results in a maximum stress acting on the last structure.

Before the demolding step starts, the pressure distribution caused by the dwell pressure is degraded. As a result of the compressibility of the polymer, the molded part is expanded and the thickness of the molded structures increases. This effect may also cause the contact stress and the friction force between mold and polymer to increase.

6.4.5 Stress Distribution during Demolding of Structures

6.4.5.1 Demolding of Ideal, Vertical Structures

As described in the section above, critical contact stress will result from shrinkage when switching from dwell pressure to velocity-controlled demolding. In this state, adhesion and the highest possible friction between substrate plate and mold can reduce shrinkage. However, shrinkage that cannot be avoided completely also results in stress inside the microstructures, which may be higher than the yield stress of the material at demolding temperature. This may give rise to the deformation or damage of microstructures. The microstructures located at the edge are especially at risk.

Representative of the large number of designs, a design with 40 systematically arranged freestanding microstructures of 100 µm height and a width of 20 µm on a residual layer of 50 µm will be used for the simulation of the demolding state. The diameter of the residual layer is in the range of 66 mm. Simulation starts with the complete filling of microcavities under dwell pressure at a molding temperature of 180°C. After cooling from the molding temperature to the demolding temperature of 90°C under dwell pressure with typical cooling times, it is changed from force-controlled dwell pressure to velocity-controlled demolding. The demolding velocity is set representatively to 1 mm/sec. Dynamic friction between mold and polymer and between polymer and substrate plate is assumed to be 0.4 and 0.5, respectively. The corresponding factor between static and dynamic friction is set to 2. As the highest contact stress occurs in the structures at the edge, the structure located at the largest distance from the center of the molded part will be observed in detail during the demolding step. To interpret the load and the resulting stress during the demolding step, a simulated demolding sequence of the selected structure is presented in Fig. 6.25. To compare the load with material properties measured in tensile experiments at a temperature corresponding to the demolding temperature, the van Mises stress is used, which can be compared to the yield stress of the material.

At the beginning of the demolding step, the maximum stress in the structure is found at the intersection between the residual layer and microstructure. This stress distribution is supported by the geometric interface between mold and polymer. Due to the sharp edges caused by fabrication processes, the interface

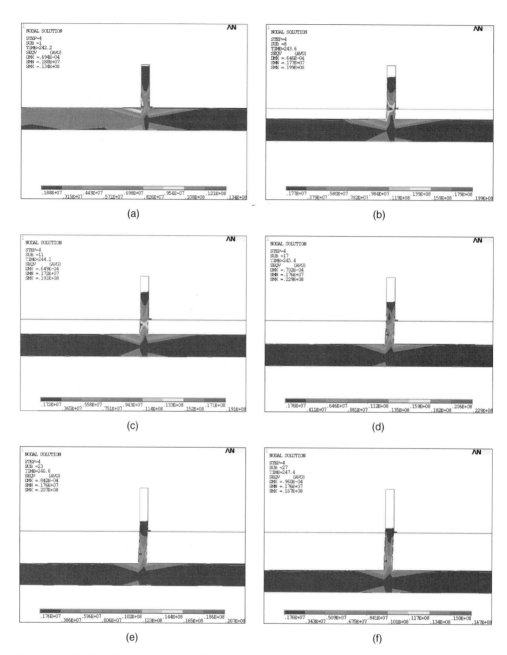

Figure 6.25 Van Mises stress inside a microstructure during the demolding step of hot embossing. The critical load during the demolding step occurs in the first part of demolding, where the highest stress is encountered at the intersection between structure and polymer.

acts like a notch, the characteristic notch stress of which may also increase the stress in this area. In this state—the beginning of demolding—the risk of tearing the structures is very high. In practice, the structures often break at the interface between the residual layer and microstructure. With further demolding, the maximum stress will be transferred to the first quarter of the structure. Now, it depends on the cross-section of the structure whether this load will cause stress that is higher than the yield stress. Continuing motion transfers the maximum stress to the surface of the structure, which also is an effect of shrinkage of the residual layer. Directly before the completion of demolding, the maximum stress is found at the top edge of the structure, which may result in the damage of the top of the microstructure, known as overdrawn edges. This sequence shows that the stress impact on the microstructures is at a maximum during the first half of the demolding step. If this critical part of the demolding step is survived, demolding of microstructures is normally finished successfully.

6.4.5.2 Demolding of Structures with Undercuts

The simulation results presented in the section above refer to ideal vertical structures and an ideal vertical direction of demolding velocity. In the practice of mold fabrication, however, the risk of undercuts cannot be excluded completely. Moreover, bending of the whole machine, the tool, and mold insert under load may cause undercuts when the polymer solidifies below a bent mold insert, with the shape corresponding to the bending line of the mold insert. After switching to velocity-controlled demolding, the elastic mold insert returns to the initial position and, depending on the bending, an angle between mold insert and molded part results during demolding (Chapter 4).

This influencing factor may be represented by typical undercuts characterized by a difference of width between the ground and the top of a cavity. Simulation refers to a difference between 0.25 μm and 1 μm for a structure 100 μm high. Analogous to Fig. 6.25, Fig. 6.26 shows the sequence of demolding for an undercut of 1 μm.

The demolding sequence shows that small undercuts already cause a significant increase in stress inside the structures, which may result in the damage or rupture of the structures during demolding. The maximum stress over the cross-section of the freestanding structure increases the risk of tear-off during the first half of demolding. Under the assumption of an undercut as described here, the maximum stress depends on the cross-section of the structure. Structures with larger cross-sections exhibit an advantageous stress distribution over a larger area, such that the maximum stress can be reduced to a value below the yield stress of the polymer. However, undercuts are critical for filigree structures with high aspect ratios and small lateral dimensions. In the second half of the demolding step, the maximum stress is encountered near the surface of the structures, which may result in shear deformation in the direction of demolding. During the last

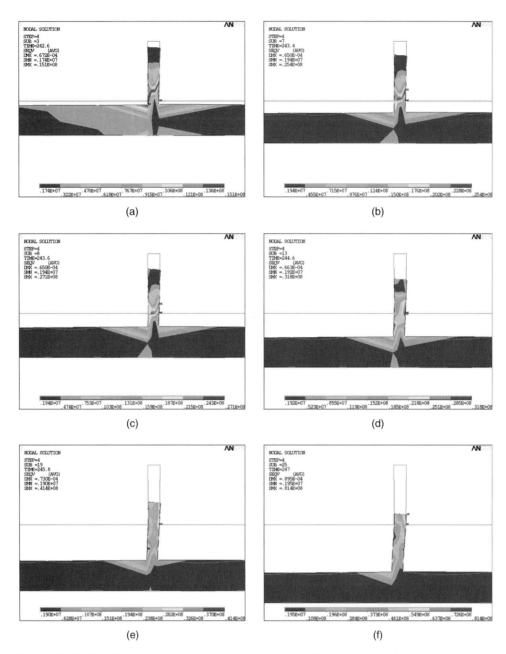

Figure 6.26 Van Mises stress inside a microstructure with an undercut during demolding. Small undercuts increase the load of the microstructures dramatically. The related stress reaches values above the yield stress of the polymer at demolding temperature, as a result of which the structure is damaged. Damage may also occur at the beginning of demolding and, in contrast to the ideal structure, at the end of the demolding step, immediately before the structure is completely demolded.

step of the demolding cycle, immediately before the structure is completely demolded, the difference between the lateral sizes of the top and the bottom of the cavity will cause a maximum stress at the top of the structure, which will result in typical deformation of the edges at the top of the microstructures. The influence of shrinkage, an insufficient adhesion of the residual layer to the substrate plate, or deformation of the top of the structures may suggest undercuts in the mold inserts. To estimate the influence of undercuts on the simulated demolding forces, all 40 arranged structures of the model described here are provided with undercuts. Undercuts of 250 nm increase the demolding force by about 100%. Undercuts of 500 nm increase the demolding force by about 600%, which may result in damage to the microstructures in most cases. This nonlinear behavior of the effect of undercuts underlines the necessity of fabricating cavities with vertical sidewalls, especially for filigree microstructures, and of using molding machines and tools with a high stiffness to prevent the bending of the integrated mold insert.

References

1. *Dynamics of Polymeric Liquids—Fluid Mechanics*, Volume 1. John Wiley, 1977.
2. M-Base. Simulation tool EXPRESS. http//www.m-base.de, 2008.
3. Moldflow. http//www.moldflow.com, 2008.
4. M. Amnes, V. Srivastava, S. P. Lele, and L. Anand. Modeling and simulation of the hot-embossing process for manufacture of microfluidic devices from amorphous polymers. *ICOMM*, (50), 2006.
5. N. Bogdanski, M. Wissen, A. Ziegeler, and H.-C. Scheer. Temperature-reduced nanoimprint lithography for thin and uniform residual layers. *Microelectronic Engineering*, 78–79: 598–604, 2005.
6. G. L. W. Cross. The production of nanostructures by mechanical forming. *Journal of Physics D: Applied Physics*, 39:R363–R386, 2006.
7. Y. Ding, H. W. Ro, T. A. Germer, J. F. Douglas, B. C. Okerberg, A. Karim, and C. L. Soles. Relaxation behavior of polymer structures fabricated by nanoimprint lithography. *ACS Nano*, 1(2):84–92, 2007.
8. A. C. Eringen and K. Okada. A lubrication theory for fluids with microstructure. *Int. J. Eng. Sci.*, 33:2297–2308, 1995.
9. E. C. Ferreira, N. M. Neves, R. Muschalle, and A. S. Pouzada. Friction properties of thermoplastics in injection molding. In *SPE Antec*, 2001.
10. Y. Guo, G. Liu, Y. Xiong, and Y. Tian. Study of the demolding process—implications for thermal stress, adhesion and friction control. *Journal of Micromechanics and Microengineering*, 17:9–19, 2007.
11. Y. Guo, G. Liu, X. Zhu, and Y. Tian. Analysis of the demolding forces during hot embossing. *Microsystem Technologies*, 13:411–415, 2007.
12. H.-C. Scheer, N. Bogdanski, M. Wissen, T. Konishi, and Y. Hirai. Profile evolution during thermal nanoimprint. *Microelectronic Engineering*, 83:843–846, 2006.
13. Y. He, J.-Z. Fu, and Z.-C. Chen. Research on optimization of the hot embossing process. *Journal of Micromechanics and Microengineering*, 17:2420–2425, 2007.
14. L. J. Heyderman, H. Schift, C. David, I. Gobrecht, and T. Schweizer. Flow behaviour of thin polymer films used for hot embossing lithography. *Microelectronic Engineering*, 54:229–245, 2000.

15. Y. Hirai, T. Konishi, T. Yoshikawa, and S. Yoshida. Simulation and experimental study of polymer deformation in nanoimprint lithography. *Journal Vac. Sci. Technol. B*, 22(6): 3288–3293, 2004.

16. Y. Hirai, S. Yoshida, and N. Takagi. Defect analysis in thermal nanoimprint lithography. *J. Vac. Sci. Technol. B*, 21(6):2765–2770, 2003.

17. Y.-J. Juang, L. J. Lee, and K. W. Koelling. Hot embossing in microfabrication. Part I: Experimental. *Polymer Engineering and Science*, 42(3):539–550, 2002.

18. Y.-J. Juang, L. J. Lee, and K. W. Koelling. Hot embossing in microfabrication. Part II: Rheological characterization and process analysis. *Polymer Engineering and Science*, 42 (3):551–566, 2002.

19. D. S. Kim, K. C. Lee, and T. H. Kwon. Micro-channel filling considering surface tension effect. *Journal of Micromechanics and Microengineering*, 12:236–246, 2002.

20. H. D. Rowland and W. P. King. Polymer deformation filling modes during microembossing. *Journal of Micromechanics and Microengineering*, 14:1625–1632, 2004.

21. H.-J. Lee, H. Wook, C. Soles, R. L. Jones, E. K. Lin, and W.-L. Wu. Effect of initial resist thickness on residual layer thickness of nanoimprinted structures. *J. Vac. Sci. Technol. B*, 23 (6):3023–3027, 2005.

22. K. F. Lei, W. J. Li, and Y. Yam. Effects of contact-stress on hot embossed PMMA microchannel wall profile. *Microsystem Technologies*, 11:353–357, 2005.

23. P. J. Leider. Squeezing flow between parallel disks: I. Experimental results. *Ind. Eng. Chem. Fundam.*, 13(4):342–346, 1974.

24. P. J. Leider and B. Bird. Squeezing flow between parallel disks: I. Theoretical analysis. *Ind. Eng. Chem. Fundam.*, 13(4):336–341, 1974.

25. T. Leveder, S. Landis, L. Davoust, and N. Chaix. Flow property measurements for nanoimprint simulation. *Microelectronic Engineering*, 84:928–931, 2007.

26. Y. Luo, M. Xu, X. D. Wang, and C. Liu. Finite element analysis of PMMA microfluidic chip based on hot embossing technique. *Journal of Physics: Conference Series*, 48:1102–1106, 2006.

27. D. S. Macintyre and S. Thoms. A study of resist flow during nanoimprint lithography. *Microelectronic Engineering*, 78–79:670–675, 2005.

28. G. Menges, E. Haberstroh, W. Michaeli, and E. Schmachtenberg. *Werkstoffkunde Kunststoffe*. Hanser Publishers, 2002.

29. K. Mohamed, M. M. Alkaisi, and J. Smaill. Resist deformation at low temperature in nanoimprint lithography. *Current Applied Physics*, 6:486–490, 2006.

30. M. Otto, M. Bender, B. Hadam, B. Spangenberg, and H. Kurz. Characterization and application of a UV-based imprint technique. *Microelectronic Engineering*, 57–58:361–366, 2001.

31. H. D. Rowland, W. P. King, A. C. Sun, and P. R. Schunk. Simulations of non-uniform embossing: the effect of asymmetric neighbor cavities on polymer flow during nanoimprint lithography. Technical report, Sandia National Laboratories, 2007.

32. H. D. Rowland, A. C. Sun, P. R. Schunk, and W. P. King. Impact of polymer film thickness and cavity size on polymer flow during embossing: toward process design rules for nanoimprint lithography. *Journal of Micromechanics and Microengineering*, 15:2414–2425, 2005.

33. H.-C. Scheer and H. Schulz. A contribution to the flow behaviour of thin polymer films during hot embossing. *Microelectronic Engineering*, 56:311–332, 2001.

34. H.-C. Scheer, H. Schulz, and D. Lyebyedyev. Strategies for wafer-scale hot embossing lithography. In *Proceedings of SPIE*, Volume 4349, page 86, 2001.

35. H. Schift, L. J. Heydermann, M. Auf der Maur, and J. Gobrecht. Pattern formation in hot embossing of thin polymer films. *Nanotechnology*, 12:173–177, 2001.

36. X.-J. Shen, L.-W. Pan, and L. Lin. Microplastic embossing process: experimental and theoretical characterizations. *Sensors and Actuators A*, 97–98:428–433, 2002.

37. D. Sus. Relaxations- und Normalspannungseffekte in der Quetschströmung. *Rheologica Acta*, 23:489–496, 1984.

38. M. Worgull and M. Heckele. New aspects of simulation in hot embossing. *Microsystem Technologies*, 10:432–437.

39. M. Worgull, M. Heckele, J.-F. Hétu, and K. K. Kabanemi. Modeling and optimization of the hot embossing process for micro- and nanocomponent fabrication. *Journal of Microfabrication, Microlithography, and Micromachining*, Special Issue Announcement NanoPatterning, 2005.

40. M. Worgull, J.-F. Hétu, K. K. Kabanemi and M. Heckele, . Hot embossing of microstructures: characterization of friction during demolding. *Microsystem Technologies*, 14:763–767, 2008.

41. M. Worgull, M. Heckele, and W. K. Schomburg. *Analysis of the hot embossing process*. PhD thesis, University of Karlsruhe, Institute for Microstructure Technology, 2003.

42. S. Y. Yang, S. C. Nian, and I. C. Sun. Flow visualization of filling process during micro-injection molding. *Intern. Polymer Processing*, 17(4):354–360, 2002.

43. D. Yao and B. Kim. Simulation of the filling process in micro channels for polymeric materials. *Journal of Micromechanics and Microengineering*, 12:604–610, 2002.

44. W. B. Young. Analysis of the nanoimprint lithography with a viscous model. *Microelectronic Engineering*, 77:405–411, 2005.

45. W.-B. Young. Simulation of the filling process in molding components with micro channels. *Microsystem Technologies*, 11:410–415, 2005.

46. Ch. Mehne. *Grossformatige Abformung mikrostrukturierter Formeinsätze durch Heissprägen*. PhD thesis, University of Karlsruhe (TH), Institute for Microstructure Technology, 2007, ISBN: 978-3-86644-185-9.

7 Hot Embossing Technique

The hot embossing technique is developed to different levels, beginning with simple manually controlled hot embossing machines used, for example, in laboratories, up to high-level machines with a high grade in automation used in industry and science. Independent from the level of complexity and automation, the fundamental components of each hot embossing system can be split into four major groups:

- the press, a frame with high stiffness with an integrated drive unit, responsible for the act of molding force and providing sensitive relative motion between mold insert and polymer
- the tool that is integrated in the molding press. The tool consists of an integrated heating and cooling unit, a vacuum unit, and an optional alignment system.
- the microstructured mold insert, which is fixed in a hot embossing tool
- the control system, which has the function of measurement, visualization, and control of process parameters. This control system is an interface between the machine and the user.

Hot embossing tools are essential components of any hot embossing system. They are characterized also by different levels of complexity and are discussed briefly in Chapter 8. The manifold of mold inserts as master for replication is discussed in Chapter 9. The following section discusses the technical aspects of hot embossing machines, with the focus on the press unit, the drive system, the control of force and velocity, and the control unit as an interface to the user.

7.1 Technical Requirements

To understand the technical solutions that are used for hot embossing machines, first the technical requirements for replication of microstructures by hot embossing need to be discussed. With this background the challenges in the development of hot embossing machines can be estimated. But first, what are the technical requirements for hot embossing? What are the technical challenges?

- The molding of microstructures on thin residual layers requires, besides low viscosity of the polymer melt, also high molding forces (Section 6.1). The requirements increase if the structured area increases and the thickness of the residual layer decreases. High molding forces are also helpful to reduce the shrinkage of molded parts, which finally will decrease the risk of damage to structures during demolding. High demolding forces require a press unit with high stiffness against bending. Depending on the molding area, the design of mold insert, and the viscosity of the polymer melt, molding forces in the range between 1 kN up to several hundred kN are necessary for successful molding. Regarding the press force, commercial machines are available in different classes, beginning at 50 kN up to 1,000 kN, allowing these machines to replicate a wide range of microstructured mold inserts.

- High forces require an adequate stiffness of the press frame. Otherwise, bending will influence the thickness of the residual layer and complicate the precise relative motion of the mold insert and substrate plate under load. Besides the stiffness of the press frame, precise parallel guidance of the crossbar is essential. Otherwise, precision during molding and demolding cannot be guaranteed. Especially the demolding of microstructures requires a parallel motion of the crossbars; otherwise, during demolding the molded structures can be deformed by forces acting on the sidewalls of the structures.

- As mentioned in the previous chapter, hot embossing is characterized by short flow paths. In particular, to fill microcavities the flow path is characterized by the depth of the cavities. To achieve also low stress inside the molded parts, the short flow paths should be supported by a low shear stress of the polymer melt during molding, which makes it necessary to achieve low molding velocities. In turn, these moderate velocities can be achieved by a precise motion of one of the two crossbars. Here, a precise drive unit and precise control of this motion is one of the challenges. Typical flow velocities are in the range of 1 mm/min or below. These velocities become more important during demolding. At this process step the motion of the crossbar has to be controlled very precisely and velocities in the range of 0.5 mm/min down to 0.1 mm/min should be achievable; otherwise, the risk of damage to filigree structures increases. It is obvious that these velocities cannot be used to open and close the tool. Therefore, the drive mechanism should guarantee besides a sensitive motion also a fast motion. Here, the bandwidth of velocity can include four dimensions (e.g., 0.1 mm/min up to 250 mm/min, hot embossing machine Jenoptic HEX03)

- Like the molding velocity, also the molding force has to be set over four dimensions. On the one side molding forces in the range of several 100 kN can be necessary; on the other side touch forces in the range of 0.1 kN or below are desired. The precise control of this large bandwidth of force is combined with the precise control of the motion of one of the two crossbars of the machine, which underlines the importance of the drive unit. The large bandwidth of forces needed for hot embossing requires a system that allows the measurement of the force precisely with a high resolution in a range over these four dimensions. For example, the resolution of the measurement system of the HEX03 is in a range of approximately 2 N by a maximum force of 250 kN.

- The control of molding force and molding and demolding velocities are supported by precise measurement of the relative distance between mold and substrate plate, respectively the position of the two crossbars. Regarding the dimensions of the structures and the velocities used for molding and demolding, the resolution of the measurement system should be set in a range of a minimum of 1 μm or below.

- If double-sided molding or position-controlled molding is required, a precise alignment system is also necessary to allow the adjustment of both mold halves relative to each other. Considering the size of microstructures, the alignment system should guarantee an adjustment in three degrees of freedom, two lateral directions, and one rotation. The precision of the alignment system and the stability of the fixation after the alignment define the quality of double-sided molding. Especially the fixation has to guarantee that during molding no further alignment has to occur. Depending on the design of the structures, typical repeatable precision in the range down to 10 μm is desired.

- Regarding the temperature-dependent flow behavior of polymer melts, the tool with an integrated heating and cooling unit has to guarantee a homogeneous temperature distribution in the mold insert and also in the substrate plate. To control the relaxation process of the polymer and the grade of crystallinity of the molded parts, it is desired to control the cooling rate. Concerning a cost-effective molding, a high cooling rate should also be achievable to obtain short cycle times.

- To guarantee a complete filling of microcavities, reducing air bubbles during molding, and to avoid oxidation of the mold inserts and substrate plates at high temperatures, mold insert and substrate plate should be isolated from the ambient pressure. These requirements can be achieved by molding under vacuum.

- To derive benefit from the molding concept of hot embossing, the process parameters should be adapted to individual needs. Therefore, a user interface is recommended, allowing the modification of the process parameters like temperatures, molding force, molding velocity, and demolding velocity in an effective and versatile way. The high potential of process variations in hot embossing will be supported by the use of a programming language.

- The advantage of hot embossing is the high grade of flexibility regarding the change of mold and polymer. To support the high flexibility, the design of a hot embossing machine and the tool should guarantee a quick change of mold inserts and polymers in terms of polymer sheets.

7.2 Technical Implementation

The requirements mentioned above can be implemented in different ways, each implementation with specific advantages and also potential disadvantages. Exemplary for the manifold of implementations, the technical concepts of two commercially available machines, Jenoptic HEX03 and Wickert WMP 1000 (Section 7.3), are explained below in detail. The focus is set on the implementation of mechanical stiffness, drive unit, and control system.

7.2.1 Mechanical Stiffness

As described above, stiffness and precise guidance of crossbars are essential properties of each hot embossing machine. To guarantee high resistance against bending at high molding forces, the cross-sections and the profiles of the two crossbars and guiding columns should be enormously enlarged. The combination of two crossbars, one fixed and one movable, and the guiding columns acts as a frame with high bending strength (Fig. 7.1).

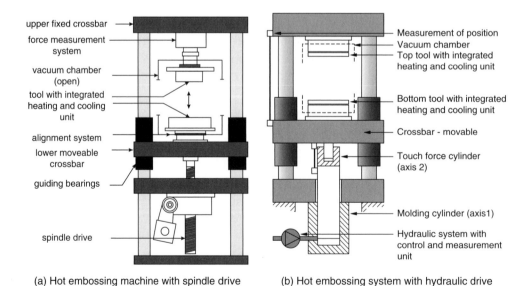

(a) Hot embossing machine with spindle drive   (b) Hot embossing system with hydraulic drive

Figure 7.1 Basic concept of an embossing frame with high bending strength. The maximum bending moment acts on the intersection between the crossbars and the guiding columns. Here, massive guiding bearings are necessary.

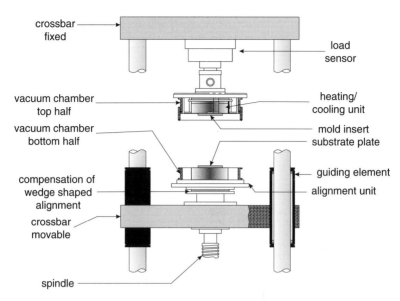

Figure 7.2 Schematic view of the guiding bearings of a hot embossing machine to compensate the high bending moments acting on the intersection between crossbars and guiding columns.

Especially the intersection between the crossbars and the guiding columns undergoes high bending moments during the act of molding forces. In parallel, a precise motion with low tolerances in guidance is a precondition for molding and especially demolding. To counter these high moments of bending, massive bearings at the intersection between crossbars and guiding columns are necessary. If the press forces increase, four column machines will split off the bending moments to four bearings and further reduce the torsion of the crossbar under load. Both commercial machines are therefore equipped with four guiding columns and massive bearings between crossbars and guiding columns (Fig. 7.2).

7.2.2 Drive Unit

To implement the requirements of the precise motion between substrate plate and mold insert, it is obvious to reduce the motion to only one crossbar. Depending on the integration of the drive unit, in most cases the bottom crossbar will be moved against the fixed crossbar on top. Two concepts of drive mechanism are implemented in commercial hot embossing machines—a mechanical spindle drive and, as an alternative for large hot embossing machines, a hydraulic drive.

7.2.2.1 Spindle Drive

Spindle drives use the principle of conversion of a rotary motion to a translational motion by a screw thread. To achieve the vertical motion the spindle has

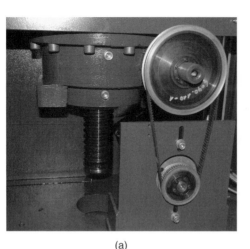

(a)   (b)

Figure 7.3 Schematic view of (a) a spindle drive and (b) a spindle for a hot embossing machine. The torque of an electric stepper motor is transferred by a belt drive to a spindle. The thread of the spindle transfers the torque into a linear motion of the crossbar. With the controlling of the stepper motor, molding velocities in the range of 0.1 mm/min up to 250 mm/ min can be achieved (hot embossing machine Jenoptik HEX03).

to be vertically mounted between the lower movable crossbar and the bottom of the hot embossing machine (Fig. 7.3). Further, the spindle has to be connected to a rotary drive, an electric motor. To achieve the precise motion of the crossbar, the rotation of the spindle has to set up in a range significantly below typical revolutions per minute of electrical drives, which makes it necessary to integrate a gear ratio. Already a belt drive with different diameters of belt pulley allows a reduction in the rotation speed. In this case the slippage has to be taken into account, especially by the transmission of large torques. The translational motion is finally determined by the slope of the spindle, the gear between the electric motor and spindle, and the revolution of the electric motor. The control of the translational molding and demolding velocity of a hot embossing machine will in this case be reduced to the control of the revolution of an electric engine. For example, a stepper motor allows the very precise control of revolutions. Depending on the concept, control of a large bandwidth and precision of revolutions per minutes can be achieved that allows the transmission of these revolutions in vertical velocities in the range of less than 0.5 mm/min and more than 200 mm/min. The advantage of mechanical spindle drives results in a compact construction. This kind of drive mechanism is already established in tensile testing machines and achieves a high precision, even at high forces.

7.2.2.2 Hydraulic Drive

If high molding forces are required, for example for large area molding, besides a large spindle drive a hydraulic drive unit can be used also as an alternative. Hydraulic drives are established drive mechanisms in macroscopic molding presses, for example to achieve the clamp force to close the molding tool during injection molding. The challenge in using this concept for micro hot embossing is to achieve the required range of molding velocities and the large bandwidth of molding forces. To fulfill these requirements the performance can be split off into two hydraulic cylinders: a first small cylinder responsible for the touch force and a second cylinder, here the main cylinder, responsible for the high molding force. The small cylinder is therefore integrated in the main cylinder (Fig. 7.4). This concept is implemented in the commercially available hot embossing machine Wickert WMP 1000.

The hot embossing machine consists of a massive frame with high stiffness. The movable crossbar is powered by a two-stage coaxial-arranged hydraulic cylinder with different diameters. Both cylinders are characterized by independent pressure generation and by an independent force and position controlling system. The measurement of force, normally measured in the flux pattern of the frame, is achieved here by the measurement of the hydraulic pressure. To eliminate external influences like the weight of the different tools or friction in the guiding

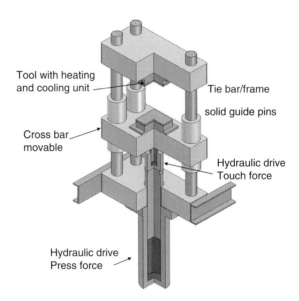

Figure 7.4 A hydraulic drive mechanism for hot embossing. To achieve the large bandwidth of force, touch force, and molding force, the mechanism is split into two cylinders. The first small cylinder is responsible for the touch force, and the second large cylinder is responsible for the molding force. The small cylinder is integrated into in the main cylinder.

elements, the force and position measurement systems are referenced by a reference step. This step has to be carried out only at the beginning of molding.

The small cylinder, called the touch force cylinder, responsible for the effecting of the touch force during heating of the polymer, provides a force in the range between 500 N and 30,000 N. The precision of the force control is typically in the range of a few 100 N; the length of stroke of this cylinder is in a range of a few millimeters with a precision of approximately 100 μm.

The main cylinder, called the molding cylinder, is responsible for the opening and closing of the tool and for the generation of the press force. The control of this molding cylinder is split into two different ranges, a coarse range for the closing of the tool and a fine range for the molding step. The coarse range is characterized by velocities in the range of 90 mm/sec, with a precision of a few decimillimeters. In the range of 50 mm before the molding tool is closed the coarse control mechanism is switched to a precise control mechanism supported by a high-resolution measurement of the crossbar position. To achieve a final high precision positioning, the velocity of the crossbar will be reduced. The achievable precision of the position is here in the range of approximately 100 μm. The maximum achievable press force of 1,000 kN can be controlled with a precision of 1,000 N.

The hot embossing process implemented by the hot embossing machine Wickert WMP1000 in combination with the basic molding tool (Section 8.4) can be illustrated in four steps in Fig. 7.5.

In detail a hot embossing cycle can be split into the following steps.

1. Insert of the polymer film and closing of the molding tool by the main cylinder 1 up to the sealing point of the vacuum chamber (Fig. 7.5(a)).
2. Evacuation of the vacuum chamber and providing of touch force by the small cylinder 2. During the providing of touch force the main cylinder 1 still remains in the position of step 1. If the basic tool with separated heating and cooling units is used, the heating plate has no contact with the cooling block because of the Belleville spring between heating and cooling plate (Fig. 7.5(b)). This air gap reduces the thermal mass of the heating block and with the implemented heating system heating rates up to 50 K/min can be achieved. Typical heating times for hot embossing processes are in the range of 120 seconds.
3. After the molding temperature is attained, the molding process starts by the switching from touch force of the small cylinder 2 to the press force of the main cylinder 1. In a first step the cylinder 2 is displaced by cylinder 1. This means that cylinder 2 under retention of the touch force is opened under the same velocity that cylinder 1 will be closed. During this switching, the position of the crossbar remains constant.
4. Velocity-controlled molding with the main cylinder 1 up to the desired press force. Under the act of the load the heating plates are pressed onto the cooling plates and generate a system with the required stiffness and

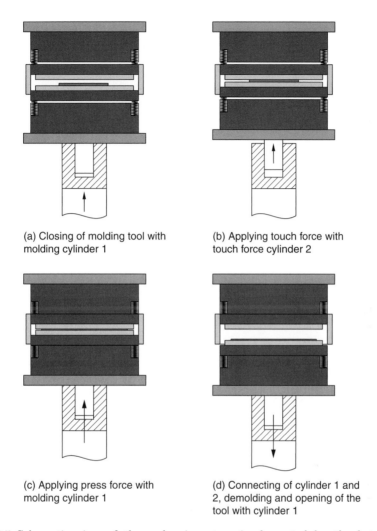

(a) Closing of molding tool with molding cylinder 1

(b) Applying touch force with touch force cylinder 2

(c) Applying press force with molding cylinder 1

(d) Connecting of cylinder 1 and 2, demolding and opening of the tool with cylinder 1

Figure 7.5 Schematic view of the embossing steps implemented by the hot embossing Machine Wicker WMP 1000 in combination with a basic molding tool.

plane-parallel configuration (Fig. 7.5(c)). The electrical heating is stopped immediately after the molding force is achieved and the heating plates are fixed by a magnetic clamping system to later provide a vertical demolding without any misalignment. In a following step the small cylinder 2 is locked with the main cylinder 1. In this configuration the load can be maintained over a selected period of time (holding time, or dwell pressure).

5. By the contact of the heating plate with the cooling plate with a large heat capacity, the mold insert will be cooled in a short period of time down to demolding temperature. Typical cooling rates for molding PMMA are in the range of 180 seconds. If the demolding temperature

is attained, the load is eliminated by a shift of cylinder 1. The control
of this demolding step requires a precise motion of cylinder 1 over a
distance of a few millimeters (Fig. 7.5(d)).

6. In the last molding step the vacuum chamber is vented and the molding
tool is opened completely. Cylinder 1 and cylinder 2 will be unlocked
and the heating plates and the cooling plates will be separated by the
springs.

Within the scope of an embossing cycle, the regulating distances of both cylinders can be programmed arbitrarily. For the closing of the molding tool, cylinder
1 can be positioned with a precision of 0.2 mm; the control of the force in this
step is not possible. The final closing of the molding tool is achieved with a precision of positioning in the range of 100 μm. The small cylinder has a total stroke
of about 50 mm and an identical precision of positioning.

7.2.3 Measurement of Process Parameters

To control the process and to get feedback over the current status of the process, the parameter has to be visualized. Therefore, a measurement of the main
process parameter molding force, temperature of mold insert and substrate
plate, and the position of mold insert and substrate relative to each other
must be determined. The measurement data should be visualized online and
stored for further analysis (Section 7.2.4). Depending on the process parameters,
different sampling times are recommended. The sampling of temperature can be
done in time steps of seconds, because of the moderate changes in time, even in
the heating and cooling state. In contrast to the measurement of temperatures, a
high-resolution sampling in the range below 0.1 second is recommended for the
measurement of force and relative position. These process parameters change
significantly with time, for example, the force during molding and demolding,
and show extreme values that are fundamental for further analysis of molding
and demolding behavior.

- Measurement of temperature: During molding the temperature
  determines the thermal behavior of the polymer. The ideal, therefore, is a measurement directly inside the polymer during molding,
  which is nearly impossible to achieve with common sensor technology. Therefore, the temperature sensors should be located in the
  mold insert and substrate plate as near as possible to the polymer.
  Typically temperature sensors, for example Pt100 resistors or
  NiCr–Ni thermocouples, are integrated under the surface of the
  substrate plate or in the plate, where the mold insert is fixed.
  At minimum one sensor in every plate is necessary. For a determination of the temperature distribution in the plates, an array of

sensors is required, especially if large areas, like 8-inch, are molded. If small areas (e.g., 4-inch) are molded, one sensor can be sufficient. Nevertheless, an array of temperature sensors will help to analyze the temperature distribution during molding, which becomes more important if semicrystalline polymers with a small molding window are molded. In this case, the temperature has to be set in a small and homogeneous range. Especially if it is necessary to determine the temperature of the polymer melt precisely, it has to be taken into account that between the polymer and the sensor two times the heat transfer will influence the measurement between the sensor and the plate, and also between the plate and the polymer. Therefore, if exact temperatures are required, during heating and cooling it is recommended to set up a short holding time at the end of the process step to reach a nearly stationary temperature state.

- Measurement of force: Regarding the range of forces during hot embossing—touch forces in the range of 100 N, molding forces in the range up to 200 kN or more, and finally demolding forces (tensile forces) in the range up to 2,000 N—the requirements on a measurement will be visible, especially when a high resolution during the touch force sequence and the demolding sequence is needed to analyze the process. Here, high-quality load sensors with this resolution are required or, alternatively, the force measurement can be split into two measurement systems, one for the small loads and a second one for high molding loads. The sensor for the lower forces has to be protected against the high molding loads. The measurement of force by load sensors, integrated in the flux pattern of the machine, on the base of strain gauge is a technical solution for every hot embossing machine. Alternatively, for hydraulically driven machines the measurement of the hydraulic pressure can give information about the molding force. Here, the resolution does not reach that of high-quality load sensors. Further, it is complicated to obtain detailed information about the demolding forces. Nevertheless, the measurement of hydraulic pressure and the reprocessing of measurement data can be used for the control of the hot embossing process.

- Measurement of relative distance: The last basic information is that of the distance between the mold and the substrate. In combination with a high resolution in time, the molding and demolding velocity can be calculated. Here, for example, touchless measurement systems like optical measurement systems are advantageous. An integration in the tools can be high risk because of the high molding temperatures. Nevertheless, these systems allow the most precise measurements. Alternatively, the position of the movable crossbar in relation to the fixed crossbar can be used as information to determine the gap between substrate plate and mold insert. If a spindle drive is used, measurement information of the drive—especially the information of the stepper motor—can be used. Here, bending of the machine, tool, and slip of the drive system possibly may not be compensated.

7.2.4 Control System

A control system is an important component of every hot embossing machine because it has the function of an intersection between the machine and the user. First, the task of a control system is to transform the process parameter entered by the user into individual machine control commands. Besides, controlling the visualization of the measured data is a second important task; through this, the user will get an overview of the process status at every moment. Further, the possibility of manual control of force, position, and temperature at every moment of the molding cycle is fundamental to avoid damaging mold inserts during a potential malfunction of the embossing machine.

To control the molding cycle and to set up the process parameter, two concepts are established: control by a programming language and control by a memory-programmable controller. Macro language programming is the most flexible solution for controlling the hot embossing process. A pool of basic functions for heating, cooling, force-controlled motions, or velocity-controlled motions can be combined into a program, which allows the adaptation of the hot embossing process to nearly all possible tasks. A further advantage is that the measured data from selected temperature sensors can be used for inquiry functions, which allow the control of the course of the program. Also different holding times, for example during the force-controlled molding step, can be integrated in this way so that the process can be very sensitively controlled. The potentials of using this concept are manifold but require a certain background of the hot embossing process. Figure 7.6 shows a simple program for a hot embossing cycle, implemented on the hot embossing machine Jenoptik HEX03.

In contrast to the machine-specific macro language control, memory-programmable controlling is state of the art for molding processes, especially in industrial use. A hot embossing machine suitable for commercial use has to be tuned to the needs of the industrial user. The control of the process should be user friendly and should allow an easy handling of all important features. Further, the control unit has to identify operating errors caused by the user to prevent damage to the molding tool and the mold inserts. Therefore, a stored program control (SPS, series-parallel-series) unit is well suited. This kind of control unit is an established process control system in industry, for example, for injection molding machines or for common presses used in the polymer processing industry. The hot embossing machine Wickert WMP1000 is equipped with an SPS control unit in combination with a graphical user interface on an integrated computer (Fig. 7.7). By the graphical user interface all process parameters and

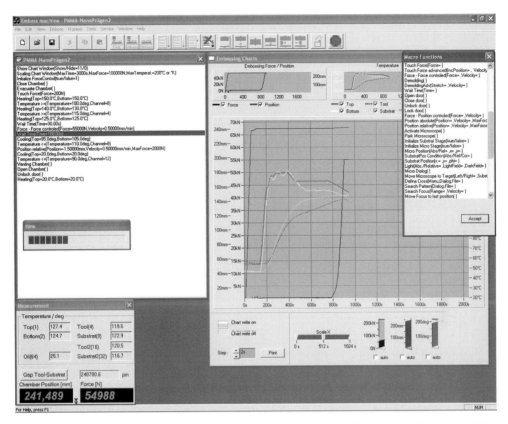

Figure 7.6 Example of a basic molding program programmed with the macro language of the hot embossing machine Jenoptik HEX03. All functions of the macro language can be combined in different ways, which allows the adaptation of the process to a wide range of tasks and the very sensitive control of the embossing cycle. Beginning by the closing of the vacuum chamber and its evacuation, the heating of substrate plate and mold insert will be started with heating commands. After the selected temperature is attained, the molding process starts with the velocity-controlled molding step up to the value of the selected molding force. Then the process switches to the force-controlled molding step, with the constant force over the determined waiting time. The cooling is initiated by the cooling command, and in parallel the heating is set down to demolding temperature. If the cooling temperature is achieved, the demolding process starts with a velocity-controlled motion of the movable crossbar. Finally, after the opening temperature is attained, the vacuum chamber opens.

process steps can be entered and will finally be executed by the machine in every molding cycle.

In contrast to a flexible control mechanism based on a programming language, this kind of control is attached to the sequence of operations of the SPS program, suitable for most hot embossing tasks. An advantage of this kind of machine control is the communication between computer, SPS, and hydraulic control unit using a standardized bus system. The hot embossing machine communicates by Profibus protocol between the individual components. This kind of

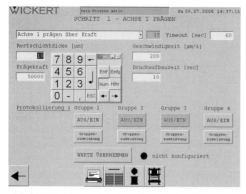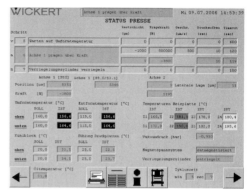

(a) Graphical user interface for entering the press force and velocity

(b) Graphical user interface for the temperature status during molding

Figure 7.7 Graphical user interface of an SPS control unit of the hot embossing machine Wickert WMP 1000. The molding process is controlled by the internal program, which can be modified and uploaded into the memory. The operation sequence is characterized by the program, and the user can enter the values for the process parameter.

communication allows the integration of additional components like molding tools with internal control units. These internal control units can be connected to the control unit of the hot embossing machine by the standardized bus communication protocols and can additionally be operated by the hot embossing machine.

Nevertheless, a memory-programmable control allows a user-friendly communication, but with the restriction that a basic program controls the molding cycle and that only the values of the parameters like temperature, force, velocity, and position can be set by the user. So the flexibility to modify the process in a wide range is lost; if not, a new modified program will be uploaded. Here, specific knowledge in memory programming is required, which can be more complex than a self-explanatory form of a macro language.

7.3 Commercially Available Machines

The market of commercially available hot embossing machines is still comprehensible. The first hot embossing machine suitable for hot embossing of high aspect ratio was developed in Karlsruhe at FZK in cooperation with Jenoptik Mikrotechnik.

7.3.1 Jenoptik Mikrotechnik

Jenoptik Mikrotechnik [2] was one of the first companies to provide a complete family of hot embossing machines. Each member of the hot embossing

(a) HEX01   (b) HEX02

(c) HEX03   (d) HEX04

Figure 7.8 The hot embossing family of Jenoptik. The machine HEX01 is a compact hot embossing system, followed by the more powerful machine HEX02 and, similar to HEX02, the hot embossing machine HEX03, with an integrated alignment system for double-sided molding. The system HEX04 is the flagship of the hot embossing machines, well suited for large-area molding. With courtesy of Jenoptic Mikrotechnik.

machine family is suited for different kinds of embossing tasks. The machine HEX01 is the smallest machine, compact and with a maximum force of 100 kN suitable for the most molding tasks, especially up to a molding area of 4 inches. If larger areas need to be molded, the machine HEX02 fulfills the requirements regarding molding forces up to 200 kN and molding temperatures up to 300°C. If further double-sided molding is required, the machine HEX03 additionally is equipped with a precise alignment system and an integrated microscope that allows the very comfortable alignment of the tool. The latest machine HEX04 is specially designed for large-area replication under a high molding force up to 600 kN. All these machines are characterized by an electrical heating unit, a convective cooling system, a spindle drive, and flexible control by a macro language. The family of hot embossing machines is shown in Fig. 7.8, and the specifications of the machines are summarized in Table 7.1.

Table 7.1 Technical Data of the Jenoptik Hot Embossing Machine Family

Press	HEX 01	HEX 02	HEX 03	HEX 04
Press force, adjustable (kN)	20, 50	200, 250	200, 250	400, 600
Press force, increment (N)	10	10	10	20, 30
Temperature inside chamber (°C)	320, 500	320, 500	320, 500	350
Molding velocities (mm/min)	0.01–600	0.01–600	0.01–600	0.01–1,000
Maximum substrate size (diameter, mm)	180	180	180	300
Maximum embossing area (diameter, mm)	150	150	150	300
Overlay accuracy alignment system (μm)	–	10 (optional)	2	2
Power consumption (kW)	9.0	14.0	16.5	31.5
Total weight (±10%) (kg)	600	1,200	1,700	4,500

7.3.2 Wickert Press

Another company who manufactures hot embossing machines is Wickert Maschinenbau [4]. This company built the first machine "MS1" used in industry for hot embossing of micro spectrometers. In 2003 a new generation of hot embossing machines was developed. The machine WMP1000 (Fig. 7.9, Table 7.2) is characterized by a hydraulic drive, a molding area larger than 8 inches, and a maximum force of 1,000 kN. The machine was developed in particular for industrial use; therefore, an automatic handling system was integrated and, in parallel, an optimized molding tool was developed (Section 8.4) to reduce the heating and cooling times, which finally resulted in a significant reduction of cycle times. Further, a user-friendly control panel was integrated allowing one to control the machine in an effective way. Nevertheless, more infrastructure is needed to operate this machine, for example, a separate room for the hydraulic pumps and a solid foundation for the heavy machine with a large overall height.

7.3.3 EVGroup

The company EVGroup [1] offers two hot embossing machines with different levels of automation [7]. The hot embossing machine EVG520HE is characterized by a semi-automatic molding process; the hot embossing machine EVG750 is, in contrast, fully automated (Fig. 7.10). This high level of automation is suited for large serial production. The machines are compatible with standard semiconductor manufacturing technologies and allow molding on substrates up to 200 mm. The hot embossing system EVG520HE includes a vacuum chamber, a drive

Figure 7.9 Hot embossing machine Wickert WMP1000. This machine is manufactured for industrial use and the molding of large series. Therefore, the machine is characterized by a fast hydraulic drive, an optimized tool for short cycle times, and a semi-automatic handling system.

Table 7.2 Technical Data of Hot Embossing Machine Wickert WMP1000, Equipped with a Basic Molding Tool

Press	Touch force	500 N–30 kN
	Molding force	50–1,000 kN
	Molding velocity	10 μm/sec–90 mm/sec
	Total weight	7,500 kg
Basic tool	Molding temperature	Max. 300°C
	Heating rate	50 kN
	Cooling rate	30 K/min
	Diameter of mold	Max. 250 mm
Advanced tool	Molding temperature	Max. 350°C
	Heating/cooling rate	50 K/min
	Diameter of mold	Max. 154 mm
	Alignment precision	≤ 5 μm

unit with a high press force up to 600 kN, and a heating system that allows the molding of a wide range of polymers. Both machines are equipped with an alignment system. The specifications of these machines are summarized in Table 7.3.

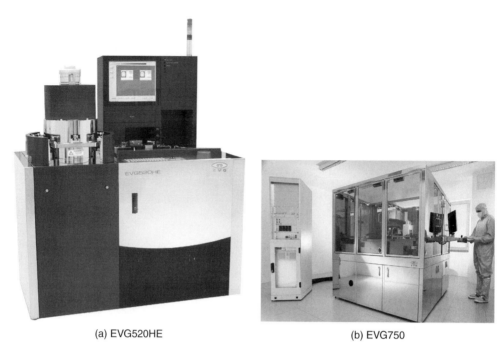

(a) EVG520HE   (b) EVG750

Figure 7.10 Nanoimprint machine EVG520HE and EVG750. The EVG520HE is semi-automated; the EVG750 is a fully automated hot embossing machine with molding forces up to 600 kN. With courtesy of EVGroup.

Table 7.3 Technical Data of the EVG Hot Embossing Machines

Press	EVG520HE	EVG750
Process capabilities	Imprinting, bonding	Imprinting, bonding
Press force, max.	60 kN	600 kN
Temperature, max.	550°C	250°C
Maximum substrate size	200 mm diameter	200×200 mm^2
Alignment system	No	Yes

7.3.4 Suess

Another category of hot embossing machines is deduced from nanoimprint machines. Nanoimprint machines are developed and optimized for the requirements of nanoimprinting. The material class of UV-curing materials is also required, like thermoplastic polymers. The company Suess [3] offers a unique nanoimprinting stepper NPS300 with hot embossing and UV-NIL capability. The requirements in precision for nanoimprint are slightly different from conventional hot embossing of structures with high aspect ratios. These kinds of machines, therefore, are optimized for the replication of structures with a

Table 7.4 Technical Data of the Suess Imprint Stepper NPS300

Press	NPS300
Press force, max.	4 kN
Press force, min.	5 N
Temperature, max.	550°C
Maximum stamp size (diameter)	100 mm
Maximum substrate size (diameter)	300 mm
Overlay accuracy alignment system	250 nm

moderate aspect ratio, mostly on substrates with a diameter up to 300 mm. Many tasks can be done by this kind of machine, but some constructive restrictions prevent its universal use as a hot embossing machine. For example, the force is limited, which makes it more complicated to mold structures with high aspect ratios on large areas. Also, the temperature range is limited, which makes it difficult to replicate high-temperature polymers. Nevertheless, with these machines excellent replications are practicable, especially in the nano range, where high-force embossing machines may cause problems regarding the precise control of molding velocity and molding force. Table 7.4 summarizes the specifications of the imprinting stepper NPS300.

Nevertheless, hot embossing is not limited to commercially available machines. There exist in laboratories a lot of modified presses used for hot embossing [5,6,8]. In general, individually developed hot embossing machines can achieve the same standards as commercially available machines. The advantage of individually equipped machines is the individual adaptation to the specific requirements and the available budget.

References

1. EvGroup. http//www.EVGroup.com, 2008.
2. Jenoptik Mikrotechnik. http//www.jo-mt.de, 2008.
3. Suess Mikrotechnik. http//www.suss.com, 2008.
4. Wickert Pressen. http//www.wickert-presstech.de, 2008.
5. D. M. Cao, W. J. Meng, and K. W. Kelly. High-temperature instrumented microscale compression molding of Pb. *Microsystem Technologies*, 10:323–328, 2004.
6. K. Idei, H. Mekaru, H. Takeda, and T. Hattori. Precise micro pattern replication by hot embossing. *JSME International Journal, Series A*, 49(1):69–73, 2006.
7. Ch. Schaefer, S. Farrens, T. Glinser, P. Lindner, and N. Roos. State of the art automated nanoimprinting of polymers and its challenges. In *7th International Conference on the Commercialization of Micro and Nano Systems (COMS)*, September 8–12, 2002.
8. X. C. Shan, R. Maeda, and Y. Murakoshi. Micro hot embossing for replication of microstructures. *Japan. Journal of Applied Physics*, 42:3859–3862, 2003.

8 Hot Embossing Tools

Apart from the molding press (Chapter 7) and the microstructured mold insert (Chapter 9), hot embossing tools are essential components of any hot embossing system. The hot embossing tool may be defined as an interface between the molding press that is responsible for applying the molding force and molding velocity and the microstructured mold insert to be replicated in polymers. Compared to macroscopic molding tools, such as tools for injection molding, where the structures are part of the tool, tools for micro replication are characterized by a reversible integration of a microstructured mold insert. This concept results from the different and incompatible fabrication processes of the macroscopic tool and microscopic structures. The tasks of a hot embossing tool are similar to tasks known from macroscopic molding tools, on the one hand. On the other hand, the embossing tool has to fulfill tasks that are specific to the molding of microstructures, like the generation of a vacuum. In detail, a hot embossing tool has to fulfill the following tasks.

- heating and cooling of the polymer film by heat conduction of the mold insert and substrate plate
- fixation of different kinds and sizes of mold inserts
- hermetic sealing of the mold insert, polymer, and substrate plate against ambient pressure
- generation of vacuum to fill the microcavities completely
- a demolding unit, which allows the vertical demolding of the embossed parts at a controlled demolding velocity
- optional alignment of both mold halves, if double-sided molding or positioned molding is desired

From these tasks, the requirements to be met by a molding tool can be deduced, including minimum requirements for molding and optional requirements for specific tasks.

8.1 Requirements on Hot Embossing Tools

- To achieve reproducible molding results over the molding area, such as a homogeneous viscosity for the complete filling of microcavities and a uniform thickness of the residual layer, a constant temperature distribution over the molding area is required. Heating of the polymer film up to the melting state is characterized by heat transfer from the mold insert and substrate plate and may be achieved, for instance, by heat flow from electrical heating elements. The heat flow that finally arrives at the surface of the mold insert or the substrate plate depends on the heat conductivity of the materials and especially on the heat transfer between the different components of the tool. Here, a good contact of the components is an important factor in determining the quality of the temperature distribution in the mold insert and also the heating time. Good heat transfer can be achieved by a high degree of evenness of the contact surfaces and avoiding any gaps in between.

- This homogeneous temperature distribution is also crucial to demolding, where inhomogeneous temperature distributions will cause anisotropy of shrinkage. Typically, convective systems are applied for cooling. For homogeneous temperature distribution, an arrangement of cooling channels is needed, which ideally guarantees a uniform heat flow over the molding area. Here, the increase in temperature of the coolant during cooling has to be taken into account (Section 8.2.2).

- Integration of mold inserts should be characterized by high flexibility and the use of uncomplicated fixation systems that allow for a quick exchange of the mold inserts. As different kinds of mold inserts are used (Chapter 9), a system is required that allows one to fix, for example, LIGA molds with a thickness of several millimeters and, with some modifications, also thin nickel shims with a thickness of 300 µm. In this respect, the creativity of construction engineers is needed. Furthermore, integration has to guarantee good mechanical contact with the heating and cooling unit to provide for an excellent heat transfer, which is fundamental to reaching adequate cycle times.

- To fill microcavities completely, vacuum is needed. Therefore, a sealing of the mold insert, polymer, and substrate plate against ambient pressure is required. To enable molding and demolding, the sealing has to ensure a relative motion between mold insert and substrate plate of several millimeters after vacuum has been established.

- A challenging task is the demolding of microstructured parts. The tool requires demolding units that, irrespective of the kind of mold insert, guarantee a demolding velocity in the range of 1 mm/min or below in a vertical direction over a distance of several millimeters under vacuum. This vertical motion is required to prevent any perpendicular forces from acting on the filigree structures. Otherwise, the risk of deformation or damage would increase significantly.

- If double-sided molding is required, the tool has to provide for a relative positioning between mold insert and substrate plate or, alternatively, a second mold. Depending on the structure size, an overlay accuracy in the range lower than 10 µm is desired. To achieve this overlay accuracy, an alignment with three degrees of freedom has to be taken into account, motion in a lateral direction (x, y), and one rotation around the center axis of the mold insert.

- An optional requirement is the compensation of uneven mold inserts. To compensate wedge-shaped mold inserts, a mechanism should be integrated to equalize these differences in evenness. With such a mechanism, a nearly homogeneous thickness of the residual layer will be achieved, which may be fundamental to further processing steps of molded parts.

8.2 Elements of Hot Embossing Tools

Most of the requirements described above can be fulfilled by a basic tool concept. Nevertheless, specific molding requirements may need individual concepts. The following section describes some technical solutions, without any claim of completeness.

8.2.1 Heating Concepts

For heating, three different physical concepts are available: heating by convection, by radiation, and by conduction.

Heating by radiation is established in macroscopic thermoforming processes, where a thin polymer foil is heated up to the glass transition range. This macroscopic thermoforming process is characterized by short heating times. It is sufficient to heat the polymer film only. Transferring this concept to hot embossing requires modifications because of the higher temperature (melting range of the polymer) and the longer molding times. Here, heating is also recommended during velocity- and force-controlled molding, which makes this concept difficult to implement because of the nearly closed gap between mold insert and substrate plate. In addition, it is difficult to control temperature during long force-controlled molding, for example, to achieve thin residual layers. Consequently, this concept is not implemented in commercially available hot embossing machines. For selected embossing processes like roll-to-roll embossing, however, this heating concept may be applicable. Seunarine et al. [9] demonstrate heating by radiation for molding low-temperature polymers like COC. A halogen lamp was used as the heating source. The heating of 0.12 g COC from 30°C up to 100°C was achieved within 45 seconds.

Compared to the heating concept based on radiation, heating by convection is a practicable concept. A heating fluid is circulated through flow channels in the heating unit, which transfer the heat to the polymer film by convection. The quality of heat transfer and temperature distribution depend on the pattern and cross-section of the flow channels, heat capacity of the fluid, and the kind of flow, laminar or turbulent (Section 8.2.2). A helical arrangement of flow channels is recommended, which guarantees an acceptable temperature distribution in the heating unit and a moderate flow resistance of the fluid. Due to the required molding temperature of more than 100°C up to 300°C, a fluid with a boiling point above the maximum achievable temperature has to be chosen. Hence, oil is a recommended heating fluid, but theoretically also water and steam can be used at high pressure. The concept of convective heating may also be used for convective cooling, whereby external heating and cooling machines heat or cool the circulating fluid to the desired temperature. Still, this concept requires external machines, and the risk of contamination by oil steam from sealing defects increases.

Another effective and comparably simple concept for heating is the concept of conduction via electrical heating elements. These elements are available with variable lengths, diameters, and performance, such that a wide range of heating concepts can be implemented. The arrangement and control of these elements determine the heat flow and temperature distribution during non-stationary and stationary heating. Heating elements are available up to several kilowatts and may be combined with other constructive solutions (Section 8.4) to make up an effective heating system, ensuring fast heating times.

The concepts of convective and conductive heating, in particular, require an optimized material concept. The material should have a high conductivity and, in parallel, a high stiffness to withstand the load during molding. For example, copper possesses excellent conductive properties ($\lambda = 372$ W/m K), but it does not sufficiently withstand high loads during molding. With a conductivity of 238 W/m K, also aluminum could be used, if the molding load were moderate. Typical steel materials with a high stiffness are mostly characterized by bad heat conductivities. For example, steel with 0.1% C has a conductivity of 52 W/m K only, and Cr-Ni steel or V2A steel reach 15 W/m K only. As a compromise between an excellent conductive material and a material of high stiffness, heat conductivity in the range of 25 W/m K can be determined. Consequently, selected materials (alloys) are required for the design of molding tools, which fulfill the contradictory requirements.

Besides conductivity, heat transfer among several components of the tool determines short cycle times. A small gap due to the unevenness of one component already can reduce heat transfer significantly. Therefore, all components should have a gap-free contact or heat transfer should be improved, for example, by the use of heat conduction pastes between uneven surfaces. Heat transfer between two components is also determined by the contact pressure between them. An increase of the contact pressure will increase the heat transfer significantly [1]. Without contact pressure, heat transfer between two even steel plates can be determined to be 2,500 W/m^2 K. If the contact pressure increases to 7 MPa, heat transfer increases to 21,000 W/m^2 K. Further increase of the contact pressure to 15 MPa results in a heat transfer of 38,000 W/m^2 K. In contrast to an increase of heat transfer with an increasing contact pressure, heat transfer will be reduced to 50 W/m² K if a gap of 500 μm exists. This significant dependence of heat transfer underlines the importance of a contact force during the heating stage of hot embossing to achieve an optimized heat transfer and to reduce cycle times. Finally, also the volume and, hence, the thermal mass of the heating unit should be minimized to an absolutely necessary level.

Figure 8.1 summarizes the technical solution of heating concepts. The concept of conductive heating and convective cooling with a minimum of thermal mass shows an efficient heating concept for micro hot embossing tools. Nevertheless, if short cycle times are required, heating systems based on radiation may applicable. Chang and Yang [4] demonstrate the performance of a convective heating system with steam, gas, and oil as working fluids, and for comparison also the heating concept by far-infrared radiation. They have shown with their equipment that a 0.2 mm thick PVC foil was heated up from 25°C up to 130°C within 30 seconds using steam heating and 25 seconds using infrared radiation. Depending on the technical implementation, these concepts can be much faster than conventional convective and conductive heating systems. Kimerling et al. [6] investigate a new rapid thermal response (RTR) tool, with rapid heating and cooling times in the range of seconds based on a high-frequency current heating (500 kHz up to 1 MHz) and a cooling by interconnected air channels embedded beneath the insert surface, about 1 mm away from the surface. Kimerling named the hot embossing process with the new embossing tool RTR embossing.

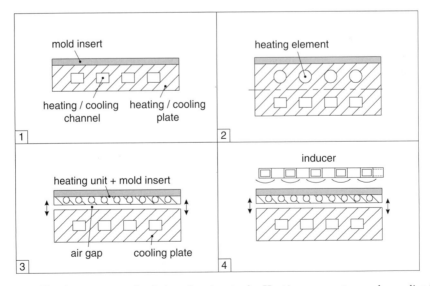

Figure 8.1 Heating concepts for hot embossing tools. Heating concepts can be realized by a pure convective heating and cooling system (1) or by a conductive heating and convective cooling system (2). A separation of cooling and heating plate by an air gap reduces the thermal mass during heating and cooling and will reduce cycle times (3). These heating systems can be supported by inductive or radiation systems (4).

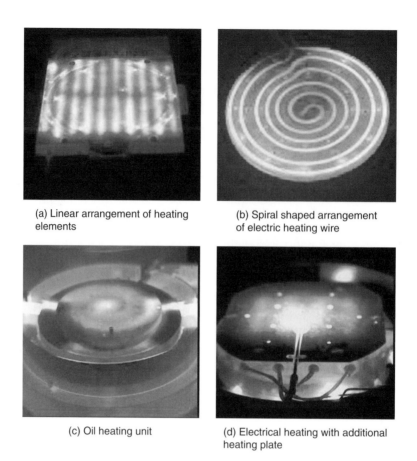

(a) Linear arrangement of heating elements

(b) Spiral shaped arrangement of electric heating wire

(c) Oil heating unit

(d) Electrical heating with additional heating plate

Figure 8.2 Thermal images of different heating concepts. The concept with a systematic arrangement of heating elements reaches a homogeneous temperature distribution over the area.

To evaluate the temperature distributions of different convective and conductive heating concepts, the temperature at the surface of a substrate plate of a hot embossing system can be visualized by thermal images. In Fig. 8.2, temperature distributions in the stationary state of a convective and a conductive heating system are compared. The concept with a systematic arrangement and control of heating elements yields the most homogeneous temperature distribution. Nevertheless, the margin areas are characterized by a lower temperature. Hence, these areas should not be part of the molding area. The molding area should be smaller than the complete area of the substrate plate.

To achieve a stationary temperature distribution in the mold insert and substrate plate, long heating times are necessary to equalize the temperature over the volume by heat conduction. If short cycle times are desired, knowledge of the temperature distribution in the polymer during heating and the time when the (nearly) stationary state is achieved will help control the molding process effectively. Measurement of the temperature inside the thin polymer foil is nearly

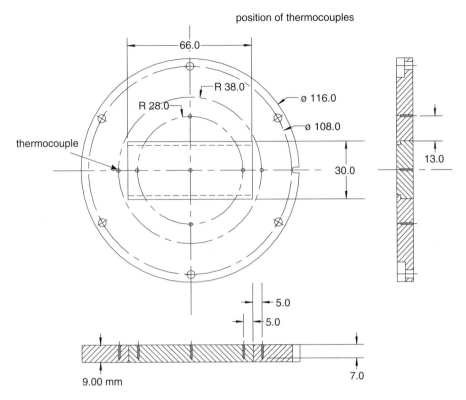

Figure 8.3 Position of sensors for the measurement of temperature distribution in a mounting plate with a mold insert. The sensors are arranged systematically over the molding area at a depth of 1.5 mm. This allows one to approximate the temperature on the surface of the plates.

impossible. Due to the comparatively small heat capacity of the polymer compared to that of the mold insert and substrate plate, it is sufficient to integrate thermocouples below the surfaces of the mold inserts and substrate plate. The systematic arrangement of these sensors is shown in Fig. 8.3. The sensors are positioned at a depth of 1.5 mm below the surface of the substrate plate and mold insert and provide a first approximation of the temperature distribution on their surface.

These measurements (Fig. 8.4) show an inhomogeneous temperature distribution during the non-stationary heating state in the mounting plate (see Section 8.2.4) with an integrated mold insert. Here, temperature differences between the center and the margin regions may be up to 10 degrees. In the stationary state, when the desired temperature is achieved, a nearly homogeneous temperature distribution will result from lateral heat conduction in the fixation plate and substrate plate at the end of the heating time. As the heating elements are controlled, the increase in temperature will decrease when the desired temperature is approached. This effect is reflected by a reduced distance of the curves measured at equidistant time intervals.

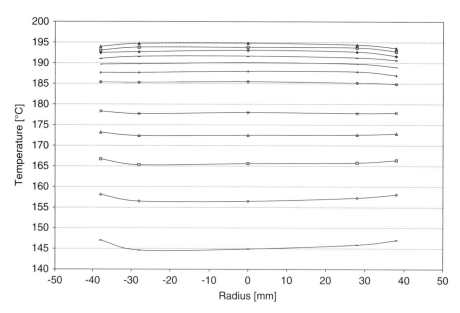

Figure 8.4 Temperature distribution measured in a mold insert integrated in a fixation plate during heating. The temperature distribution is nearly homogeneous in the stationary state. In the non-stationary state during heating, differences of up to 10°C appear between the center and the margin regions. In the microstructured mold insert, a nearly homogeneous temperature distribution can be achieved. The different curves are measured in time steps of one minute.

8.2.2 Cooling Concept

Compared to the different concepts of heating, cooling of the polymer melt can be reduced to a concept by convective cooling. Similar to heating by a thermal fluid, a fluid can be used for cooling the mold insert and substrate plate. A cooling fluid is circulated through flow channels in the cooling units. The achievable cooling rate depends on different influencing factors:

- heat capacity of the mold material
- heat capacity of the cooling fluid
- viscosity of the cooling fluid
- geometry (cross-section) of the flow channels
- flow velocity of the cooling fluid
- temperature difference between the in-flowing and out-flowing fluid

Convective heat transfer depends on the flow behavior. In the case of laminar flow, heat transfer is much smaller compared to cooling by a turbulent flow of the cooling fluid. To estimate the effectiveness of a cooling system, calculation of the Reynolds number and Nusselt number is recommended for a given design. The fundamentals of the estimation of convective heat transfer under stationary conditions for a helical cooling channel (Fig. 8.5) are described below.

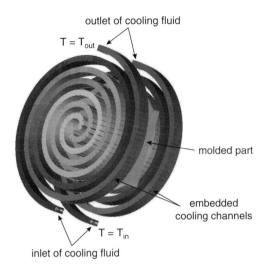

outlet of cooling fluid

$T = T_{out}$

molded part

embedded
cooling channels

$T = T_{in}$

inlet of cooling fluid

Figure 8.5 Helical design of a cooling system. This design is suited for cooling the molding area and is characterized by a streamlined shape without sharp corners.

The heat transfer coefficient α can be calculated by:

$$\alpha = Nu\frac{\lambda}{d} \tag{8.1}$$

where Nu is the Nusselt number, λ is the heat conductivity of the cooling fluid (temperature-dependent), and d is the diameter of the flow channel.

The Nusselt number depends on the kind of flow, laminar or turbulent,

$$Nu_{lam} = 0.664\left(Re\frac{d}{L}\right)^{1/2}Pr^{1/3} \tag{8.2}$$

$$Nu_{turb} = \frac{0.125\xi(Re - 1000)Pr}{1 + 4.49\sqrt{\xi}(Pr^{2/3} - 1)}\left[1 + \left(\frac{d}{L}\right)^{2/3}\right] \tag{8.3}$$

$$\xi = (1.82 \ log \ Re - 1.64)^{-2} \tag{8.4}$$

where Re is the Reynolds number, d is the diameter of the flow channel, L is the length of flow channel, and Pr is the Prandtl number.

The Prandtl number is a characteristic material property and defines the ratio between the thickness of the thermal boundary layer and the boundary layer of the fluid

$$Pr = \frac{\nu}{a} \tag{8.5}$$

where v is the kinematic viscosity and a is the temperature conductivity.

For a given design of flow channels, the kind of flow has to be determined first by the calculation of the Reynolds number,

$$Re = \frac{\varrho v d}{\eta} = \frac{v d}{\nu} \tag{8.6}$$

where η is the dynamic viscosity and d is the diameter of the flow channel.

If the flow channel is characterized by a rectangular cross-section, the diameter d has to be substituted by the hydraulic diameter d_h,

$$d_h = \frac{2ab}{a+b} \tag{8.7}$$

where a is the length of sidewall in direction a and b is the length of sidewall in direction b.

Flow in pipes changes from laminar flow to turbulent flow at a critical Reynolds number of $Re_{crit} = 2,300$. If a helical design is used, the critical Reynolds number increases depending on the ratio between the diameter d_h of the flow channel and the diameter D_{coil} of the coil.

$$Re_{crit} = 2300 \left(1 + 8.6 \left(\frac{d_h}{D_{coil}} \right)^{0.45} \right) \tag{8.8}$$

A significant increase of the efficiency of heat transfer can therefore only be achieved by a turbulent flow of the cooling fluid. As mentioned above, oil is a recommended cooling fluid. Compared to water, oil has a high thermal stability and a high boiling point, but also a higher viscosity that results in a low Reynolds number at a given volume flow. For the helical arrangement of cooling channels, a Reynolds number in the range below 300 was determined (volume flow 4.28 l/min, cross-section of the flow channels 5×12 mm^2, material properties of BP Olex WF 0801 at 20°C). Under these boundary conditions, flow is in the laminar state. For effective cooling, a Reynolds number of at least 4,000—better 10,000—is recommended [1]. To achieve this value, the diameter of the flow channels, velocity, and viscosity of the cooling fluid can be changed. Low viscosity of the fluid, in combination with high flow velocities at an optimized cross-section, will increase the Reynolds number up to turbulent flow. Nevertheless, it has to be taken into account that the necessary pumping power P will increase with the third dimension of the flow velocity ($P \propto v^3$). Consequently, pumps of high performance will be required. As an alternative, water as a cooling fluid has the advantage of a comparatively low viscosity, which makes it suitable for turbulent flow even in flow channels of smaller diameter. Small cross-sections allow the arrangement of a high density of cooling channels in the cooling unit. The advantages of high cooling rates using water are compensated by the risk of steam formation in the case of a potential failure of the cooling flow. Consequently, the risk of tool damage increases. To avoid this risk, use of oil is recommended.

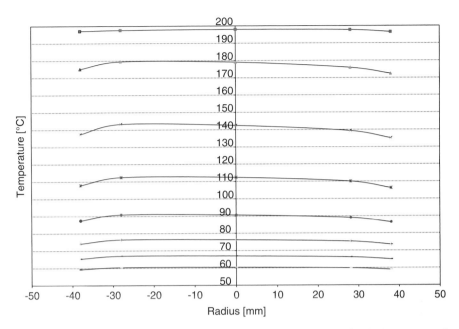

Figure 8.6 Measured temperature distribution during cooling state. At the beginning of cooling from 200°C to finally 60°C, the temperature differences along the radius of the mounting plate decrease. Because of the decreasing heat flow with a lower temperature, the temperature homogenizes. The different curves are measured in time steps of one minute.

The boiling point of selected cooling oils is in the range above the molding temperatures, which makes these cooling fluids safe. The disadvantage is the lower heat capacity c_p (approx. 1–2 kJ/kg K compared to 4.19 kJ/kg K for water) and the higher viscosity (kin. viscosity of BP Olex WF0801 approx. 16.5 mm^2/ sec compared to 1 mm^2/sec for water at 20°C), which makes it challenging to achieve Reynolds numbers in excess of 2,300.

Similar to the measurement of the temperature distribution in the mounting plate and mold insert during heating, the temperature distribution during cooling can be determined (Fig. 8.6). During the non-stationary cooling state at the beginning of the cooling cycle an inhomogeneous temperature distribution over the radius is measured. With further cooling the temperature distribution homogenizes over the radius. At demolding temperature a nearly homogeneous lateral temperature distribution is achieved.

8.2.3 Alignment Systems

Alignment systems are recommended in the case of double-sided molding or positioned molding. As mentioned above, an overlay of two mold halves in lateral dimensions (x–y) and rotation is necessary to theoretically achieve a perfect

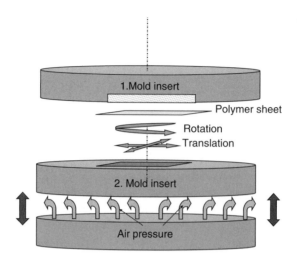

Figure 8.7 Alignment system for hot embossing tools. It is based on the hover cushion, which reduces friction. With mechanical elements and piezo-elements, an independent alignment in lateral dimensions (x, y) and rotation around the center axis can be achieved.

alignment. It is sufficient to align only one mold half. In practice, the lower mold half only will be shifted and rotated in filigree steps. Consequently, the lower mold half has to be separated from the lower movable crossbar of the machine. This is achieved by a hover cushion (Fig. 8.7). By this principle, friction is nearly eliminated and the lower mold half can be shifted and rotated with relatively small forces. To control the accurate movement in the micron range, piezo-elements are suitable. For a coarse adjustment, mechanical elements like screws can be used.

The design and principle of an alignment system can be explained using an implemented system as an example (Fig. 8.8). This system combines a hover cushion, solid joints, and piezo-elements [5].

In an air-bearing plate, 16 bearing elements are arranged. Each of these elements is supplied with air via a nozzle. The package of table plate and thermostating plate floats on these bearing elements and is guided by a solid guiding unit. The latter is installed symmetrically to the main axes of the table and connected with the air-bearing plate via two solid bearings. Between the solid bearings, the guiding frame is suspended from four solid joints. The frame is moved along the x-axis by a piezo-actuator and an opposite spring unit. By means of another four solid joints, the table plate is attached to the guiding frame. Using two piezo-actuators and two spring units fixed to the guiding frame, the table plate may be moved along the y-axis when the actuators are deflected in the same direction or rotated around the vertical axis when the direction of deflection is not the same. As a result of such a table design, shifts along the x-axis are decoupled mechanically from the movements along the y-axis. This is advantageous, as no transverse shift occurs between the table and the actuator(s) of the other axis when shifting the table.

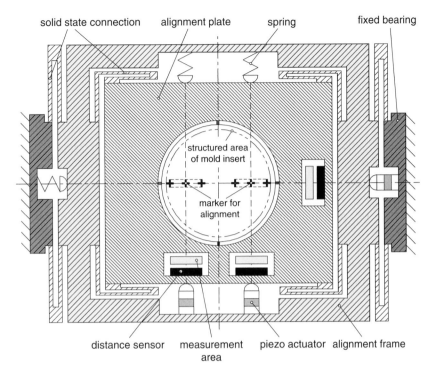

Figure 8.8 Arrangement of guiding elements, sensors, and actuators of an implemented alignment system.

Functioning of the positioning table is represented schematically in Fig. 8.8. It shows the solid guiding system, the table plate, the driving units (consisting of the piezo-actuators and the opposite spring units), as well as the sensors and measurement areas of the tool. In addition, the structured area of the mold insert with the adjustment marks is visible.

The mold inserts are installed in the tool halves in a pre-adjusted manner using clamping rings and alignment pins, such that an offset of less than ± 50 μm is expected between the mold insert structures. This offset is determined by a first-test embossing step and by measuring the molded part at an external measurement station. For this purpose, two adjustment marks each must be provided in both mold inserts. By embossing, they are transferred to the molded part. The adjustment marks are produced on the mold insert in addition to the functional structures, or they may be part of the function-bearing structure.

At the external measurement station, for example, a microscope with reflected light and transmitted light or an instrument for double-sided wafer inspection, the adjustment marks on the front and rear sides or on the top and bottom surfaces of the molded part are analyzed simultaneously and the offset is determined with the help of, for example, digital image processing techniques. After the values measured by the distance sensors during first embossing have been stored as reference values, the offset determined is converted into correction values for the table position in x- and y-direction and the necessary rotation around the vertical axis of the table. These correction values are input in the positioning control system. By another test embossing step and measurement of the molded part, overlay accuracy of the structures is checked and, if necessary, another correction is made. Due to the air bearing, the table is lifted by the bearing gap width of about 15 μm, and the target position is approached by three bearing control circuits. After the position has been reached, the table is put down again. If applicable, the position deviation resulting from the tilting of the table is corrected when putting down the table. For this purpose, another positioning process is performed with the necessary allowance.

Having reached the best possible overlay of the mold insert structures, the target distance values found by the three sensors are controlled automatically prior to each embossing cycle and approached again, if required. Determination of the offset between the mold insert structures by a test embossing step and by an external measurement of the molded part has the advantage that all influences acting on the molded part, such as machine deformation due to the embossing forces, are also acquired and corrected. Moreover, an external measurement station is required for quality assurance purposes. Hence, a measurement unit does not have to be integrated in the machine.

8.2.4 Integration of Mold Inserts

The integration of mold inserts in molding tools has not yet been standardized. On the one hand, molding tools are individual developments. On the other hand, mold inserts are characterized by different sizes, materials, and fabrication processes, which also requires tailored solutions in the end. These individual solutions should ensure that the mold inserts can be exchanged easily and in a short time. Furthermore, the thermal contact has to be guaranteed over the whole contact area. Gaps between mold inserts and tool reduce the thermal contact significantly and may also be responsible for the bending of mold inserts under load, which may result in the damage of mold inserts if brittle materials like silicon are used. The evenness of mold inserts and the corresponding integration system should be as high as possible. Evenness also is required, if a homogeneous thickness of the residual layer is desired. Here, a compensation for wedge-shaped mold inserts can be implemented by a compensation system. For example, the principle of two wedge-shaped disks that can be twisted relative to each other is a suitable system that is already part of commercially available hot embossing systems (Jenoptik).

Irrespective of the individuality of fixation, some general examples are presented here and suggestions are made for different kinds of mold inserts. The boundary condition is given by the maximum molding area defined by the size of the heating and cooling unit. On this area a plate, here an adapter plate with tapped holes, typically acts as ground plate for any fixation. On this

Figure 8.9 Fixation of a mold insert fabricated by mechanical machining. The systematic arrangement of holes at a defined pitch diameter allows the fixation of the mold insert to an adapter plate. Mold inserts with diameters from 4 inches up to 8 inches can be fixed.

plate the tapped holes may be arranged at variable pitch diameters, which allows one to fix several mold inserts of different sizes with screws only.

If a mold insert is to be produced by mechanical machining, for example, a raw material in the form of a disc can be used. This disc may be provided first with holes arranged at a defined pitch diameter. In this way, the complete disc can be fixed with screws to an adapter plate (Fig. 8.9).

This kind of fixation can also be used to integrate mold inserts with smaller lateral dimensions, for example, electroplated LIGA mold inserts. Here, the disc has the function of a mounting disc. In this case, a window has to be cut into the disc instead of the microstructured area. In this window, an electroplated mold insert can be integrated. The tolerances of the window and the mold insert should be very narrow; otherwise, a gap between the mounting disc and the mold insert will occur. These gaps have the function of an unwanted microcavity that will be filled during molding and increase the demolding force during demolding. Therefore, an interference fit between mounting disc and mold insert is recommended. Additionally, the mold insert should be positioned by a circulatory step to fix it in vertical direction against the tensile forces during demolding. The step sizes of the mold insert and mounting disc have to correspond to each other to achieve an even surface of both components (Fig. 8.10). The method of electro-discharge machining is recommended for the structuring of the window and the shape of the mold insert.

This method can also be used in a second step to integrate several mold inserts into a large (e.g., 8-inch) mounting disc. In this way, large-area mold inserts can be assembled easily by the integration and arrangement of smaller mold inserts, for example, LIGA mold inserts. The advantage of this assembly technology is

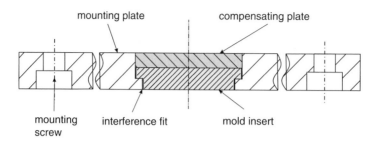

Figure 8.10 Fixation of a LIGA mold insert. The mold insert is integrated in the mounting disc via a previously cut window. The mold insert is fixed in the disc window by graded sidewalls. To compensate the differences in thickness of mounting disc and mold insert, combinable plates of different thicknesses in the lateral size of the mold insert can be fixed to the rear of the mold insert until an even rear surface is achieved.

the fabrication of large mold inserts by mold inserts that cannot be fabricated in such large dimensions. This kind of assembly, however, requires precise fabrication of the windows and the corresponding shape of the mold inserts (Section 9.3).

The fixation method described above requires a thickness of mold inserts of several millimeters. Electroplated nickel shims (Chapter 9), by contrast, are characterized by a thickness in the range of 300–500 µm and diameters typically amounting to 4 inches or 6 inches. Due to these large lateral dimensions compared to the thickness, such mold inserts are flexible and, hence, the fixation method described above is not suitable. To avoid any bending of the mold insert during demolding, a holohedral fixation is recommended, especially when high demolding forces are expected. An established technique in nanoimprinting technology, for example, is the fixation of shims by the use of a vacuum chuck [2]. This technology fails in hot embossing, however, because vacuum is needed to fill the microcavities and the pressure difference necessary for the fixation is lacking.

Gluing a nickel shim onto a flat mounting disc with different gluing materials and techniques is an acceptable approach (Fig. 8.11). The disadvantage is a permanent fixation. A separation may result in the damage of the shim. Gluing has to be prepared very carefully. Apart from a homogeneous thickness of the glue, an even and clean surface of the substrate is necessary. Any particle between mounting disc and shim will deform the nickel shim during molding and also replicate these deformations in the polymer.

A process of gluing on a flexible intermediate layer was developed by Bründel [3] (Fig. 8.12). The advantage of this flexible layer is the equalization of unevenness and the compensation of deformations caused by enclosed particles during gluing.

For small nickel shims (e.g., 4 inches) with microstructures requiring low demolding forces, fixation by a clamping mechanism is suitable. As in the case of fixation by gluing, an even and clean mounting disc is recommended.

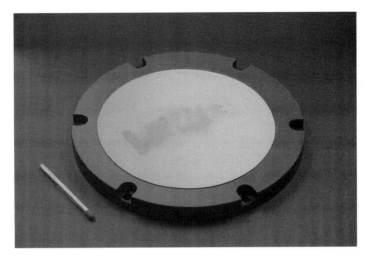

Figure 8.11 Fixation of a nickel shim on a rough substrate plate by gluing. The gluing area has to be clean and even; otherwise, the nickel shim may be deformed by enclosed particles.

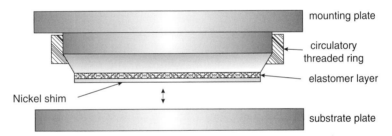

Figure 8.12 Fixation of a nickel shim by gluing on a flexible elastomeric layer. This layer equalizes unevenness and enclosed particles.

The clamping mechanism may consist of a simple circular frame to clamp the shim in the margin region. The frame is fixed by screws to the mounting disc (Fig. 8.13). Such a frame, however, eliminates the contact between the microstructured shim and the substrate plate without any modifications of the substrate plate. For molding, the area of the substrate plate therefore has to be modified, such that the size of the substrate plate is reduced to a geometry that is similar to the geometry enclosed by the frame.

8.2.5 Demolding Systems

A demolding system has to provide for a precisely controlled vertical movement. Demolding can be achieved by three different concepts: adhesion of the residual layer to a rough substrate plate, use of ejector pins, and use of

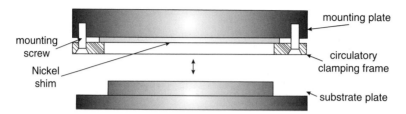

Figure 8.13 Fixation of a nickel shim by clamping. A simple circular frame can be used to clamp a nickel shim in the margin regions. To obtain contact between mold insert, polymer, and substrate plate, the geometry of the substrate plate has to be modified, such that it corresponds to the area enclosed by the frame.

pressure-assisted demolding with the option of an advanced demolding concept for double-sided molding.

8.2.5.1 Adhesion to Rough Substrate Plates

A simple, but effective concept for single-sided molding is the use of a substrate plate with a high adhesion to the surface. This high adhesion can be achieved, for example, by a rough surface with undercuts in the nano range, which are fabricated by sand blasting or lapping of the surface. During molding, a high adhesion between the residual layer and the rough surface will occur. Typically, it is significantly higher than the demolding forces. In this case, demolding of the structures is achieved by the precise relative motion between mold insert and substrate plate, which is generated by the motion of the movable crossbar of the hot embossing machine. This concept is recommended because of its simplicity. To obtain a good adhesion, however, the substrate plates have to be cleaned and any residues from the fabrication processes have to be removed. During molding, adhesion decreases, such that the substrate plates have to be exchanged after several molding cycles. It is also recommended to use a different substrate plate for every polymer. Otherwise, adhesion might decrease significantly.

8.2.5.2 Ejector Pins

Ejector pins are state of the art in the demolding of macroscopic molded parts, especially in the injection molding process. This concept may be transferred to micro replication processes. From the sizes of mold inserts and their fabrication processes, it is obvious that ejector pins cannot be positioned easily into the mold insert. Therefore, only an arrangement around mold inserts is recommended. The ejector pins act at the circumference of the residual layer. For mold inserts with a small microstructured area, the ejector pins transmit the

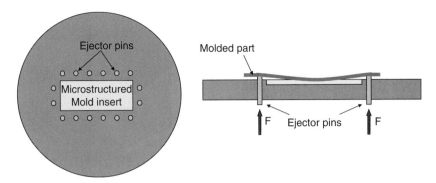

Figure 8.14 Principle of demolding by ejector pins. The ejector pins may be arranged around microstructured areas only.

demolding forces nearly in a vertical direction, such that microstructures can be demolded vertically. Nevertheless, the thickness of the residual layer has to be large enough to transmit the forces; otherwise, the residual layer will break if higher forces at the ejector pins are needed for demolding. If the molding area increases, the situation will change unfavorably. If the ejector pins are arranged around a mold insert only, bending during demolding cannot be avoided and will result in forces on the microstructures that are perpendicular to the demolding direction. Demolding large structured areas on a thin residual layer may therefore result in problems due to the ejector pins arranged at the circumference. This concept has to be modified, such that the ejector pins are arranged systematically over the molding area, also inside or between the mold inserts. Ejector pins are suited if several small mold inserts are combined into a large mold insert. In this case, ejector pins may be arranged between the individual mold inserts. The use of ejector pins requires a high technical expenditure and the area where mold inserts can be arranged cannot be changed easily. Hence, this system is characterized by a lower flexibility, but it allows for effective demolding of microstructured molds over a large area (Fig. 8.14).

8.2.5.3 Air Pressure–Assisted Demolding

Air pressure–assisted demolding is particularly suitable for double-sided molding. The principle is shown in Fig. 8.15. To initiate demolding, a small gap between mold insert and molded part is necessary, into which air pressure is injected. The air pressure separates the molded part from the mold insert by aerostatic pressure acting on an extending gap. This principle is supported by a fixation of the residual layer at the circumference. This residual layer acts as a kind of seal against the ambient pressure. If air pressure is injected into a small circular gap, the molded part is demolded from the mold insert, beginning in the margin region and proceeding to the center of the mold insert. The injected air

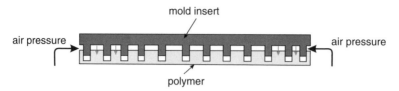

Figure 8.15 Principle of air pressure–assisted demolding. Air pressure is injected into a small gap between mold insert and molded part, and the structures are demolded by aerostatic pressure.

applies an aerostatic pressure between the mold insert and molded part, and the molded part is removed nearly vertically.

The concept of air pressure–assisted demolding can be used for demolding parts molded on both sides. Here, an adapted solution is required. The advanced concept for double-sided demolding is presented in Section 8.6.2. A detailed description can be found in Dittrich et al. [5].

8.3 Simple Tool for Hot Embossing

A minimal configuration for a tool can be implemented by the symmetrical arrangement of the following components (Fig. 8.16):

- heating plate with heating elements or convective heating system
- cooling unit with typical convective heat transfer
- vacuum chamber, implemented by a simple circular elastomeric seal
- fixation system for mold insert and substrate plate, implemented by screws

Already with this minimal configuration most of the embossing tasks can be done. Nevertheless, the quality of the tool manufacturing, the arrangement of the heating elements, and the design of the cooling channels will define, besides the quality of the temperature control unit, the obtainable replication results. Because of the direct connection of heating and cooling unit, a lot of thermal mass has to be heated and cooled at every cycle, which results in comparably long cycle times for these kinds of molding tools. The reduction of thermal mass is, therefore, an important task in the development of hot embossing tools and is implemented by an advanced basic tool (Section 8.4).

Independent of the long heating and cooling times, this concept allows, in combination with an effective temperature control unit, the achievement of stationary temperatures over long times. Especially if thin residual layers are desired, a constant temperature during the force-controlled embossing state is required. This kind of universal tool is well suited for a large bandwidth of

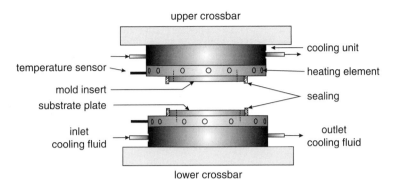

Figure 8.16 Schematic view of an elementary molding tool. The tool consists of a heating and cooling unit and a circulatory elastomeric ring for sealing against ambient pressure. Substrate plate and mold insert can be mounted by screws.

experimental tasks and small series in which the replication of prototypes and feasibility studies are in the foreground of interests.

8.4 Basic Tool for Hot Embossing

The requirements and concepts underlying the design of tools for hot embossing have been described above. An example of a hot embossing tool with an optimized heating and cooling concept was developed at the Forschungszentrum Karlsruhe. In this case, thermal mass is reduced, while the stability and evenness of the surfaces needed for molding microstructures on large areas of a thin residual layer are maintained (Fig. 8.17).

Functioning of this tool is illustrated in Fig. 8.18.

To reduce the heated masses to the largest possible extent, the hot and cold areas of the tool are separated thermally. Hence, such a tool is divided into a heating plate and a cooling block (Fig. 8.18). In the basic state of the tool, both functional units are insulated thermally by an air gap produced with the help of springs (Fig. 8.18(a)). This air gap is kept when melting the polymer. The contact force is generated by the springs only (Fig. 8.18(b)). Due to thermal insulation, the relatively thin heating plate and the mold insert can be heated up rapidly.

In the displacement- and force-controlled embossing process, the molding force presses the heating plate onto the massive cooling block, which results in a mechanically stable set-up, by means of which homogeneously thin residual layers may be produced even on large areas (Fig. 8.18(c)). As soon as the heating plate is in contact with the cooling block, heat is removed from the heating plate to the comparably large and permanently cooled cooling block. As the cooling block acts like a heat sink, the heating plate and the mold insert can be cooled down rapidly. Subsequently, the components can be demolded (Fig. 8.18(d)).

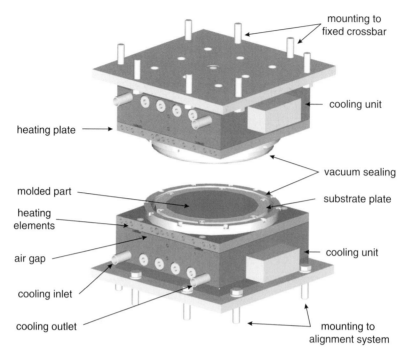

Figure 8.17 Schematic view of a basic molding tool with reduced thermal mass. The tool is characterized by a symmetrical top and bottom architecture. The electrical heating unit is characterized by a reduced thermal mass, separated from a convective cooling unit by air gaps.

Except for the vacuum chamber, the tool halves are designed symmetrically and consist of a water-cooled cooling block and a heating plate each. The heating plate is lifted off the cooling block by pre-stressed disc spring packages. Mold inserts of 250 mm in diameter can be fixed on the heating plate; the maximum molding temperature is 300°C. In this tool, the heating plate can be clamped to the cooling block using a magnetic clamping system. A tolerance-free opening movement of the hot embossing machine thus allows for offset-free demolding. As in conventional hot embossing tools, substrate plates roughened by lapping or sand blasting may be used for demolding. The demonstration tool is presented in Fig. 8.19.

For precise temperature control, the heating plates of the basic tool are divided into four zones each, which are controlled separately and may be set to various temperatures. In this way, a highly homogeneous temperature distribution can be achieved in the mold insert. In the hot embossing process, three different temperatures can be input for each tool half: The molding temperature is the temperature to which the heating plates are heated, and the embossing temperature is the temperature at which the embossing force starts to build up. At the demolding temperature, demolding of the embossed component starts. The molding temperature is measured directly in the individual zones

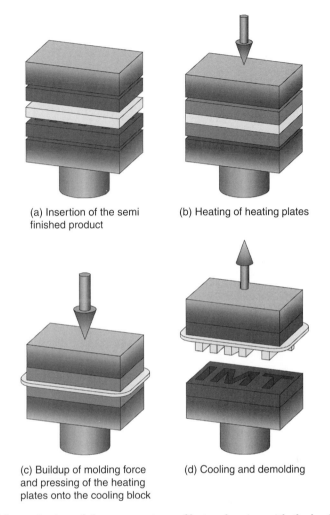

(a) Insertion of the semi
finished product

(b) Heating of heating plates

(c) Buildup of molding force
and pressing of the heating
plates onto the cooling block

(d) Cooling and demolding

Figure 8.18 Schematic view of the process steps of hot embossing with the basic molding tool.

of the heating plates; the embossing and demolding temperatures are measured in the mold insert and substrate plate, respectively. As the mold insert cools down quickly when the embossing force is generated, it may be reasonable to select a molding temperature far above the embossing temperature. As a result of the thermal inertia of the heating plate and the cooling block, cooling of the polymer melt then is slowed down slightly. This allows for the fabrication of very thin components of low stress.

The heating concept based on a spring was also implemented by Schift et al. [8] for the molding of wafer-type substrates. A clamped stack of stamp and substrate was preassembled in an alignment system. The clamped stack was not in contact with the heating plate because of the spring system. The gap was closed by the acting of force, pressing the stack onto the heating plate. After embossing

Figure 8.19 Lower half of a basic molding tool with a moldable surface area of up to 250 mm diameter.

the force was set to a low value, which resulted in a separation of the stack from the heating plate caused by the springs. The molded part was cooled and demolded manually.

8.5 Basic Tool for Industrial Applications

Based on the concept described in the previous section a hot embossing tool for industrial application was developed [7]. This tool can be used in a wide range of different embossing systems independent of the existence of a sensitive touch-force control mechanism. The heating plate with the integrated heating elements is separated from the cooling block by a number of springs between the heating and cooling unit. If the tool is closed by motion of the crossbars of an embossing machine, the touch force will be generated by these springs. The quality and the value of the touch force is a function of the number of springs, the arrangement, the spring constants, and the distance the springs will be pressed.

Compared to the basic tool described above, during demolding the heating plate is not fixed by a magnetic fixation system. The lack of this fixation system allows the arrangement of the flow channels near the surface of the cooling block, which guarantees a quick and efficient heat transfer from the heating plate. The desired guidance of the mold and substrate plate during demolding can be achieved by a connected guidance of the top and bottom heating plates.

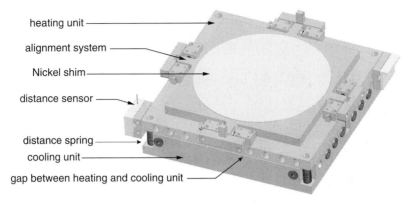

heating unit

alignment system

Nickel shim

distance sensor

distance spring

cooling unit

gap between heating and cooling unit

Figure 8.20 Molding tool for industrial applications. The fast heating and cooling tool is optimized for use in a wide range of presses without precise touch-force control.

To measure external influences like bending of the crossbars under load or thermal strains, both heating plates are connected with distance sensors. These sensors allow the measurement of the gap between mold insert and substrate and also allow, therefore, the determination of the thickness of the residual layer precisely. If an array of these sensors is integrated, the tilt caused by flow behavior or eccentric position of the polymer foil can be determined. Figure 8.20 shows in a schematic view the construction of one-half of this tool.

8.6 High-Precision Tool for Double-Sided Molding

On the one hand, the hot embossing tool described above is well suited for the cost-efficient molding of parts molded. On the other hand, this tool can also be used with an alignment system for double-sided molding. In this case, demolding has to be done manually at one of the microstructured mold inserts, which will increase the risk of damaging the structures. A new concept for double-sided high-precision molding was developed by the Forschungszentrum Karlsruhe (Fig. 8.21, [5]). This concept also includes an advanced heating concept to reduce cycle times and especially a demolding concept for double-sided molding. Compared to the basic tool described above, the high-precision tool developed differs as follows:

- During the complete process, the semifinished product is fixed between two clamping plates and can be moved by the latter and by another servo-electric drive independently of the opening and closing movement of the embossing press.
- Together with the mold insert, the clamping plates that fix the semifinished product form a shearing-edge sealing.

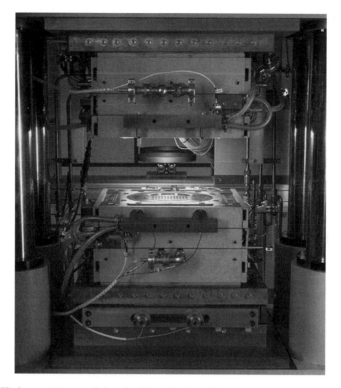

Figure 8.21 High-precision tool for double-sided molding mounted into the hot embossing machine of the type Wickert WMP 1000.

- Separate vacuum and pressurized gas supplies in the upper and lower tool allow for an independent evacuation of the cavity and a separately adjustable pressurized gas loading of the upper and lower side of the semifinished product. In this way, pressurized gas-supported demolding can be achieved.
- An alignment system based on the concept of an air bearing and flexible joints is integrated.

8.6.1 Construction

The principle of this tool and the components can be illustrated in Fig. 8.22.

The tool halves each consist of a tool frame of constant temperature, a tool insert of minimum mass that is subjected to thermal cycling, and a cooling block kept at constant temperature. During heating, the tool insert is thermally insulated from the cooling block. In this way, short heating times are reached. During the embossing and cooling phase, the tool insert is coupled thermally to the cooling block such that heat can be removed rapidly. As a result of the

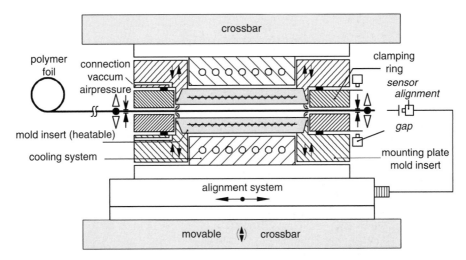

Figure 8.22 Schematic view of high-precision tool for double-sided molding. Basic components of this tool are both mold halves with clamping ring for the clamping of the polymer foil, a demolding system with ejector pins and air-pressure assistance, and an alignment system. This tool is further enabled for the use of polymer film on coils.

short times needed for heating and cooling the tool during embossing, cycle time is reduced considerably.

The tool halves are additionally provided with plates, by means of which the semifinished product is fixed during heating and cooling and shrinkage is avoided. By a demolding drive, the fixing plates can be moved precisely and independently of the tool opening movement. This allows the application of the semifinished product to the mold inserts or to separate it from them for demolding. Via separate vacuum and pressurized air connections in the upper and lower tool, independent evacuation of the cavity or controlled separate pressurized gas supply to the upper or lower side of the semifinished product is ensured. In this way, the large-area molded parts that have been microstructured on both sides can be demolded easily.

The positioning table is equipped with an air bearing and flexible joints. Consequently, positioning is not influenced by friction or play of the joints. Precise orientation of the tool halves is achieved by piezo-actuators driving the table and high-resolution sensors integrated in the tool frame.

A detailed view of the tool is shown in Fig. 8.23.

Both mold halves are nearly identical and are assembled by a tool frame with three functional levels (A, B, C) and the mold insert (level D).

The basic level A consists of a clamping plate or ground plate (1), the core of the tool (2), the alignment system (3), and the distance frame (4). A two-layer heat-insulating plate (5) isolates the clamping plate against the temperature-equalization plate. Circulatory insulation strips (6) reduce the heat flow to the environment. Centralized on the ground plate or clamping plate, the core

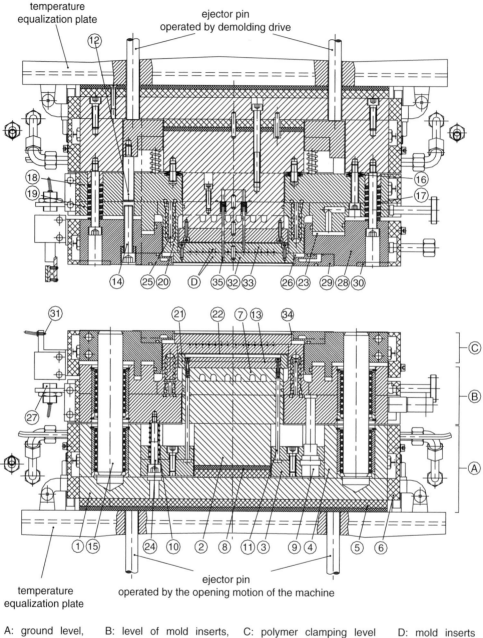

Figure 8.23 Detailed construction principle of high-precision tool for double-sided molding.

A: ground level,   B: level of mold inserts,   C: polymer clamping level   D: mold inserts
1: ground plate 2: core 3: ejector system 4: distance frame 5: insualting plate 6: insulating strip
7: cooling plate 8: insulationg plate 9: ball bearing 10: spring 11: ejector pin 12: setbolt
13: second ejector unit 14: threaded bolt 15: ball bearing 16: Belleville spring 17: compression spring
18: carrier plate 19: annular piston plate 20: clamping ring 21: distance plate 22: isolation gap
23: cylinder 24: screw 25: inner raceway 26: ring channel 27: distance sensor 28: carrier plate
29: clamping ring for polymer film 30: distance strip 31: distance sensor 32: mold insert
33: heating plate 34: gap for evacuation and venting 35: temperature sensor

of the tool is mounted. This core consists of a cooling plate (7) and two intermediate plates. It guides temper oil, conductors, and sensor connections to the cooling plate and to the mold insert (D). The core is also isolated by a two-layer insulation plate (8). The alignment system (3) encloses the core and is activated eccentrically by an ejector bolt. The system is guided by ball bearings (9) and is supported by springs (10) against the mold-insert carrier level (B). The alignment system consists of six ejector pins (11) and eight set bolts (12). The ejector pins act directly on the molded part or can be coupled to a second alignment system, which can be arranged between cooling plate (7) and mold insert (D). This second alignment system is exchangeable and allows the individual adaptation of the position of the ejector pins to the desired geometry of the mold insert. The set bolts (12) act over adjustable threaded bolts (14) on the polymer clamping level (C) so that these levels can be shifted simultaneously to the shift of ejector pins. In the edges of the distance frame (4) are arranged four high-precision ball bearings, responsible for the guiding of the mold-insert carrier plane (B) and the polymer clamping plane (C). Packages of Belleville springs (16) support the mold insert carrier plane (B) against the ground plane (A). Also, prestressed compression springs (17) support the polymer clamping level (C) against the ground plane (A).

The mold-insert carrier plane (B) consists of a carrier plate (18), an annular piston plate (19), and a clamping ring (20) responsible for the fixation of the mold insert. In the ground state described here, the mold-insert carrier plane (B) is lifted by the Belleville springs (16) in a way that between the mold insert plane (D) and the distance plate (21) of the second ejection unit an isolation gap of approximately 1 mm occurs (22). This gap will close if the cylinder (23) is activated by air pressure. The ejection system, connected to the carrier plate at the end position by screws, follows this shift so that no relative shift between ejector pins and mold insert occurs. The clamping ring is tempered with the aid of an inner ring (25) by two oil heating circuits. By the ring channel (26) the cavity can be evacuated and vented. With the help of the distance sensor (27), the molding gap during embossing and the relative demolding distance between molded part and mold insert during demolding can be determined.

The polymer clamping plane (C) consists of a carrier plate (28), a clamping ring for the polymer sheet (29), distance strips with friction lining (30), and distance sensors (31) for the determination of the relative position between both mold halves. The guiding pins of the high-precision ball bearings are clamped in the carrier plate. During the embossing state both mold halves are supported against each other by the distance strips. The frictionally engaged connection protects both mold halves against lateral shift. The polymer film, the semifinished product, is clamped between the clamping rings. During the heating clamping is still active up to the demolding state, in which finally the molded film is lifted from the mold insert. With an outer diameter of 258 mm between the clamping rings, polymer foils with a maximum width of 250 mm can be clamped.

The mold insert (D) is mounted inside the mold insert zone, clamped by the clamping ring and fixed in this mold insert plane. The mounting zone is defined by the inner diameter of the polymer clamping ring and the geometry of the second ejection unit. The polymer clamping ring has, in this case, an inner diameter of 150.4 mm so that it is possible to use also polymer-coated silicon wafer as a semifinished product. The maximum thickness of a mold insert is in the range of 24 mm. This large range allows the mounting of mold inserts (32) with a connected heating plate (33) or thin shim mold inserts with an adapted distance plate (21).

The clamping ring for the semifinished product is characterized by a circulatory groove, reducing the shrinkage during cooling by form-fitting. Further, the task of the clamping ring is to seal the cavity against penetration of polymer melt into the evacuation and venting gap (34).

8.6.2 Operation Principle

The operation principle can be illustrated for the main steps of hot embossing: heating, molding, and demolding.

8.6.2.1 Heating

In Fig. 8.24, the process steps required for heating the semifinished foil are represented schematically. The initial situation (A) shows the tool area close to the cavity with the cooling blocks, mold inserts, and clamping rings for the semifinished product, and the semifinished foil supplied. The cooling blocks are set to a constant temperature far below the demolding temperature of the semifinished plastic product. While the semifinished product is heated, the mold inserts that have been lifted off the cooling blocks and, thus, separated thermally, are heated up to molding temperature. At the beginning of the heating of the semifinished product, the temperature of the mold inserts exceeds the glass-transition temperature of the plastic. The clamping rings are set to a constant temperature slightly below the glass-transition temperature. The semifinished foil possesses room temperature, as it is done without external preheating.

- The semifinished product is fixed between the clamping rings as a result of the closing movement of the tool and moved toward the structured surface of the upper mold insert up to the gap width s_{heat} (step 1).
- By evacuating the upper partial cavity and loading the lower partial cavity with pressurized gas, the semifinished product is deflected and pressed onto the surface of the mold insert (step 2).

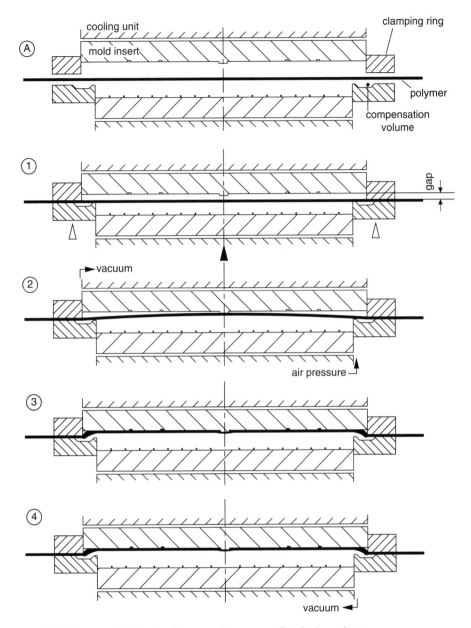

Figure 8.24 Concept for the heating of polymer semifinished products.

As a result of the contact, the semifinished product is heated and completely nestles against the surface of the mold insert due to thermal elongation.

- Heating is completed (step 3) when the semifinished product is in contact with the mold insert up to the edges. Elongation and shrinkage of the semifinished product result in an excess of material at the edges, which is dependent on the extent to which elongation is inhibited by the mold insert structures. During later process steps, this is compensated by the compensation volume of the clamping rings.

- As soon as the mold insert has reached molding temperature, the lower partial cavity is evacuated (step 4).

8.6.2.2 Molding

The process steps required for the molding of the semifinished product are displayed in Fig. 8.25. The initial situation (A) shows the cavity with the mold inserts heated to molding temperature, the clamping rings heated to demolding temperature, and the semifinished product contacting the surface of the upper mold insert and heated to molding temperature.

- The mold inserts are applied to the cooling blocks. In this way, cooling of the mold inserts is started (step 1). By the closing movement of the tool, the clamping rings are moved synchronously with the upper mold insert, which prevents the semifinished product from being stretched at the edge of the cavity.
- The tool is further closed, as a result of which the sealing edge of the lower clamping ring is moved against the surface of the upper mold insert and blocks material flow toward the outside (step 2).
- Then, the tool is closed, such that the mold insert structures are immersed into the plastic feedstock and filled (step 3).
- Finally, closing force is increased up to embossing force, such that the structure edges are molded and the melt is compressed to compensate shrinkage (step 4).
- To compensate volume contraction of the melt during cooling, the closing force is maintained until the glass-transition temperature of the plastic is reached. When the temperature drops below the glass-transition temperature, closing force is reduced to a small residual force, as material flow does not take place any longer. Shrinkage of the molded part is then inhibited by the mold insert structures. At the circumference, shrinkage is prevented by the fixation of the molded part between the clamping rings. As the molded part remains fixed during demolding from the mold inserts, defects of the molded part that may result from the release of energy-elastic forces at the end of demolding in principle are expected to be reduced significantly.

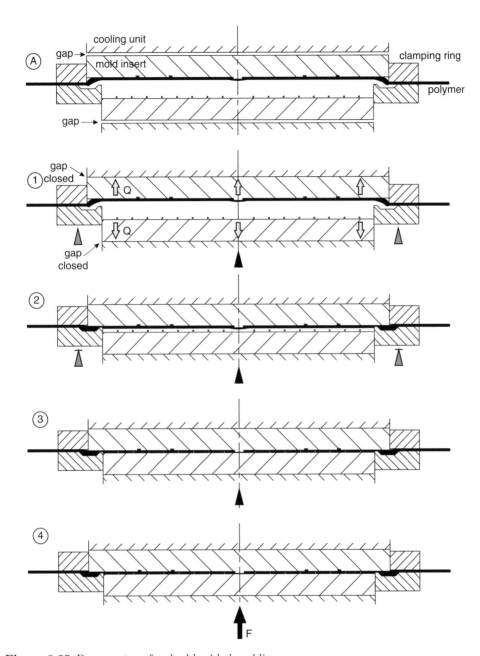

Figure 8.25 Process steps for double-sided molding.

8.6.2.3 Demolding

The initial situation (A) shows the closed cavity after molding. The mold inserts are cooled down to demolding temperature. Closing force is reduced to a small residual force (Fig. 8.26).

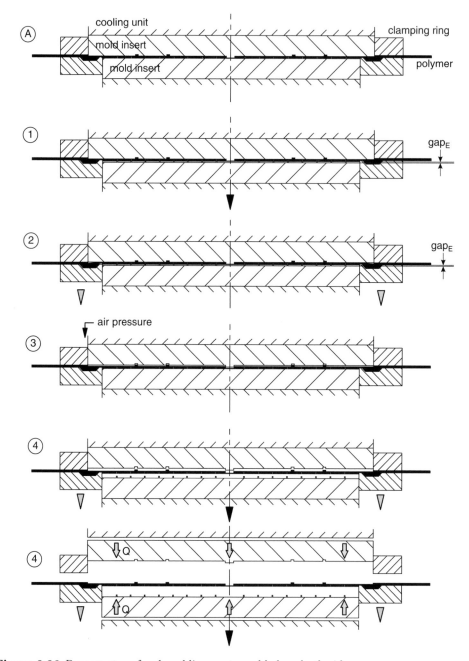

Figure 8.26 Process steps for demolding parts molded on both sides.

- The tool is opened up to the gap width s_{demold}. Due to higher adhesion forces of the molded part in the upper old insert, the breakaway torque between molded part and lower mold insert is overcome first (step 1). The gap width s_{demold} is preferably chosen to be smaller than the structural height of the lower mold insert, such that the molded part that is removed from the upper mold insert by pressurized gas in the following step nestles against the lower mold insert and is supported by the latter. In this way, plastic deformation or stretching of the molded part is prevented even in the case of high pressure.

- The clamping rings are moved by the same distance s_{demold}. Along its circumference, the molded part is peeled from the upper mold insert. To ensure favorable force input into the part, the demolding force is introduced via the strengthened area in the compensation volume. This area compensates the differences in the demolding behavior of the molded part by bending (step 2).

- The wedge-shaped gap generated between the molded part and the upper mold insert is loaded by pressurized gas. The propagating pressurized gas front detaches the molded part radially from the mold insert (step 3).

- After the breakaway torques between the molded part and both mold inserts have been overcome, the tool is opened in the following steps and the molded part is demolded completely from the mold insert structures. Then, the mold inserts are lifted off the cooling blocks and heated up again (steps 4 and 5).

References

1. Moldflow reference handbook. http//www.moldflow.com, 2003.
2. L. Bendfeldt, H. Schulz, N. Roos, and H.-C. Scheer. Groove design of vacuum chucks for hot embossing lithography. *Microelectronic Engineering*, 61–62:455–459, 2002.
3. M. Bründel. *Herstellung photonischer Komponenten durch Heissprägen und UV-induzierte Brechzahlmodifikation von PMMA*. PhD thesis, University of Karlsruhe (TH), 2008, ISBN: 978-3-86644-221-4.
4. J.-H. Chang and S.-Y. Yang. Development of fluid-based heating and pressing systems for micro hot embossing. *Microsystem Technologies*, 11:396–403, 2005.
5. H. Dittrich, M. Heckele,and W. K. Schomburg. *Werkzeugentwicklung für das Heissprägen beidseitig mikrostrukturierter Formteile*. PhD thesis, University of Karlsruhe (TH), Institute for Microstructure Technology, FZKA Report 7058, 2004.
6. T. E. Kimerling, W. Liu, B. H. Kim, and D. Yao. Rapid hot embossing of polymer microfeatures. *Microsystem Technologies*, 12:730–735, 2006.
7. C. Mehne, R. Steger, P. Koltay, D. Warkentin, and M. P. Heckele. Large-area polymer microstructure replications through the hot embossing process using modular moulding tools. In *Proc. IMechE Part B: Journal Engineering Manufacture*, 222:93–99, 2008.

8. H. Schift, S. Bellini, J. Gobrecht, F. Reuther, M. Kubenz, M. B. Mikkelsen, and K. Vogelsang. Fast heating and cooling in nanoimprint using a spring-loaded adapter in a preheated press. *Microelectronic Engineering*, 84:932–936, 2007.

9. K. Seunarine, N. Gadegaard, M. O. Riehle, and C. D. W. Wilkinson. Optical heating for short embossing cycle times. *Microelectronic Engineering*, 83:859–863, 2006.

9 Microstructured Mold Inserts for Hot Embossing

For every replication process a mold or master is necessary to copy the structures of the mold into a molding material. Because of the different structuring methods of classical mechanical mold fabrication in the macroscopic range and the structuring methods of microstructures, the mold has to be split into the tool (Chapter 8) and the mold insert with a microstructured surface. In theory every microstructured surface can be used as a mold insert. The precondition is that the mold material and the microstructures will withstand the temperature and mechanical load during molding. Nevertheless, for successful molding and especially demolding, the mold insert has to fulfill the following requirements.

- The yield stress of the mold material at maximum molding temperature has to be significantly higher than the load effected by the molding force.
- To avoid any bending and to guarantee the best possible evenness of the mold, the residual stress inside the mold, caused by the fabrication process, should be reduced to a minimum.
- The mold material should show chemical resistance to the polymer.
- A high heat conductivity of the mold material will reduce heating and cooling times.
- For cost effectiveness, the lifetime of the mold should be extended over many cycles.
- To support successful demolding, the surface roughness, especially of vertical sidewalls, should be reduced to an unavoidable minimum.
- Demolding angles are advantageous because they facilitate demolding. In contrast, undercuts prevent successful demolding of microstructures. Even small undercuts in the submicron range can increase demolding forces significantly (Section 6.4.5).

Regarding the requirements, especially the requirement of high yield stress, it is obvious that mold inserts fabricated in metals are well suited. The technique of microstructuring of metals is therefore essential for mold fabrication. But also glass or polymers like UV-transparent PDMS or high-temperature resistant PEEK can be used for selected replication tasks. Nevertheless, regarding the lifetime of a mold insert and the high stiffness, molds fabricated of metals are mostly used for replication. Therefore, this section of mold insert fabrication is focused on the fabrication technology and the properties of metal mold inserts.

An overview of the different mold fabrication processes is shown in Fig. 9.1.

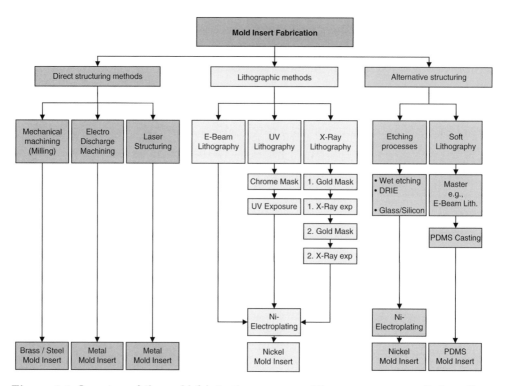

Figure 9.1 Overview of the mold fabrication processes. The processes are split into direct structuring methods, lithographic methods, and alternative methods like etching processes or PDMS casting.

The structuring processes can be split off into two groups: (1) direct structuring methods, like mechanical machining, electric discharge machining (EDM) or laser structuring [7]; and (2) the field of lithographic methods, like E-beam lithography, UV-lithography, and for structures with high aspect ratio, X-ray lithography. All lithographic processes require the step of electroforming to obtain a metal mold insert. Each structuring method has different characteristics and is therefore suited for different kinds of applications.

9.1 Direct Structuring Methods

Not every design requires the quality of lithographic processes for mold fabrication. Direct structuring methods are an alternative technique. These methods have a limitation in size of structured areas of some lithographic processes. Therefore, direct structuring processes are well suited for large microstructured areas. Structure sizes in the range of 30 μm can be achieved, but the surface quality does not achieve the quality of lithographic processes. With selected methods and selected machines it is possible to achieve optical qualities of surfaces, but the requirements and therefore the complexity of the whole process increase significantly.

9.1.1 Mechanical Micro Machining

Mechanical micro machining includes the methods of micro milling, micro drilling, and micro grinding. The fabrication of microstructured molds with conventional machines is limited because of the accuracy of the machines and the limitation of integrating micro tools. Micro cutting processes are therefore not just a miniaturization of the conventional cutting technology [4,28]. For mold fabrication, specialized machines and processes are developed corresponding to the requirements of micro machining.

The precision of the microstructuring process should be in the range significantly below the minimum lateral structure size of a mold insert. The precision of the machine and the process results finally in the surface roughness of the mold [20]. Therefore, high-precision machines are developed with geometric and thermal stability of the tool guide and clamping chuck. To achieve a high grade of evenness of the structured area, the centering and the position of the work piece has to be exactly defined. To achieve this accuracy, ultraprecise and adjustable clamping systems with an integrated measurement system are part of such machines. Further, the work piece is clamped during the whole structuring process to avoid inaccuracies on the evidence of multiple adjustments.

To achieve exact geometries with a high surface quality, the tools for microstructuring should have well-defined cutting edges. Therefore, tools with cutting edges of natural diamonds or hard metal are common. The sidewalls with the best quality are achieved by diamond tools; the rounding of the cutting edges is in the range of nanometers. Further, diamonds show a maximum of hardness, a low friction coefficient, and a high heat conductivity. The smallest diameter for diamond tools is in the range of 200 µm; hard metal tools are available down to a diameter of 30 µm (Fig. 9.2). Structuring in these dimensions requires adapted methods for the removal of chips; otherwise the risk of generating burr increases. The use of a suitable material combination between tool and work piece is essential, as well as the use of lubricant, cooling liquids, or air pressure. Finally, the cutting depth has to be limited. To achieve high structures the contours have to be structured in several steps with an increase in cutting depth [7,9]. The technique of micro machining is also suitable for the manufacturing of mold inserts with different levels of height and can further combine with a lithographic process step to achieve an electroplated mold insert with different levels of height [6].

Which materials can be used for structuring by micro machining? As mentioned above, well suited are materials with high stiffness and high heat conductivity. In praxis metals like copper, aluminum, brass, or steel can be used. Using iron alloys it is important to take into account that diamond tools should not be used because of the interaction between diamond and iron that results in signs of

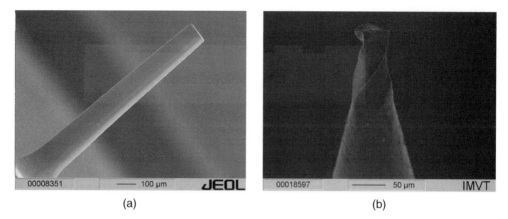

(a)   (b)

Figure 9.2 (a) Diamond tool and (b) hard-metal tool for micro machining. Diamond tools show the best available hardness and geometry, with rounding of the cutting edges in the nanometer range. They are available down to 200 µm. Hard-metal cutting tools are available down to approximately 30 µm.

wear of the diamond. Besides metals, polymers and ceramics are also suitable. Using these materials the cutting speed has to be adapted to avoid damage of them [8].

The advantage of the fabrication of mold inserts by micro milling is, besides the large bandwidth of materials, the possibility to structure large areas. If the same structures can be fabricated by lithographic processes or by micro milling, the structures can be produced, normally, more cost effectively. Compared to lithographic processes where only vertical sidewalls can be fabricated, micro machining allows the fabrication of three-dimensional structures. In a simple case demolding angles can be integrated using milling tools with cone-shaped profiles. Well suited for micro machining are microfluidic designs [13,16] with feature sizes in a range down to 50 µm or housings for microsystems like pumps or sensors. Finally, an example will show typical mold inserts for hot embossing (Figs. 9.3–9.5).

9.1.2 Laser Structuring

Lasers are a flexible tool for micro machining. They are suited for structuring metals and polymers and can be used to fabricate mold inserts for hot embossing [11,27,32]. The removal of material can be achieved physically by laser ablation. The Nd:YAG laser, for example, is a rapid and efficient tool for micro machining, even into materials that are hard to machine such as hard metals or ceramics. A short pulse duration leads to high-peak power intensities and removes material via vaporization. To fabricate micro features the laser beam is focused below 50 µm. The focus size therefore determines the size of the microstructures

Figure 9.3 Microfluidic mold insert for a dispensing well plate, fabricated by micro machining of steel. With the cones, through-holes can be molded.

Figure 9.4 Microfluidic structures structured by micro machining. The structure size is in the range of 50 μm; the height is in the range of 100 μm. The length of these structures can achieve the range of one hundred millimeters.

that can be achieved with this technology. A fundamental issue of laser structuring is the incompatibility of high ablation rates and a smooth surface with low roughness. High ablating rates results in mechanical damage of the surface and heating of the mold material. To achieve uniform removal and to improve surface quality, laser pulses and traces are overlapped in the range of 50–80%. The depth of ablation is limited by the energy of the beam, the material, and the desired surface quality. Furthermore, to reach a compromise between processing speed and adequate surface quality, the machining is split into a rough and a finish machining. After finish machining typical roughness can be achieved in

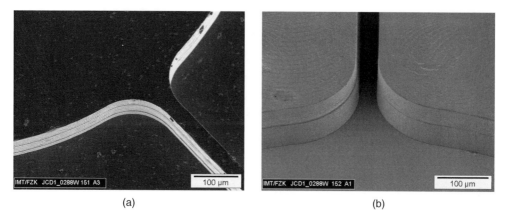

Figure 9.5 SEM picture of a mold insert fabricated by micro machining (a) and the corresponding replication in PMMA (b). The roughness of the sidewalls is typically in the range of 200 nm. The grooves at the vertical sidewalls fabricated by milling are oriented perpendicular to the demolding direction. The microgrooves from the machining process are also replicated in polymers.

the range of 1 μm, which is significantly higher than the roughness of molds fabricated by micro milling with diamond tools. Finally, to achieve high structures in a mold, respectively depth, material has to be removed layer by layer in several processing steps.

9.1.3 Electric Discharge Machining (EDM)

Electric discharge machining (EDM) is a structuring method referring to the ablating of material by electrical erosion. The tool with the function of an electrode and the work piece are embedded in a dielectricum like water. Between both electrodes—here the tool and the work piece—a difference in potential with a frequency larger than 100 kHz is applied, which results in sparks. These sparks are responsible for the erosion of material. The dielectricum water functions as a cooling liquid and is responsible for the transport of ablated material.

Depending on the form of the tool as electrode, different kinds of structuring methods can be achieved (Fig. 9.6). If the tool has the shape of a wire, a kind of cutting of the mold material can be achieved. If a wire is used it has to be renewed continuously, which makes it necessary to use wires on coils. A structured metal electrode, for example from the LIGA process, can be used to structure materials by abating the electrode into the material. With this method different kinds of structures, like holes up to complex geometries, can be fabricated. The structured electrode can consist of only one structure or multiple structures, allowing the fabrication of structures of microstructured mold inserts in

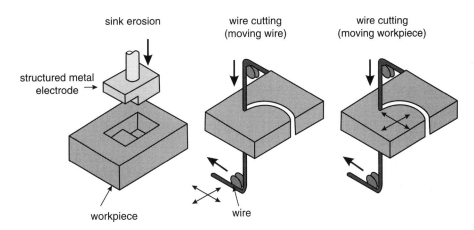

Figure 9.6 Structuring methods of EDM: sink erosion with electroplated microelectrodes (left); wire cutting (center); wire cutting with movable work pieces (right).

parallel [2,3]. Finally, the work piece can also be moved by a control unit during the erosion. If, for example, a wire is used, different structures can be cut like a fret saw. This principle of structuring allows the use of every material that is characterized by electric conduction; suitable are, therefore, a wide range of metals and selected semiconductors like silicon [31].

The advantage of this technology is the moderate force acting on the work piece during structuring and the use of a wide range of metals. Even hardened steel can be structured. Electrodes can be produced down to several micrometers; for example, wires with a thickness of 30 µm are available. Sinking electrodes can be produced by micro electroforming with dimensions in the range of 25 µm. Conversely, because of the erosion process the surface roughness is much higher than the surface roughness of mold inserts fabricated by micro milling. Typical surface roughness is in the range of 300 up to 800 nm (R_a). Therefore, the process of EDM is suitable to fabricate microstructured mold inserts with structures for microfluidic applications or housings (Fig. 9.7).

9.1.4 Non-conventional Molds—Alternative Methods

As mentioned at the beginning of this chapter, every structure can be used as a mold if the named requirements are fulfilled. Structured metals are well suited, but with some restrictions other structuring concepts and materials are also suitable.

Common processes are etching processes to structure glass or semiconductors like silicon. Wet etching of silicon is suitable to fabricate structures with low and high aspect ratios. A disadvantage is the non-rectangular shape formed due to the crystallographic planes. Alternatively, dry etching avoids the non-rectangular shapes. For example, deep reactive ion etching (DRIE) can be used to fabricate structures with high aspect ratios and vertical sidewalls. Compared to

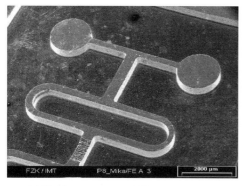

(a) Fluidic structure, mold insert fabricated by EDM

(b) Detailed view of the surface of the EDM mold insert

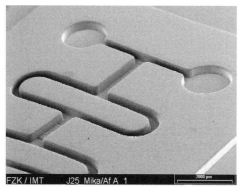

(c) Inpolymer replicated mold insert

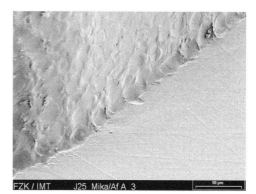

(d) Detailed view of the surface roughness in the replicated structure

Figure 9.7 Microfluidic structures produced by EDM [12]. Because of the erosion process, the surface roughness is in the range up to 800 nm.

wet etching the roughness of the surface increases and can cause high demolding forces. In combination with the material properties of silicon, the risk of damage to the mold and molded structures increases. However, silicon is a very brittle material and the risk of damage to the mold during embossing is much higher compared to molds fabricated in metals or elastic polymers. A precondition for the use of brittle material like glass or silicon in a hot embossing machine is a precise and even integration of this mold insert into a molding tool. Local stress concentration, for example as a result of an uneven substrate, or high molding pressure can damage the structures. The expected lifetime of a pure silicon mold insert in conventional hot embossing cycles is relatively short. Therefore, it is recommended to copy the etched structures via electroforming to a nickel mold insert, which allows the significant increase of its lifetime. Nevertheless, for some processes, like UV nanoimprint, UV-transparent mold inserts have to be used. Here, etched quartz glass is a suitable material [26]. Compared to hot embossing of polymers, the stress effected on a mold insert during nanoimprint with UV curable polymer is moderate so that the lifetime of brittle mold inserts will increase. Independent of the brittle behavior, also mold deformation of silicon molds during embossing can be observed. Lazzarino et al. [19] demonstrate mold deformation of a silicon-etched 4-inch mold insert. They found that the residual thickness of the final replicated parts is also not uniform. They observed some degradation of mold pattern and also recommend dummy patterns around the nanoimprinted zones (see Section 9.5).

As an alternative to microstructured mold inserts of metal or silicon, also polymers as mold materials are suitable. An established polymer in the field of nanoimprint is PDMS (polydimethylsiloxane), a polymer with elastic and UV-transparent properties [25]. PDMS is therefore a favorite material for UV-nanoimprinting. To obtain a microstructured mold from PDMS, first a microstructured master is required. This microstructured master can be produced theoretically by all known structuring methods. The replication from a master can be done by simple casting and is a comparably short process. The curing of PDMS can be achieved in approximately 30 minutes. Because of the elastomeric behavior the PDMS stamps can be easily demolded form the master by peeling [21].

In general the disadvantage of using flexible molds can be seen in the reduced accuracy in lateral dimensions of molded parts. The structures of the PDMS mold insert can be deformed by high molding pressure, which can result in inaccurate shapes of molded parts. It is obvious that polymer molds do not have the lifetime of metal molds. The replication of structures with PDMS molds is not limited to the UV-nanoimprint process. PDMS shows a temperature resistance up to 250°C, which allows one to mold thermoplastic polymers with a molding temperature below, for example, PMMA, PC, or COC. Molding with PDMS molds is also called soft-embossing [18,30].

The use of polymers as mold inserts is not limited to PDMS. Theoretically every polymer that has a significantly higher glass transition or melting temperature as the polymer material that should be structured is suitable for use as a mold. Appropriate for mold materials are especially semicrystalline high-temperature polymers like LCP or PEEK, fluoropolymer films [17], or thermosets [29]. Semicrystalline high-temperature polymers show a temperature resistance up to 340°C, which allows the molding of a wide range of polymers with lower molding temperature. The structuring of the mold inserts can be done by mechanical machining, laser structuring, or by a previous molding step from a metal master. Nevertheless, the lifespan of polymer molds is limited and the surface quality decreases with an increase in the number of replications (Fig. 9.8).

9.2 Lithographic Structuring Methods

Lithographic structuring methods include the processes of UV lithography, X-ray lithography, and E-beam lithography. The resolution that can be achieved

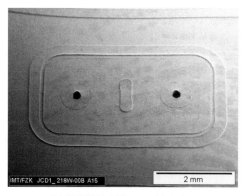

(a) PEEK-Mold insert with holes

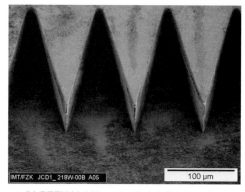

(b) PEEK-Mold insert with details of a micro spectro meter

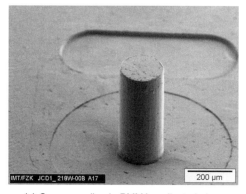

(c) Corresponding in PMMA replicated pins

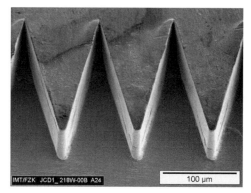

(d) Corresponding in PMMMA replicated structures

Figure 9.8 Microstructured mold insert of PEEK, fabricated by a previous hot embossing step. The structures were replicated in PMMA and PC. Compared to mold inserts from metal, the polymer mold inserts show a decrease of quality with an increase of replication cycles [10].

is a function of the wave length of the electromagnetic waves. Therefore, each of these lithographic structuring methods is characterized by it is own qualities and is suited for different kinds of structures, lateral dimensions, and aspect ratios. Independent of the lithographic process, the post-process step of electroforming is required to obtain a mold insert of metal.

9.2.1 Electroforming of Mold Inserts

Electroforming allows the fabrication of microstructured nickel mold inserts from materials typically not appropriate for use as mold inserts. For example, brittle microstructured silicon or microstructured already replicated polymers can be copied into a mold insert. Further, microstructures fabricated by micro

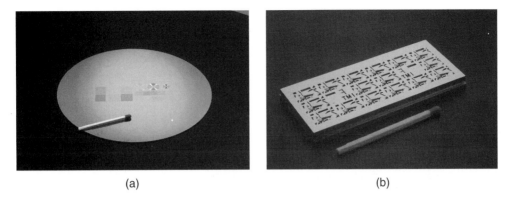

(a)   (b)

Figure 9.9 Different types of electroformed mold inserts for hot embossing. Two different types can be distinguished: (a) Ni shims, characterized by a thickness of several hundred micrometers and diameters up to 6 inches and (b) typical LIGA mold inserts with a thickness of approximately 5 mm and lateral dimensions of 28×66 mm^2.

machining in soft metals like copper or brass can be copied into hard alloys like Ni or Fe-Ni. The process of the inverted copy by electroforming can also be used to fabricate complex structures that cannot be produced directly. Some kinds of structures can be easily fabricated in the inverse form by milling or drilling; for example, narrow rip can be easily produced by mechanical machining. The inverted forms are narrow grooves that are complicated to fabricate by mechanical machining.

In practice, two kinds of electroplated mold inserts for hot embossing can be distinguished (Fig. 9.9). Mold inserts with a thickness of only several hundred micrometers (e.g., 300–500 µm), called shims, mostly fabricated from a structured silicon wafer or a wafer with structured resist, are recommended if the depth of the cavities and the aspect ratios are in a low range. Nevertheless, many kinds of microstructures can be fabricated by this technology, for example, optical structures like waveguides or small pitches for CD or DVD replication. Especially the replication of CDs is a common example for the use of shim technology. Depending on the structured wafer, 4-inch or 6-inch shims are common. For replication, these shims have to be fixed onto an even substrate plate in a hot embossing tool (Section 8.2.4). The advantage of Ni shims is the relatively short processing time of one or two days compared to the second kind of mold inserts, inserts with a thickness of approximately 5 mm. These molds refer to the LIGA technology with deep cavities and structures with high aspect ratios and have, after cutting to a defined shape by EDM, typical lateral dimensions in the range of 28×66 mm^2. These molds are fixed into a hot embossing machine by an individual mounting system (Section 8.2.4). The advantage of these mold inserts refers to an uncomplicated handling and mounting and especially a higher stability against deformation under molding conditions.

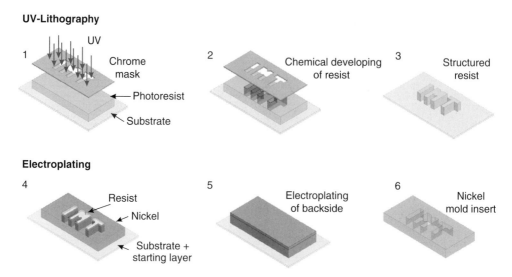

UV-Lithography

Figure 9.10 Schematic illustration of mold fabrication by UV lithography. Through a chrome mask a negative photo resist (e.g., SU8) is exposed by UV light. After developing the exposed photo resist, a microstructured mold insert can be fabricated by electroplating.

9.2.2 Mold Inserts Fabricated by UV Lithography

UV lithography is one of the established processes in microelectronics for structuring UV-curable photo resist. Typical thickness of the layer of the resist is in the range of only a few micrometers, which is sufficient for the task. But this technology shows also the potential to structure UV-curable polymers of more than $100\,\mu m$ (Fig. 9.10). In further process steps the structured photo resist can be copied into a nickel mold insert or nickel shim by electroplating. Today with spin coating, resists of a thickness of several hundred micrometers are available. Besides the fabrication of thick photo resists, the lithography step of these thick resists is a challenge to obtain structures with vertical sidewalls over the height of the resist. Depending on the quality of the mask surface used, roughness in the range of 40 nm can be achieved. The resolution that can be achieved depends, among other things, on the wavelength of the UV beam used. Typical structure sizes are below 5–$10\,\mu m$; typical heights of structures are in the range of 50–$200\,\mu m$. Because of the surface quality, this technology is well suited to produce mold inserts with optical components (Fig. 9.11).

9.2.3 Mold Inserts Fabricated by X-ray Lithography

The process of mold fabrication by X-ray lithography was first developed at Forschungszentrum Karlsruhe in the 1980s [1,5]. The process is split into the main steps of E-beam lithography, mask fabrication by electroforming, X-ray

Figure 9.11 Nickel mold insert (4-inch shim) with optical components fabricated by UV lithography. The mold inserts consist of several structured areas. The structure sizes are in the range between 16 μm and 10 mm; the height of structures is in the range between 1.8 and 16 μm [33].

lithography, and electroforming of the mold insert (Fig. 9.12). Referring to the process steps, the name of this process is LIGA (a German acronym for *Lithographie, Galvanik, Abformung*). In general, the mask needed for UV or LIGA processes is fabricated by E-beam lithography.

Because of the beam characteristic of X-rays, molds can be fabricated characterized by structure heights larger than 1000 μm with vertical sidewalls. The resolution allows one also to fabricate structures down to the submicron range. This process is therefore well suited for the fabrication of microstructures with high aspect ratios. Because of the parallel beam, vertical sidewalls of the structure are a further characteristic of LIGA molds. Depending on the quality of the mask, surface roughness below 40 nm [24] can be achieved. Optical components are therefore one of a wide range of applications of this technology. Figure 9.13 shows two mold inserts fabricated by the LIGA process.

9.2.4 Mold Inserts Fabricated by E-beam Lithography

E-beam lithography is one of the interfaces between a CAD design and a structured part and is therefore a fundamental step in microstructuring. This process is characterized by an electron beam with a focus down below 10 nm, which allows one to structure a photo resist in high resolution but with limited depth. Therefore, this process is used for the fabrication of masks for UV and X-ray lithography. Compared to the processes of UV and X-ray lithography, the process of E-beam lithography is a sequential processing step. This results in comparatively long processing times. Nevertheless, because of the high

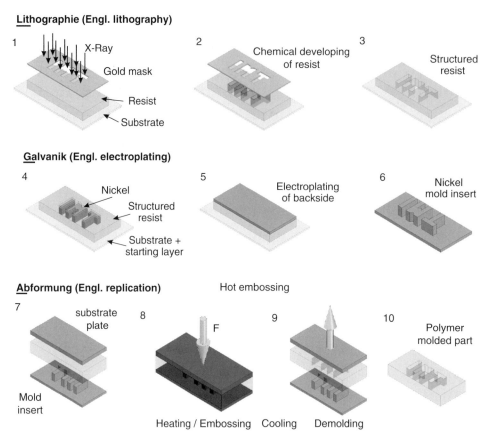

Lithographie (Engl. lithography)

1 X-Ray / Gold mask / Resist / Substrate

2 Chemical developing of resist

3 Structured resist

Galvanik (Engl. electroplating)

4 Nickel / Structured resist / Substrate + starting layer

5 Electroplating of backside

6 Nickel mold insert

Abformung (Engl. replication)   Hot embossing

7 substrate plate / Mold insert

8 F

9

10 Polymer molded part

Heating / Embossing   Cooling   Demolding

Figure 9.12 Schematic illustration of the different process steps of the LIGA process. The process consists of the four main steps of E-beam lithography, mask electroplating, X-ray lithography, electroplating of mold insert, and finally replication.

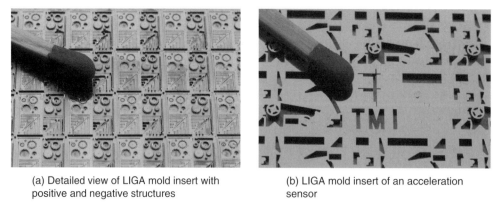

(a) Detailed view of LIGA mold insert with positive and negative structures

(b) LIGA mold insert of an acceleration sensor

Figure 9.13 Examples of mold insert structures, fabricated by the LIGA process. The structures are characterized by vertical sidewalls and high aspect ratios, down to the submicron range.

(a) Structured field   (b) Detailed view of the structures

Figure 9.14 Mold insert (shim) fabricated by E-beam lithography. The structured field of E-beam lithography is limited because of the sequential processing and the maximal shift of the alignment system. The structured size in this example is an area with a diameter of 1 mm. The rough surface on top of the structure corresponds to the roughness of the adhesion layer of the photo resist and starting layer for electroplating.

resolution, E-beam lithography can also be used for mold fabrication, especially for structures in the nanometer range, where already a typical structure height in the range between 1 μm and 5 μm results in structures with comparatively high aspect ratios. Instead of gold electroforming for mask fabrication, the mold inserts are fabricated by nickel electroforming. Because of the moderate structure height, the fabrication of nickel shims with a thickness of only several hundred micrometers (e.g., 300–500 μm) is sufficient (Fig. 9.14).

9.3 Assembled Molds

The molding of large series requires, besides short cycle times, also large batch sizes that can be implemented by large microstructured areas. The areas that can be structured by the processes named above are limited or cause problems and costs if the area of the structured field increases over the process's typical structured areas. An implementation of a large area mold is the assembling of large molds based on single structured mold inserts. The single mold inserts can be arranged in a space-efficient way, so that identical mold inserts or mold inserts with different structures can be replicated in a single molding step. Important for an arrangement is the homogeneous thickness and evenness of the assembled mold insert, which requires a precise assembling and, if required, an individual alignment of each mold insert. Therefore, the requirements for a clamping system for the mounted mold inserts are high; here a precise evenness and a gap-free fixation system is required. An example for the integration of mold inserts into a large mold is shown in Fig. 9.15. Five rectangular

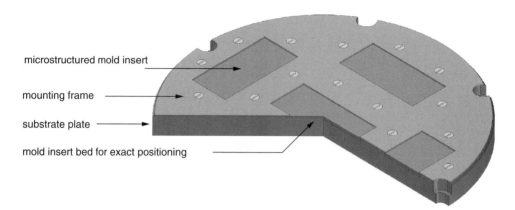

microstructured mold insert

mounting frame

substrate plate

mold insert bed for exact positioning

Figure 9.15 Arrangement of single mold inserts into a large mold. The single nickel shims are arranged in a sink and clamped by a frame. A precise clamping system and a homogeneous thickness of each mold insert are required to obtain an equal height of the large mold.

nickel shims are mounted into a clamping frame. To achieve a homogeneous evenness of the final mold, the mold inserts are mounted into a sink and fixed finally by a large frame. The precondition here is the identical thickness of each mold insert and the precise fabrication of the sink in the ground plate [23].

Another implementation is the concept of modular mold inserts, fabricated by mechanical machining. In this case, microstructured modules are assembled into a large mold insert by sequencing of single modules. Each module can be structured individually or in the same way. Also, distant modules without any structures are applicable. This concept requires a ground plate and a clamping unit to fix the modules. This modular mold requires a precise fabrication of the even sidewalls of each module; tolerances will result in a gap between each module. During molding, polymer melt can flow into the gap and will cause high demolding forces during the demolding step, which can damage structures or deform the whole molded part. An example for a modular mold is shown in Figs. 9.16 and 9.17. This mold was used for the molding of dispensing well plates (DWP), with the aim to mold through holes [22].

9.4 Mold Coatings

As mentioned before, the demolding of filigree structures with high aspect ratios without defects is one of the challenges in micro replication. The reduction of adhesion and friction between mold and polymers during demolding is therefore essential for successful demolding. Besides the reduction of roughness of the mold surface and the reduction of shrinkage by the selected setting of the process parameters, mold coatings are an established method to reduce demolding forces and the load on the molded part during demolding.

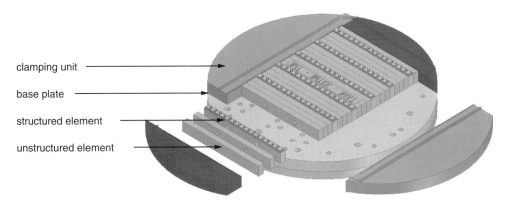

clamping unit

base plate

structured element

unstructured element

Figure 9.16 Mold insert, assembled by modular units, fabricated by mechanical machining. The modules are arranged on a base plate and fixed by a lateral and circulatory clamping unit. The sidewalls of each unit have to be fabricated very precisely to avoid any gaps between the modules.

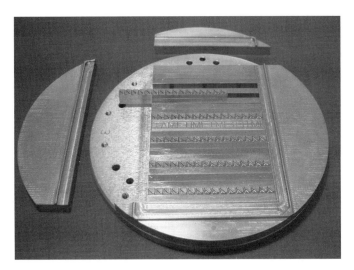

Figure 9.17 Modular mold inserts for dispensing well plates (DWP). The mold inserts are fabricated by mechanical machining of steel.

Coatings are well used in macroscopic molding. In principle, deposition technology like PVD or CVD can also used for microstructured molds. Nevertheless, the thickness of the coating has to be compared to the structure size. Typical coating layers are in the range below 50 nm down below 10 nm. If the structure size decreases, especially in the submicron range, the coating can eliminate the quality of the profile of the structures. Therefore, the coating of microstructured molds is limited or only reasonable if the size of the structure is significantly larger than the thickness of the coating.

The lifetime of mold coatings was systematically analyzed by Jaszewski et al. [14,15]. They investigated the degradation of anti-adhesive properties of protective PTFE films during hot embossing and identified possible failure causes that reduce the lifetime of these protective films. In this work, nickel shims with flat and structured surfaces were covered with two sorts of fluorinated polymer films. One film was deposited by microwave discharge; the second film was deposited by ion sputtering. Both films were about 5 nm thick and had a good spatial homogeny over an area of a few tens of square microns. The analysis of the coating (film homogeny, structure of nickel/polymer interface, depth-resolved chemical composition) after several molding steps was done by X-ray photoelectron spectroscopy (XPS). The microstructures used in this experiment were diffractive optical elements with low aspect ratios. The PTFE-covered shims were replicated by hot embossing into polycarbonate (PC) and polymethylmethacrylate (PMMA) at different molding temperatures up to 205°C. Jaszewski et al. found that the effect of molding temperature and the total embossing time are crucial parameters. The diffusion rate of the polymer film depends on the temperature; more fluorine is lost at higher temperatures and during longer embossing. It was also found that a temperature over 190°C led to the destruction of most of the C–Ni bonds at the interface. Therefore, if PTFE films are used, the embossing temperature chosen has to be as low as possible (175°C for PC and 140°C for PMMA) and the total embossing time should be minimized. Also, PDMS can be used as an anti-adhesion coating. Lee and Kim [18] fabricated a flexible mold using an ultraviolet-curable prepolymer. The surface was coated with a thin layer of PDMS to impart an anti-adhesion property.

Another common method in macroscopic replication processes is the use of a release agent. The deposition of a release agent is done typically by spraying with coarse nozzles, so that a thick layer of release agent is deposited on the surface. Here, the risk of filling fine microcavities with a release agent increases significantly. Because of the nearly incompressible behavior of the release agent, the filling of microcavities with polymer melt can fail because the release agent cannot be removed completely before molding. Further, the surface of the molded part is contaminated with the release agent, which makes it necessary to clean the molded parts before the next process step, for example, before thermal bonding can be initiated. Spray coating with a release agent is therefore only suited for relatively large microstructures without any further process steps. Otherwise, a cleaning of the molded part is recommended.

Not every design and not every material makes it necessary to use a mold coating. Some polymers are mixed with additives that have the function of a release agent. Therefore, it is recommended to use selected polymers. Common polymers like PMMA, PC, and PSU are available with additives supporting the demolding. Nevertheless, there exist polymers that are difficult to demold, for example, semicrystalline high-temperature polymers like LCP or PEEK. They show a high adhesion on microstructured molds that will increase with the surface roughness. Here, the risk of damage to the structures during demolding increases significantly. The problem becomes more important, as this kind of polymers shows an excellent resistance against most chemicals, which makes it nearly impossible to clean microstructured molds, especially when residue of polymers sticks in microcavities. Here a coating is recommended; otherwise a mold insert can be damaged by the first molding.

9.5 Design of Microstructured Molds

Independent of the further application that will determine the kind of structures on mold inserts, simple design rules can be defined regarding the aspects of mold filling, shrinkage compensation, and reduction of load on structures during demolding.

9.5.1 Mold Filling

The filling of a microcavity depends on the cavity size and the related flow length. Here, the design of the cavity is determinant, but mold filling is also a function of the pressure of the polymer melt during the molding state. During mold filling by the velocity-controlled molding state, the maximum pressure is located in the center of the mold (Section 6.4.2). Therefore, the microcavities should be arranged in a way that the cavities that require the highest filling pressure, like holes or grooves with high aspect ratios and small cross-sections, are arranged in the center of the mold insert. To improve the pressure distribution during molding, a frame around the structured area can be helpful to homogenize pressure distribution. This homogeneous pressure will help to fill microcavities in the margin regions and will further improve the shrinkage and warpage behavior of the molded part (Fig. 9.18).

9.5.2 Compensation of Shrinkage

Another aspect is the difference of shrinkage between mold insert and polymer that results in differences in lateral dimensions of both components. This difference of shrinkage of the molded part is marked by molding process, process parameters, material-combination mold polymer, and finally the point of time of measurement of the molded part. Therefore, definitive values about shrinkage cannot be found in literature; mostly intervals of shrinkage values are presented. Nevertheless, if a part needs to be molded with defined lateral dimensions, the lateral dimensions of the mold have to be corrected related to the expected shrinkage. It is obvious that this individual correction can theoretically only be optimized for a specific polymer and process. Here, experimental results based on systematic measurements will help to find individual values. In a first

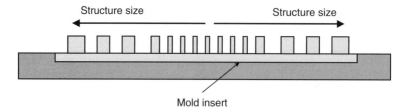

Figure 9.18 Arrangement of microstructures on mold insert. Regarding the pressure distribution during molding, the highest pressure is located in the center of the mold. This helps to fill the cavities that require the highest filling pressure. Because of the increasing influence of shrinkage with an increasing area of the structured filed, structures with larger cross-sections should be arranged in the margin regions of the mold insert. Filigree structures should be arranged as near as possible to the center of the mold.

step, the p-v-T diagram of a polymer and the characteristic process data can help to estimate the volume shrinkage.

Shrinkage and the anisotropy of shrinkage result also from the differences of pressure distribution during molding. If the pressure distribution over the molded area can be homogenized, the anisotropy in shrinkage will be reduced, which results in a decrease of the warpage of the whole flat molded part. Therefore, the design of mold inserts should consider any additional structures, eliminating the typical pressure distribution achieved by free-flow fronts. Here, a circulatory shoulder or structures creating a nearly closed mold will help to homogenize the pressure during the velocity- and especially force-controlled molding state. Especially during cooling under dwell pressure, a homogeneous pressure profile can be achieved. Nevertheless, if such a circulatory shoulder or frame is used, the volume of the enclosed material has to be determined precisely; otherwise, if the volume of the polymer cannot compensate the contraction during cooling, sink marks will occur [34,36].

9.5.3 Reduction of Demolding Forces

As described in Chapter 4, typically molded parts shrink in the direction of the center of the mold, which results in high contact stress between the molded part and the mold inserts at the vertical sidewalls. Regarding this effect of difference in shrinkage, simple recommendations can be deduced.

First, the molding area and the thickness of the residual layer will determine, besides the process parameters and the polymer material, the values of shrinkage. If the molding area increases, the shrinkage, especially in the margin region, will effect higher normal forces on the sidewalls of the structures, which results in higher demolding forces for the structures arranged in this area. Therefore, if filigree freestanding structures have to be replicated, it is recommended to reduce the molding area and to arrange the structures in the center of the mold.

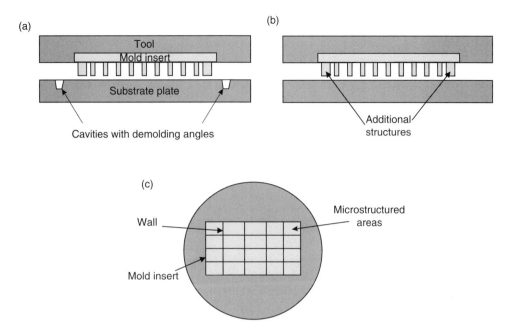

Figure 9.19 Additional structures for protection against high contact stress of the microstructures during demolding. (a) Additional structures integrated as circulatory cavities in the substrate plate; (b) additional structures integrated as circulatory structures in the mold insert; and (c) systematically arranged sidewalls as part of the mold insert will homogenize the pressure distribution inside the separated areas during molding.

Regarding the direction of shrinkage toward the center of the mold, the sidewalls whose direction is perpendicular to the center-oriented direction will effect the highest contact stress. If possible, the arrangement of structures should consider this tendency. For example, if long freestanding walls need to be replicated, a star-shaped arrangement of cavities or groups of cavities can be helpful. Depending on the design, other arrangements can be reasonable. The aim is to reduce the areas of sidewalls perpendicular to the direction of shrinkage of the residual layer.

The process simulation in Section 6.4.4 shows that the structures arranged in the margin regions will be affected by the highest contact stress. These structures absorb the majority of contact stress and protect the neighboring structures oriented to the center of the mold insert. A simple but effective method to reduce contact stress and demolding forces is, therefore, the integration of additional structures, especially in the margin regions. Regarding the simulation results, these structures will protect the structures adjacent to the center-oriented structures in a way that the normal forces, acting on the outer structures, will be absorbed by the additional structures. The additional structures should be not be filigree, and if possible, demolding angles will help to reduce further demolding forces. In a simple case these additional structures can be implemented by a circulatory cavity or with demolding angles around the micro-structured area.

Additional structures can also be integrated into the microstructured areas by the implementation of additional walls (Fig. 9.19). Besides the circulatory arrangement around the microstructured area, also a separation of microstructured areas in separate fields is imaginable. The systematic arranged walls will protect the filigree structures inside the mold against high contact stress and will further guarantee a more homogeneous pressure distribution during molding. The arrangement of the additional walls can also form a closed die, which can effect better replication accuracy and a reduction of mold-related stress [35].

References

1. *Mikrosystemtechnik für Ingenieure.* ISBN 3-527-29405-8. VCH Verlagsgesellschaft, 1997.
2. D. M. Cao, J. Jiang, W. J. Meng, J. C. Jiang, and W. Wang. Fabrication of high-aspect-ratio microscale Ta mold inserts with micro electrical discharge machining. *Microsystem Technologies,* 13(5):503–510, 2007.
3. D. M. Cao, J. Jiang, R. Yang, and W. J. Meng. Fabrication of high-aspect-ratio microscale mold inserts by parallel μEDM. *Microsystem Technologies,* 12(9):839–845, 2006.
4. B. Denkena, H.-W. Hoffmeister, M. Reichstein, S. Illenseer, and M. Hlavac. Micro-machining processes for microsystem technology. *Microsystem Technologies,* 12:659–664, 2006.
5. E. W. Becker, W. Ehrfeld, P. Hagmann, A. Manera, and D. Muenchmeyer. Fabrication of microstructures with high aspect ratios and great structural heights by synchroton radiation lithography galvanoforming and plastic molding (LIGA process). *Microelectronic Engineering,* 4:35–56, 1986.
6. J. Fahrenberg, Th. Schaller, W. Bacher, A. El-Kholi, and W. K. Schomburg. High aspect ratio multi-level mold inserts fabricated by mechanical micro machining and deep etch X-ray lithography. *Microsystem Technologies,* 2:174–177, 1996.
7. J. Fleischer and J. Kotschenreuther. The manufacturing of micro molds by conventional and energy-assisted processes. *I. Journal Adv. Manuf. Technol.,* 33(1–2):75–85, 2006.
8. T. Gietzelt, L. Eichhorn, and K. Schubert. Material and micromachining aspects of manufacturing micromolds for replication techniques. *Advanced Engineering Materials,* 8 (1–2):33–37, 2006.
9. T. Gietzelt, L. Eichhorn, and K. Schubert. Manufacturing of microstructures with high aspect ratio by micromachining. *Microsystem Technologies,* 14(9–11):1525–1529, 2008.
10. C. Guichard. *Use of Plastic Mold Inserts in Hot Embossing.* Master's thesis, Université de Franche Comté—Science et Techniques, 2003.
11. T. Hanemann, W. Pfleging, J. Hausselt, and K.-H. Zum Gahr. Laser micromaching and light induced reaction injection molding as suitable process sequence for the rapid fabrication of microcomponents. *Microsystem Technologies,* 7:209–214, 2002.
12. D. Herrmann. *Herstellung gedeckelter Mikrokapillarstrukturen.* Master's thesis, University of Karlsruhe (TH), Institute for Microstructure Technology, 2000.
13. M. L. Hupert, W. J. Guy, S. D. Llopis, H. Shadpour, S. Rani, D. E. Nikitopoulos, and S. A. Soper. Evaluation of micromilled metal mold masters for the replication of microchip electrophoresis devices. *Microfluid Nanofluid,* 3:1–11, 2007.
14. R. W. Jaszewski, H. Schift, P. Gröning, and G. Margaritondo. Properties of thin anti-adhesive films used for the replication of microstructures in polymers. *Microelectronic Engineering,* 35:381–384, 1997.

15. R. W. Jaszewski, H. Schift, B. Schnyder, A. Schneuwly, and P. Gröning. The deposition of anti-adhesive ultra thin teflon-like films and their interaction with polymers during hot embossing. *Applied Surface Science*, 143:301–308, 1999.

16. W.-C. Jung, Y.-M. Heo, G.-S. Yoon, K.-H. Shin, S.-H. Chang, G.-H. Kim, and M.-W. Cho. Micro machining of injection mold inserts for fluidic channel of polymeric biochips. *Sensors*, 7:1643–1654, 2007.

17. D.-Y. Khang, H. Kang, T.-I. Kim, and H. H. Lee. Low-pressure nanoimprint lithography. *Nano Letters*, 4(4):633–637, 2004.

18. N. Y. Lee and Y. S. Kim. A poly(dimethysiloxane)-coated flexible mold for nanoimprint lithography. *Nanotechnology*, 18:415303, 2007.

19. F. Lazzarino, C. Gourgon, P. Schiavone, and C. Perret. Mold deformation in nanoimprint lithography. *J. Vac. Sci. Technol. B*, 22(6):3318–3322, 2004.

20. H. Li, X. Lai, C. Li, J. Feng, and J. Ni. Modelling and experimental analysis of the effects of tool wear, minimum chip thickness and micro tool geometry on the surface roughness in micro-end-milling. *Journal of Micromechanics and Microengineering*, 18:025006, 2008.

21. C. Khan Malek, J.-R. Coudevylle, J.-C. Jeannot, and R. Duffait. Revisiting micro hot embossing with moulds in non-conventional materials. *Microsystem Technologies*, 13(5): 475–481, 2006.

22. C. Mehne, R. Steger, P. Koltay, D. Warkentin, and M. P. Heckele. Large-area polymer microstructure replications through the hot embossing process using modular moulding tools. *Proc. IMechE Part B: Journal Engineering Manufacture*, 222:93–99, 2008.

23. Ch. Mehne. *Grossformatige Abformung mikrostrukturierter Formeinsätze durch Heissprägen.* PhD thesis, University of Karlsruhe (TH), Institute for Microstructure Technology, 2007.

24. M. Mekaru, T. Yamada, S. Yan, and T. Hattori. Microfabrication by hot embossing and injection molding at LASTI. *Microsystem Technologies*, 10:682–688, 2004.

25. J. Narasimhan and I. Papautsky. Polymer embossing tools for rapid prototyping of plastic microfluidic devices. *Journal of Micromechanics and Microengineering*, 14:96–103, 2004.

26. H. Niino, X. Ding, R. Kurosaki, A. Narazaki, T. Sato, and Y. Kawaguchi. Imprinting by hot embossing in polymer substrates using a template of silica glass surface-structured by the ablation of LIBWE method. *Applied Physics A*, 79:827–828, 2004.

27. W. Pfleging, W. Bernauer, T. Hanemann, and M. Torge. Rapid fabrication of microcomponents—UV-laser assisted prototyping, laser micro-machining of mold inserts and replication via photomolding. *Microsystem Technologies*, 9:67–74, 2002.

28. G. M. Robinson, M. J. Jackson, and M. D. Whitfield. A review of machining theory and tool wear with a view to developing micro and nano machining processes. *Journal Mater. Sci.*, 42(6): 2002–2015, 2007.

29. N. Roos, H. Schulz, L. Bendfeldt, M. Fink, K. Pfeiffer, and H.-C. Scheer. First and second generation purely thermoset stamps for hot embossing. *Microelectronic Engineering*, 61–62:399–405, 2002.

30. A. P. Russo, D. Apoga, N. Dowell, W. Shain, A. M. P. Turner, H. G. Craighead, H. C. Hoch, and J. N. Turner. Microfabricated plastic devices from silicon using soft intermediates. *Biomedical Microdevices*, 4:277–283, 2002.

31. A. Schoth, R. Forster, and W. Menz. Micro wire EDM for high aspect ratio 3D microstructuring of ceramics and metals. *Microsystem Technologies*, 11(4–5):250–253, 2005.

32. P. P. Shiu, G. K. Knopf, M. Ostojic, and S. Nikumb. Rapid fabrication of tooling for microfluidic devices via laser micromachining and hot embossing. *Journal of Micromechanics and Microengineering*, 18:025012, 2008.

33. A. J. Waddie, M. R. Taghizadeh, J. Mohr, V. Piotter, Ch. Mehne, A. Stuck, E. Stijns, and H. Thienpont. Design, fabrication and replication of microoptical components for educational purposes within the network of excellence in micro optics (NEMO). In *SPIE Proceedings Series No. 6185*, 2004.

34. M. Worgull, M. Heckele, and W. Schomburg. Large scale hot embossing. *Microsystem Technologies*, 12:110–115, 2005.

35. C.-H. Wu and C.-H. Lu. Fabrication of an LCD light guiding plate using closed-die hot embossing. *Journal of Micromechanics and Microengineering*, 18:035006, 2008.

36. D. Yao and R. Kuduva-Raman-Thanumoorthy. An enlarged process window for hot embossing. *Journal of Micromechanics and Microengineering*, 18:045023, 2008.

10 Hot Embossing in Science and Industry

This last chapter presents some applications in which hot embossing and related process injection compression molding play an important role for their fabrication. The number of applications increases continually, which makes it impossible to describe all of them. Here, selected examples will be presented to show the manifold of this replication technology. Regarding the different aspects of commercialization, like cost effectiveness and the size of the series of replicated structures, the applications presented here will be split into two different groups. The first group will describe applications that are fabricated in small series or exist only as prototypes. Here, the aim is to show the manifold and the flexibility of this replication process. Because of this development characteristic, these kinds of applications will be classified here in the group of scientific applications. Conversely, large series of parts are commercially replicated by hot embossing and injection compression molding, for example, Fresnel lenses or compact discs. Here, large series, a high grade of automation, and short cycle times are the dominant factors for the cost-effective fabrication of a large number of parts. Compared to the development characteristic, the replication technology for these kinds of applications is typically optimized for cost effectiveness. Therefore, they will be classified here in the group of industrial applications.

Nevertheless, a lot of applications are still in development and may be transferred into mass production.

10.1 Requirements for Hot Embossing in a Scientific Environment and Industry

During development of prototypes, mostly a high grade of flexibility regarding the technology and the set-up of process parameters are in the foreground of interest. The requirements of hot embossing machines in prototype fabrication can be summarized in a compact way:

- a high grade of flexibility regarding the integration and fixation of different kinds of mold inserts
- a variable process control to set up the process to individual requirements. Here, process control by a programming language appears to be the most flexible solution.
- a wide temperature range to allow the molding of different kind of polymers and, further, to allow the characterization of molding windows

- a modular system to allow a comparatively easy adaptation of individual tools and mold inserts
- a measurement system to help analyze the process. It is essential for the further development of the process.

Besides the use of hot embossing in science and prototype fabrication, the process also shows the capability of use in industry for the replication of medium series. Here, the requirements on the hot embossing machines are different. The requirements are mainly oriented toward cost-effective fabrication of molded parts. In the foreground of interest, therefore, is a high grade in automation in combination with short cycle times. Industrial machines do not have the flexibility of laboratory machines; often they are developed and optimized for a certain product. Therefore, a robust system with an efficient process control is desired. Nevertheless, for industrial applications the hot embossing process is, because of the lack of industrial machines, not in as common use as injection compression molding technology. This technology is well developed for large series and optimized for short cycle times. Independent of the technology, the molding cycle, the compression, or the embossing step is similar. Therefore, for industrial use the applications fabricated by injection compression molding can also be taken into account.

10.2 Micro-optical Devices

Micro-optical devices are one of the main applications in microsystem technology. The devices that can be replicated by hot embossing are manifold, beginning with micro-optical components like lenses, mirrors, optical benches, or waveguides, up to microsystems like micro spectrometers, DFB-laser systems, optical switches, fiber connectors, photonic crystals, or anti-reflection films [41,46]. Because of the requirements regarding structure sizes, surface quality, and accuracy of lateral distances, the mold inserts for hot embossing are typically fabricated by lithographic processes. In the following section the replication of micro-optical components is discussed regarding the aspects of replication in polymers. It is obvious that a limited number of examples can be presented here.

10.2.1 Lenses

Representative of the replication of microlenses, the replication of an educational kit shall be described. With the frame of the Network of Excellence in Micro Optics (NEMO), an educational kit has been developed regarding fundamental optical applications of refractive and diffractive optical elements [48] (Fig. 10.1). The mold insert is fabricated by lithography processes and copied to nickel shims for replication by hot embossing. Structure sizes are in the

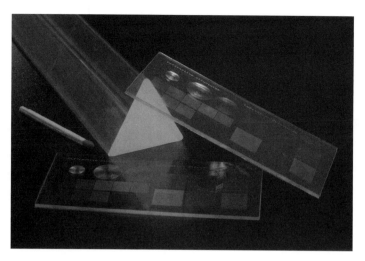

Figure 10.1 Replicated optical diffractive and refractive structures of an educational kit within the frame of the NEMO project. The replicated structures are illuminated by laser light and show the elementary function of optical elements.

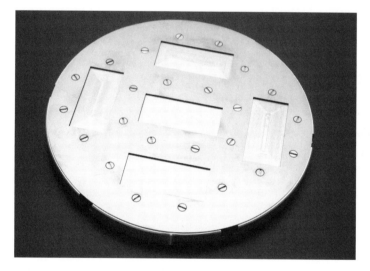

Figure 10.2 Arrangement of electroplated mold inserts for parallel replication.

range of the wavelength of visible lights; the aspect ratio is typically in the range of one. The structures are illuminated by laser light and display the elementary function of optical elements. Therefore, the replication requires a transparent molded part, which is fulfilled by the use of an amorphous polymer like PMMA and a polished substrate plate. Because of the low aspect ratios, the demolding can be achieved by air pressure without the use of a rough substrate plate. Using a mold with an integration of five identical inserts a small series of 10,000 parts have already been replicated by hot embossing (Fig. 10.2).

Besides the optical elements presented above, the mold inserts for convex microlenses can be fabricated and replicated in different ways. Moon et al. [35] showed the replication of a nickel mold insert by micro compression molding of an array of micro lenses with a diameter of 36–96 μm and a radius of curvature of 20–60 μm. A surface roughness of the molded parts of only 4 nm was achieved. Zhang et al. [52] presented a micromachining technique using an etched glass mold for microlens fabrication. The glass master was fabricated by wet isotropic etching and was treated by plasma to form an anti-adhesion layer. The hot embossing technique was used to replicate the microlenses with a diameter of 200 μm into PMMA and COC. The fabrication of microlenses by gas-assisted micro hot embossing was presented by Chang et al. [9]. A silicon mold of 300 × 300 holes with a diameter of 150 μm fabricated by photolithography and deep reactive ion etching was replicated into a polycarbonate film with a thickness of 180 μm at 150°C under a pressure load between 10 40 kg/cm^2 and 40 kg/cm^2. The surface measured roughness was in the range of 4 nm. Chang et al. [10] also presented a UV roller embossing process for the fabrication of microlenses. Based on the microstructured polymer film fabricated by gas-assisted hot embossing, a thin flat mold was fabricated by electroforming of nickel and was wrapped onto a cylinder to form the roller. A UV-curable resin was coated on a glass substrate and structured by roller embossing. The curing was achieved by UV light radiation at the rolling zone. The rolling speed was in the range between 0.5 mm/sec and 2 mm/sec and a molding pressure between 1 kg/cm^2 and 2 kg/cm^2. The array of 100 × 100 microlenses were characterized by a diameter of 100 μm and a sag height of 21 μm.

10.2.2 Optical Gratings

Optical gratings are characterized by systematically and homogeneously arranged structures with a size in the range of the wavelength of light. The replication of optical gratings, therefore, will always refer to the replication of structures in the submicron range. The molds are typically fabricated by lithographic processes. Depending on the application, the aspect ratio of the structures can vary, which can cause a challenging replication, especially if high-aspect ratio structures on large areas are required (Fig. 10.3). Optical gratings can also be part of optical microsystems, like the micro spectrometer by which the physical effect of diffraction the spectrum of light can be analyzed (Fig. 10.4; see also Section 10.2.6). Optical gratings are also suitable for the replication of grating-based optically variable devices using the effect of image switching and color movement. Hot embossing offers the possibility of incorporating optically variable devices, for example, into documents [29].

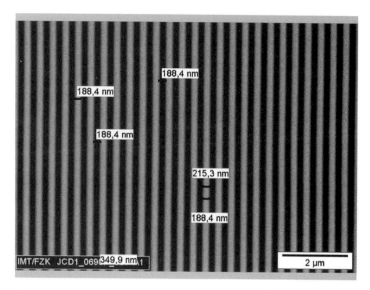

Figure 10.3 Molded optical gratings on an area of 70 × 70 mm². The structures are characterized by a lateral size of 200 nm and a length of 70 mm. The height of the structures is in the range of 600 nm.

Figure 10.4 Molded optical grating of a micro spectrometer. The grating in a Rowland arrangement is part of a vertical sidewall of the spectrometer. The step size of the grating is about 200 nm; the height of the structure is in the range of 120 μm.

10.2.3 Optical Benches

Optical benches can be components of micro-optical systems like spectrometers or sensors. Because of their tasks of an accurate positioning of optical components, optical elements like reflective mirrors with different angles can also be part of these benches. Therefore the requirements on mold fabrication and replication are ambitious. The optical components require a low surface roughness, which results in fabrication by lithographic processes. To integrate and fix optical components like lenses, the structure height of the benches has to be set typically in a range of several hundred micrometers. This requirement typically results in mold fabrication by X-ray lithography. Further, the interaction of different optical components often requires accurate distances between the molded structures. Here especially, the compensation of shrinkage of the polymer has to be taken into account by a correction of lateral geometries during design generation. The replication of optical benches with high aspect-ratio structures, vertical sidewalls, a low surface roughness, and high accuracy in lateral geometries reduces the molding window and requires a precise setting of the process parameters to reduce the demolding force and avoid damaging structures during demolding. Representative for a manifold of optical benches, an optical bench for a Fourier transmission spectrometer [44] (Fig. 10.5) and an optical bench for a distance sensor [4] are presented in Fig. 10.6.

10.2.4 Optical Waveguides

For optical interconnection, the replication of polymeric waveguides opens a new field of applications. Depending on the application, monomode or multimode waveguides with different sizes are required. Lateral dimensions at approximately

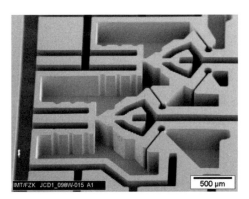

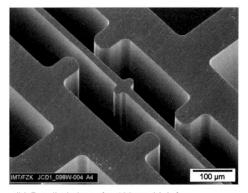

(a) Optical bench of a fourier transmissions spectrometer

(b) Detailed view of a 420 µm high free standing structure

Figure 10.5 Molded optical bench of a Fourier transmission spectrometer. The height of the structures is 420 µm; structure details are in a range down to approximately 20 µm.

Figure 10.6 Molded optical bench of a distance sensor. The sensor was developed within the frame of the FEMOS project. The height of the structures is 750 µm; part of the structures have the function of optical elements like reflecting mirrors.

6 µm [7] up to 500 µm [34] are fabricated. Sufficient for most applications is an aspect ratio in the range of one but typically a guiding path over several millimeters or centimeters. The design refers typically to freestanding rectangular shapes without any additional supporting structures, which makes it necessary to reduce internal stress inside the structures to avoid any deformation of the shape of waveguides. Several techniques allow the modification of the refractive index of the material, for example, the UV radiation of PMMA, which makes this material suitable for the replication of waveguides. Representative of a manifold of polymer waveguides, replicated rectangular waveguides and further optical splitters are presented in Fig. 10.7.

The technique of hot embossing is well suited for the fabrication of optical waveguides. Ulrich et al. [47] fabricated optical waveguides by hot embossing in 1972. They used a glass fiber as an embossing die and replicated the shape into PMMA. The replicated grooves were filled by a material with a higher refractive index. The dimensions of the waveguides were 7 µm wide and about 3.5 µm deep. The attenuation was in a range between 2 4 dB/cm and 4 dB/cm at 633 nm. Lehmacher and Neyer [30] used a two-step hot embossing process to integrate polymer optical waveguides into printed circuit boards. The inverse waveguide structures with a cross-section of 100 µm × 100 µm were replicated into polycarbonate. The core was fabricated by pressing a high-refractive index foil under heat into the grooves. Finally, a low-refractive substrate foil was laminated onto the waveguides. Krabe and Scheel [26] used hot embossing technology for the development of electrical optical circuit boards (EOCB). In this approach the fabrication of an SMD-compatible system with electrical and optical functionality was presented. The fabrication of optical waveguides

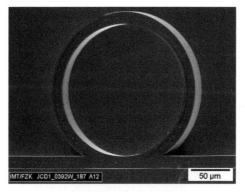

(a) Linear and circular molded polymer wave guides

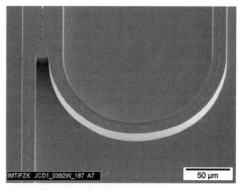

(b) Wave guide splitter, linear and circular

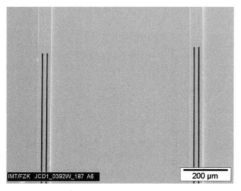

(c) Linear wave guide splitter

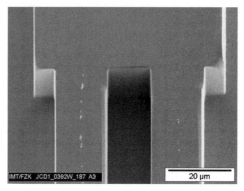

(d) Detailed view of wave guide splitter

Figure 10.7 Molded optical waveguides with a cross-section and height of approximately 6 μm [7]. The electroplated nickel shim mold was fabricated by UV lithography and replicated into PMMA. The refractive index of the molded waveguides was modified after replication by UV radiation. The basic optical waveguide elements allow further development of an interaction between waveguides for subsequent applications.

also refers to the replication of the inverted structure with a cross-section of 60 μm × 60 μm into COC. In a second step the molded grooves were filled with a core material with a higher refractive index that will be thermally cured. Mizuno et al. [34] replicated large-core waveguides with a cross-section between 100 μm and 500 μm in an ultraviolet-cured epoxy resin and achieved a low propagation loss of 0.19 dB/cm at 650 nm. Yoon et al. [51] fabricated multimode waveguides with a cross-section of 43 μm and a propagation loss of 0.2 dB/cm at 850 nm. The use of hot embossing for the fabrication of polymeric optical waveguides in thermosetting polymers was presented by Choi [13]. The waveguide patterns in the silicon master had a quadrangle shape, with 7 μm (width) × 7 μm (height) × 15.8 mm (length). The pitch of each channel was 127 μm. The calculated propagation loss was about 0.67 dB/cm at a 1,550 nm wavelength. Waveguides for millimeter waves were fabricated by hot embossing technology by Sammoura et al. [39]. The shape of the hollow waveguides was 2.54 mm × 1.27 mm, optimized for 95 GHz systems. Over a length of 2.54 mm a minimum insertion loss of –07 dB at 92.5 GHz was achieved. Chen and Jen [11] used the Taguchi method to find the most significant parameter for hot embossing of polymer splitters in PMMA. They found that, in their case, the molding and demolding temperature showed the most significant influence on the quality of the molded part. Kim et al. [22] used the hot embossing process to replicate a passive alignment structure for the optical coupling of multichannel polymeric planar lightwave circuits to optical fibers. The replication corresponded to a simultaneous replication of waveguide channels and micro pedestals in a single replication step. An alignment accuracy of 0.5 µm was achieved. Chien and Chen [12] investigated the molding of light-guiding plates by double-sided hot embossing. The plates were characterized by pyramid arrays on the top surface and micro prisms on the bottom surface.

10.2.5 Photonic Structures

Photonic structures as representative nanostructures can be successfully replicated by hot embossing or, with a similar terminology, by thermal nanoimprint. In this case, the process is identical. Schift et al. [40] fabricated mold inserts of photonic structures by electron-beam lithography. This master was copied via NIL, which enables varying fill factors and aspect ratios. Using those stamp copies, polymeric photonic structures with an aspect ratio of two were successfully replicated. Kim et al. [23] replicated a cubic array of pillars into a polymer film by hot embossing. Pillars with a diameter of 67 nm and a pitch of 200 nm were successfully replicated. A method for nanoimprint stamp fabrication of photonic crystals was presented by Kouba et al. [25]. The stamps made of silicon or nickel were fabricated by electron-beam lithography and advanced dry-etching techniques. With this technology smallest feature sizes below 50 nm and aspect ratios more than 5 were fabricated. The stamps were evaluated by imprinting into polymers.

10.2.6 Micro Spectrometer

At the Institute for Micro System Technologies at the research center of Karlsruhe a micro spectrometer was developed by the use of LIGA technology [28,37]. By this technology all components of the spectrometer—like the connection component to optical waveguides, optical gap, grating, and reflecting mirrors—can by integrated and adjusted in a precise arrangement on a mold insert. This allows the reduction of fabrication and assembling steps and will eventually reduce the cost of high-precision micro spectrometers. These kinds of spectrometers can be used to analyze different ranges of wavelengths, beginning from the UV, to the visible range, up to the infrared range of wavelength. The UV- and visible-range spectrometer can be used in the range of 380–780 nm; the infrared spectrometers are developed in different classes, 600–1100 nm, 1100–1950 nm, 1550–1750 nm, and 3200–5200 nm.

Typical lateral dimensions of these kinds of micro spectrometers are in the range of 45 mm × 22 mm and a total thickness of 2.5 mm. The light that needs to be analyzed is guided and launched by optical waveguides into the spectrometer (Fig. 10.8). Further the light is split off into the spectral colors by a concave optical grating, arranged in a Rowland arrangement. The concave-curved form of the grating has the function of a concave mirror, which maps the image of the connection gap onto a detector array. After reflection and diffraction at the grating, the light is guided by a waveguide system to an output mirror with an angle around 45 degrees. The spectral light is reflected by this mirror to a detector array and can be analyzed by an electronic system.

The launching element that connects the waveguide and the micro spectrometer corresponds to the diameter of the common waveguides between 50 μm and 300 μm. The height of the waveguide of a spectrometer has to be higher than the thickness of the waveguide to avoid any blooming. For example, the height of the optical element will be 130 μm if a waveguide with 125 μm is used, or the height of the optical elements will be 340 μm if optical waveguides with a 330 μm diameter is used. This height determines the height of the optical grating (Fig. 10.4) and also the waveguide, which guides the light inside the spectrometer system. Two different kinds of waveguides can be distinguished that will determine the fabrication and replication process during fabrication: the guiding of light by total internal reflection in a polymer sheet or the guiding of light in a hollow waveguide supported by Fresnel reflection.

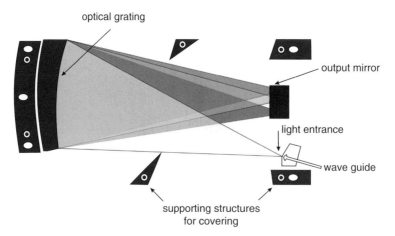

Figure 10.8 Ray trajectory of a micro spectrometer. The light is launched by optical waveguides and split off into the spectral colors by a concave optical grating arranged in a Rowland arrangement. The spectral light is guided to an output mirror and reflects to a detector array.

10.2.6.1 Polymer Waveguide Spectrometer

The polymer waveguide spectrometer refers to the light guiding inside a polymer film with the refractive index n_1 (Fig. 10.9). To achieve a total internal reflection, the waveguide has to be covered by layers with a lower refractive index (n_2, $n_3 < n_1$). The replication of these kinds of spectrometers by hot embossing requires an adapted process (Fig. 10.10). In a first step, the waveguide with the fiber connector, the optical grating, and the mirror with a material with a refractive index n_1 has to be replicated without any residual layer. This requires a selected volume of polymer film. Because of the lack of the residual layer, demolding can be critical. Therefore, after molding the waveguide layer in a second step, the cover layer with a lower refractive index is welded on the backside. For optimal welding, the polymer materials should consist of the same material (here, PMMA) but with a different refractive index. The set-up of the process parameters, especially the welding temperature, has be to done very carefully to avoid a melting and therefore a mixing of both polymers, which would destroy the effect of light guiding. After the welding step, the combination of both polymers can be demolded manually. In the next steps, the sputtering of reflective

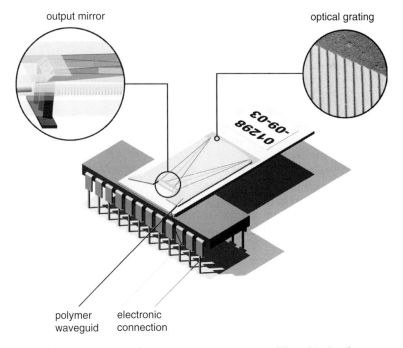

Figure 10.9 Polymer waveguide micro spectrometer. This kind of spectrometer is characterized by a waveguide inside a polymer layer by total internal reflection achieved by a combination of a polymer layer with a refractive index n_1 covered by polymer layers with a lower refractive index.

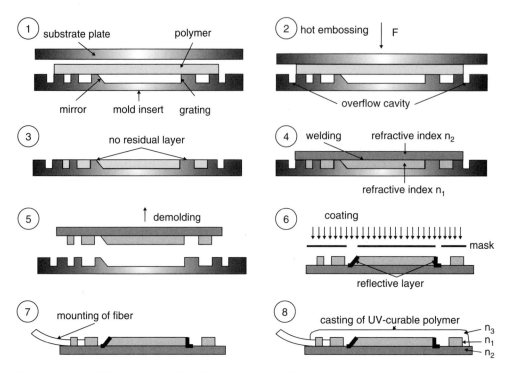

Figure 10.10 Molding steps for the replication of a micro spectrometer. Initially, the waveguide with the optical and connecting elements is embossed without a residual layer. One of the cover layers is welded during a following embossing step. After the welding step, the bond is demolded and the mirror and grating is coated with a reflective layer. Finally, the fiber is mounted and the system is encased by a UV-curing layer.

layers and the mounting of fibers follow. Finally, the system is topped by a final polymer layer to cover the optical system.

The advantage of polymer waveguide spectrometers is the lack of loss during the reflection of light inside the waveguide. The disadvantage is the absorption loss inside the waveguide, which makes it impossible to use this kind of spectrometer in the infrared range. This kind of spectrometer is therefore well suited for the analysis of light in the UV and visible range.

10.2.6.2 Hollow Waveguide Spectrometer

Spectrometers based on hollow waveguides are characterized by a covering of metallic surfaces that guide the light inside the spectrometer by Fresnel reflection (Fig. 10.11). In this case, the absorption loss is a function of the wavelength, the reflecting angle, and the polarization of light. These kinds of losses can be reduced to a few percent. Inside the waveguide the absorption losses are determined by the absorption of air, which makes it possible to use this kind of

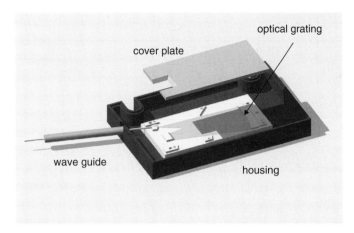

Figure 10.11 Hollow waveguide micro spectrometer. This kind of spectrometer is characterized by a waveguide inside a hollow space and Fresnel reflection on the top and bottom of the spectrometer.

spectrometer for a wide range of wavelengths. Finally, because of the lack of the polymer waveguide, replication can be done in one step, which is more cost effective.

Independent of the light-guiding system, the defect-free replication of the high-aspect ratio grating integrated in an arrangement of optical components is the challenging task of the replication process. Hot embossing is, in this case, well suited because of the short flow distances and the vertical demolding of the molded parts, especially the optical grating.

10.3 Microfluidic Devices

Microfluidic systems are a part of life science technology and diagnostic and therapeutic biomedical engineering. Passive micro components like capillary micro channel structures and so-called wells, reservoir areas, or miniaturized sample chambers can be part of micro total analysis systems (μ-TAS) or lab-on-a-chip systems [15]. Representative of passive microfluidic systems are, for example, capillary electrophoresis chips. Active microfluidic components like pumping systems or valve systems are mostly part of complex total analysis systems.

10.3.1 Capillary Analysis Systems

The functioning of a capillary electrophoresis system can be explained principally by Fig. 10.12. The system consists of two intersecting micro channels with wells at the beginning (buffer) and at the end (waste). The shorter channels

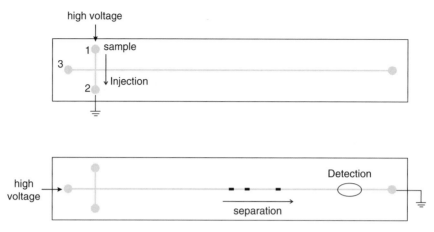

Figure 10.12 Principle of a capillary electrophoresis system. The principle refers to an intersection of two micro channels, where a small volume of the sample fluid is separated and injected into the long fluid channel with a buffer fluid inside. During the flow the sample volume will be separated into its components which can be detected at the end of the flow path. The injection and the flow are supported by a high voltage difference between the buffer and the waste.

contain the sample material that needs to be analyzed; the longer channels contain a buffer solution. To achieve flow of an injection of the sample fluid in the buffer fluid at the intersection point, a difference in potential has to be obtained by electrodes integrated in the wells. By electric switching the sample volume located in the intersection area can be injected into the longer separation channel. In this channel the plug is separated into the components, depending on molecule size and electric charge [14]. Characteristic of microfluidic structures are grooves typically with an aspect ratio in the range of one and a flow path of up to several centimeters, which can be arranged in a meandering shape to achieve compact systems. For an easy handling of those structures, it is recommended to arrange these CE-systems to standardized platforms, for example, micro titer plates with an area of 125 mm × 85 mm. On this area, for example, 96 CE-systems could be integrated (Fig. 10.13).

The cross-section of those microfluidic channels is characterized here in a range of 50 μm × 50 μm, an aspect ratio in the range of one. Considering the structured area, the mold fabrication method of micro machining is recommended. In this case, the mold insert is fabricated by micro machining in brass (Fig. 10.14). By this technology an exact geometrical design of the capillary intersection can already be achieved. The radius of the corners at the intersection depends on the channel cross-section and the required cutting tool; in this case, it is 25 μm.

Regarding the fluidic structures with aspect ratios of one and their arrangement on a large area with a thin carrier layer, replication by hot embossing is, besides the injection compression process, a suitable replication method.

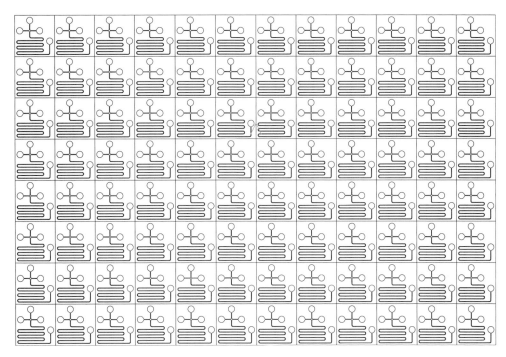

Figure 10.13 Arrangement of 96 CE-systems with a meandering flow channel. This shape allows the integration of CE-systems into established handling systems like micro titer plates [14,15].

The advantage of these methods is the short flow paths and the reduced stress inside the molded parts, which reduces the warpage of the fluidic parts. A moderate difference in shrinkage and a reduced warpage is advantageous for the next process step of UV, thermal, or ultrasonic covering. CE-systems are typically replicated in PMMA by hot embossing. Not only PMMA, but also COC or PEEK are recommended materials because of their chemical resistance, biocompatibility, and electrical properties. To obtain capillary systems after molding, the microfluidic channels have to be closed by a cover process. For this purpose, cover plates with through-holes are positioned and covered, for example, by UV-supported bonding. Figure 10.15 shows a molded and covered CE-system with integrated conducting paths that apply a potential difference.

The fabrication of microfluidic applications by hot embossing is manifold. Kricka et al. [27] showed the fabrication of polymer microchips with micro channels (100 μm wide, 40 μm deep) of varying designs. The chips were fabricated in polymethylmethacrylate by a hot embossing process using an electroformed tool produced starting with silicon chip masters. Koesdjojo et al. [24] used a two-stage embossing technique where, in a first step, a CE structure mold was fabricated in aluminum by micro machining. This master was replicated in polyetherimide (PEI) by hot embossing. This replicated part was used as the mold insert and was replicated in PMMA, at about 100°C—a significantly lower

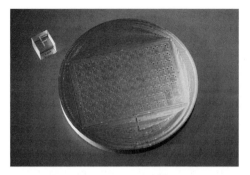

(a) CE-mold insert with 96 CE-systems

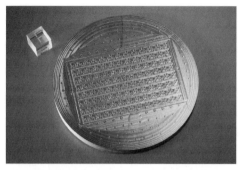

(b) CE-mold insert with the corresponding holes for the coverplate

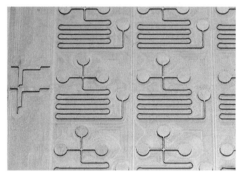

(c) Detailed view of the mold insert with a capillary system

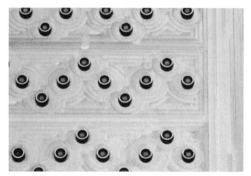

(d) Detailed view of the structures for molding holes in the cover plate

Figure 10.14 Mold insert for the replication of an arrangement of 96 CE-systems. The mold was fabricated by mechanical machining.

(a) Microtiterplate with 96 CE-systems

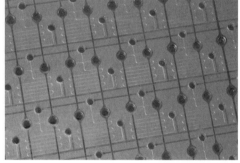

(b) Electrical connections of CE-systems

Figure 10.15 Micro titer plate with 96 CE-systems [14,15].

glass-transition temperature than PEI (about 216°C). The covering was done by solvent welding. To avoid damaging the micro channels during bonding, they were filled with water that formed a solid sacrificial layer when frozen. Ng and Wang [38] fabricated CE structures by roller embossing. A 50 µm–thick nickel mold fabricated by electroplating was wrapped around a roller that could be heated up over 170°C. By this technology microfluidic structures with feature sizes down to 50 µm and depths of 30 µm were successfully replicated within a tolerance of about 2% with regard to the lateral dimensions. This technology shows the potential for mass production but it has to be taken into account that structures with a high aspect ratio cannot be replicated without any deformation of structures during demolding. This roller technology is therefore well suited if the depth of the fluidic channels is relatively small and the aspect ratio is below one.

Besides the replication of channels in the micro range, the dimension of channels can decrease to the nano range. Thamdrup et al. [45] fabricated nano channels to analyze DNA molecules. The major advantage of nanofluidic structures is, in this case, the probing and investigating of single DNA molecules on an appropriate length scale. This allows the extraction of information that is not accessible in a bulk solution, coiled-up formation. The two-level Ormocomp stamp with micro and nano channels was fabricated by a process chain of E-beam lithography, UV lithography, and reactive ion etching. The stamp was coated with a nonstick coating by molecular vapor deposition and finally replicated in PMMA by thermal imprint.

Finally, especially for the fabrication of micro- or nanofluidic passive fluidic systems, hot embossing technology is well suited for the additional process step of thermal bonding to cover the microfluidic channels. In this case, for example, the molded part can remain in the hot embossing machine and will be covered with a polymer film by an additional welding step. Heckele et al. [18] described the bonding technologies for microfluidic systems and, further, the integration of conducting paths by embossing technique [17]. The challenge of replication and bonding of replicated nanochannels was described by Abgrall et al [3]. In this paper the replication of nanostructures in PMMA under the use of silicon wafer as mold and the additional thermal bonding technique was presented by an array of sealed planar nanochannels with a depth of 80 nm and low AR ranging from 0.008 to 0.05.

10.3.2 Microfluidic Pumps

Besides the passive microfluidic elements presented above, active elements correspond to microfluidic systems like pumps or valves. Here, the contribution of hot embossing is the replication of the housing with micro features responsible for the working of complex systems. The designs of such housings are manifold. Typically for the functioning of microfluidic systems, the feature size is in

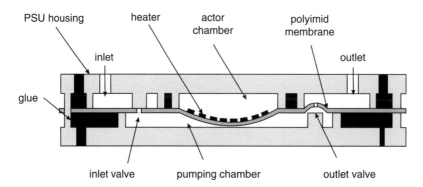

Figure 10.16 Schematic view of a micro pump. The pumping effect refers to a cyclic heating of gas in an acting chamber and the resulting oscillation of a polyimide membrane.

a range in which mold inserts can be fabricated by micro mechanical machining. Here, the batch sizes can be increased to large molding areas in the range of 8 inches.

Representative of a large number of pumping principles that may be realized in microsystem technology, one of the principles of a micro pump developed at Forschungszentrum Karlsruhe [32,33] can be explained by Fig. 10.16.

The pumping effect relates to a closed actuator chamber with a cycled heated polyimide membrane. The heating of the membrane causes an expansion of the gas in the actuator chamber and presses the membrane into the pumping chamber. With cyclic heating and an appropriate arrangement of inlet valve and exhaust valve, a pumping effect can be achieved. The schematic view illustrates the components: top housing, membrane, and bottom housing. Therefore, two different mold inserts are required. In this case the mold inserts were fabricated by mechanical machining of brass. Twelve housings are integrated in a mold insert (Fig. 10.17). The mounted micro pump with fluidic and electrical intersections is shown in Fig. 10.18.

A polymer-based vortex micro pump fabricated by hot embossing was described by Lei et al. [31]. The pumping principle relates in this case to a circumrotating impeller inside the pump. The pump molded in optically transparent and biocompatible material was characterized by a pump rate of 2.5 ml/min. Also in this case the task of hot embossing was the replication of the housing.

10.3.3 Microfluidic Valves

A precise dosing of small amounts of liquids and gases can be achieved by micro valves. A microfluidic polymer valve with a small dead volume in the range of 6 nl and a response time faster than 1 msec was presented by Shao et al. [42]. The top and bottom housings and the integrated fluid channels of this valve can be replicated by hot embossing, which further allows a low-cost batch fabrication.

Figure 10.17 Mold insert for micro pumps. The mold insert was fabricated by mechanical machining. Twelve pumps can be replicated in a batch.

Figure 10.18 Assembled micro pump.

The use of hot embossing as a replication technology for microfluidic systems can also be explained well by the fabrication of a membrane-based magnetic inductive micro valve [21]. The function principle is shown by Fig. 10.19. The micro valve is normally closed and the valve membrane is near a labile equilibrium in the opened state, so once switched only a little power is consumed to keep the valve open. Removing the acting force closes the valve automatically. The valve membrane is subject to mechanical stress, giving it a convex shape and thereby sealing the valve seat. Inside the membrane there is a gold conductor, and two permanent magnets with yoke are placed outside of the valve. A Lorentz force acts on the membrane if an electric current flows through

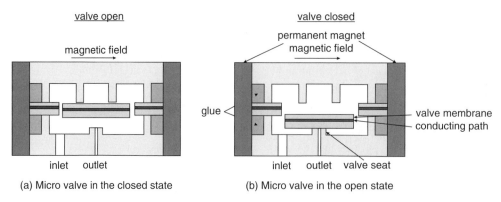

Figure 10.19 Scheme of a micro valve based on the magnetic inductive principle. The Lorentz force achieved by a current flow in the pre-stressed membrane lifts the membrane from the valve seat and opens the valve.

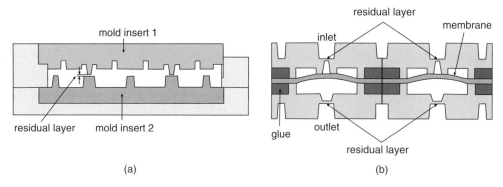

Figure 10.20 (a) Schematic view of double-sided molding of a micro valve housing. (b) The residual layer in the inlet and the outlet is removed by further process steps like penetration with a pin.

the conductor path. This force lifts the membrane from the valve seat and opens the valve.

For the fabrication of the housings of a micro valve, the process variation of double-sided hot embossing has been used to achieve through-holes for the inlet and outlet of the valve (Fig. 10.20). Therefore, both mold inserts have to be positioned by an alignment system. The embossing process is optimized such that a thin residual layer between both pins is achieved. This thin residual layer is easily removed by further process steps, like penetration with a needle. The material used for the replication in this figure was PSU.

The mold inserts are fabricated by mechanical machining of brass (Fig. 10.21 (a)). The final dimension of the polymer valve is in the range of $1.8 \times 12 \times 1.5$ mm^3 without the magnets (Fig. 10.21).

(a) Mold insert of the micro valve

(b) Assembled micro valve with fluidic and electric connections

Figure 10.21 Mold insert for the molding of the micro valve fabricated by mechanical machining in brass. The pin in the middle of the rectangular structures is the structure for the outlet of the valve. The polymer part of the valve has a dimension of $1.8 \times 12 \times 1.5$ mm^3.

10.4 Further Applications

10.4.1 Micro Needles

Another application for hot embossing technique is in the medicine, in particular for drug delivery. Micro needles are one of the minimally invasive drug delivery systems entering the body through the skin. Therefore, the outer layer of the skin (*stratum corneum*) with a typical thickness in a range between 10 µm and 20 µm hast to be pierced. The thickness of this layer determines the minimum length of the needles. The biocompatibility of selected polymers and the fabrication of an array of micro needles for drug delivery make this application well suited for a polymer replication processes. One of the requirements is a hollowness, which makes it necessary to integrate a fluid channel inside the polymer needles.

A suitable fabrication method is the replication of micro needles by double-sided, positioned hot embossing. In this case, a cone-shaped needle on the one side hits a cone-shaped hole on the other side of a two-sided mold insert (Fig. 10.22). The accuracy that can be achieved regarding the homogeneous thickness of the side-walls depends here on the overlay accuracy of both mold halves. The advantage of this method is that only a thin residual layer on the tip of the cone has to be disrupted. This can be easily done, for example, by laser structuring. The fabrication of the mold inserts—the positive cone and the negative cone-shaped hole—can be done by mechanical machining (Fig. 10.23). The mold material used here was brass.

Finally, the residual layer of the needle has to be eliminated. In this case the residual layer is, as well as the carrier layer of the needle, also located in the tip of cone. To achieve a fluid channel through the needle, the residual layer has

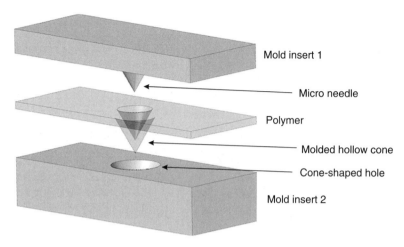

Figure 10.22 Double-sided molding of micro needles. To achieve hollow, cone-shaped needles, double-sided positioned molding of a cone-shaped needle into a cone-shaped hole is required.

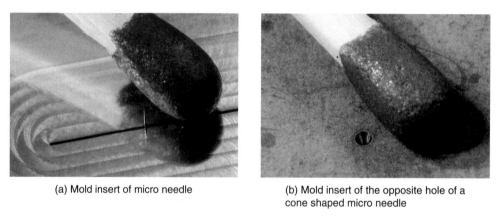

(a) Mold insert of micro needle
(b) Mold insert of the opposite hole of a cone shaped micro needle

Figure 10.23 Mold inserts for the replication of a cone-shaped micro needle.

to be removed. A molded micro needle with a fine hole in the tip is shown in Fig. 10.24.

Another method for the fabrication of micro needles arrays was presented by Han et al. [16]. They fabricated an in-plane micro needle by UV lithography and copied these needles into an out-of-plane needle array. The third step was the fabrication of an inverted PDMS mold insert, which was replicated into polymer by hot embossing. Moon et al. [36] investigated the fabrication of micro needles by hot embossing with a successive deep X-ray exposure. Initially, tetrahedral intermediate structures were fabricated by two-step X-ray lithography. These structures were then used to fabricate a mold insert by electroplating or casting. In the following step these mold inserts were replicated in PMMA by hot embossing and finally structured into micro needles by X-ray lithography.

(a) Molded micro needle　　　　(b) Micro hole on top of a micro needle

Figure 10.24 Molded micro needles with a through-hole on top of the cone-shaped needle.

10.4.2 3D Structures in Terms of Micro Zippers

As mentioned in the previous chapters, the replication of structures with undercuts causes damage to the structures during demolding. This damage refers to replication in thermoplastic polymers that show a nearly inelastic behavior. In contrast, if an elastic material is selected as molding material, the demolding of structures with undercuts can be achieved. Bogdanski et al. [6] showed the replication of structures with undercuts with the elastic material PDMS. These structures can further be used, for example, to prepare polymeric zippers. In this work, undercut structures were fabricated in a Si-wafer by a wet-etching processes in KOH using silicon dioxide as a highly selective etching mask. By this method two kinds of three-dimensional (3D) structures in terms of V-shaped grooves on an area of 5×5 cm^2 and dovetail-shaped undercuts on an area of 20×5 mm^2 were fabricated. The depth of the structures was in a range of 20 μm; the width of the undercuts varied from 0.5 μm up to 10 μm. The V-shaped structures were replicated in PMMA; the dovetailed-shaped structures were replicated in elastic PDMS. In this case, the residual layer had to be thick enough to guarantee a successful demolding of the structures. If the thickness of the residual layer is in the range of the structure height, the structures tear off at the intersection between residual layer and structures. This example shows that with an elastic molding material hot embossing is not limited only to structures with vertical sidewalls or demolding angles.

10.4.3 Comb Drive, Acceleration Sensor

A further application is the prototype fabrication of comb drive structures [50]. These kinds of structures are characterized by a large number of freestanding

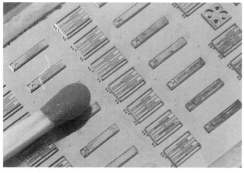

(a) LIGA-mold insert of an acceleration sensor   (b) Detail of replicated acceleration sensor

Figure 10.25 Mold insert of an acceleration sensor fabricated by the LIGA process. The mold insert is replicated in polymer onto a sacrificial layer. After electroplating, the sacrificial layer is removed and the oscillating metallic mass between two condensor plates measures the acceleration by the change of the capacity of a condensor.

vertical sidewalls on a large area. The precise arrangement of the structures requires low stress in the structures to avoid any deformation that can result in contact with comb structures, with the effect of a malfunction of the comb drive.

The principle of an electric condenser with an oscillating mass between two condenser plates can be used for the application of a micro acceleration sensor [33]. This principle requires metal structures and can be fabricated by electroplating of previously in-polymer replicated and metalicized inverted structures. Therefore, the structures are first replicated by hot embossing of a nickel mold, fabricated in this case by X-ray lithography and electroplating. These replicated structures contain the inverted structures of the condensor plates and the oscillating mass on a sacrificial layer for the second step of electroplating. These structures are characterized by sidewalls with high aspect ratios (Fig. 10.25).

10.4.4 Lotus Structures

Submicrostructured surfaces allow modifying the behavior of polymer films or components. Especially in microfluidics a lotuslike characteristic is requested for many applications. Structure details with a high aspect ratio are necessary to decouple the bottom and the top of the functional layer. Unlike the stochastic methods, patterning with a LIGA-mold insert it is possible to structure surfaces very uniformly or even with controlled variations (e.g., with gradients). The realization of lotuslike patterns on surfaces requires precise control of the structuring method. The manufacturing method of such structured surfaces should also enable mass fabrication, for example, by polymer replication [19,49].

The liquid repellency of a surface is principally governed by a combination of its chemical nature (i.e., surface energy) and, in the case of stochastic surfaces, by its topography at the micro scale (i.e., surface roughness). Although flat, low surface energy materials can often exhibit high water contact angles, this is normally not sufficient to yield the repellency of super-hydrophobic surfaces. In order to obtain this, the difference between the advancing and the receding contact angle (contact angle hysteresis) must be minimal. Effectively, contact angle hysteresis is the force required to move a liquid droplet across a surface. In the case of little or no hysteresis, very little force is required to move a droplet; hence, it rolls off easily. The boundary conditions to achieve super-hydrophobic surfaces are described by Cassie and Baxter [8]. The contact angle can be influenced by the fractional area f, that is, the total area of solid-liquid interface in a unity of geometrical plane area parallel to the surface. In this example, honeycomb structures were replicated over a broad range of geometric parameters. To compare different wetting behaviors, two different designs were investigated and manufactured. The pitch of the microstructured pattern was constantly maintained (pitch = 4 µm), whereas the diameter of the combs and the thickness of the separating walls varied. The two patterns had the following geometrical characteristics: wall thickness = 1,000 nm and comb diameter = 3,000 nm (opening ratio $f = 0.25$), and wall thickness = 400 nm and comb diameter = 3,600 nm (opening ratio $f = 0.10$). The height of the structures was 4 µm, resulting in an aspect ratio of the sub-micro walls between 4 and 10. The height of the structures (and therefore the aspect ratio) was chosen in order to obtain super-hydrophobic surfaces and also to allow demolding after replication by hot embossing without damaging the separating walls. Regarding the boundary conditions named above, the fabrication of lotus structures can be illustrated by Fig. 10.26. The process chain consists of the process steps of E-beam lithography, mask fabrication, X-ray lithography, electroplating of the mold insert, and finally replication by hot embossing.

10.5 Industrial Applications

10.5.1 Compact Disc Replication

A popular example of mass replication is the compact disc (CD) [20] or digital versatile disc (DVD) fabrication [43]. In 2004 about 25 billion discs were fabricated [5]. CD-ROM production peaked in 2001 with about 14 billion discs. Nowadays, DVD production increases strongly. Regarding these large numbers of pieces, the challenge for disc production is to guarantee an effective machine throughput and to keep the production cost on a low level, independent of the kind of format (CD or DVD) and the size of digital structures.

As mentioned in the introduction, the fabrication of old-time records is an impressive example of the commercial use of the embossing process.

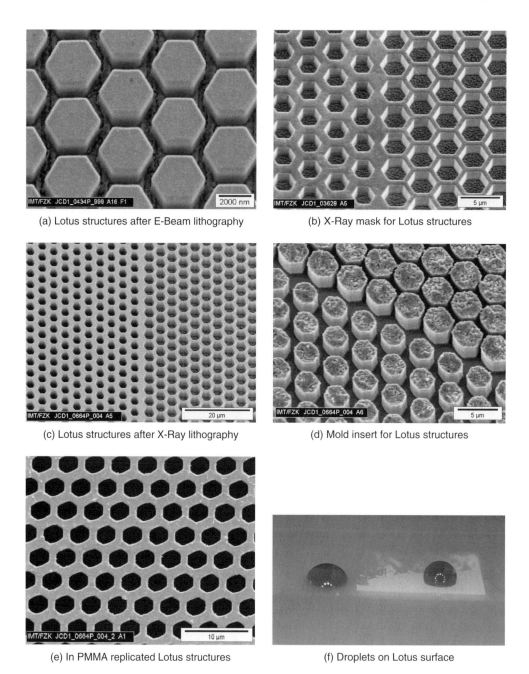

(a) Lotus structures after E-Beam lithography

(b) X-Ray mask for Lotus structures

(c) Lotus structures after X-Ray lithography

(d) Mold insert for Lotus structures

(e) In PMMA replicated Lotus structures

(f) Droplets on Lotus surface

Figure 10.26 Process chain for the manufacturing of controlled lotuslike surfaces. The process steps correspond to the LIGA process, E-beam lithography, mask fabrication, X-ray lithography, mold electroplating, and replication. The advantage of this process for the fabrication of lotuslike structures is the potential to fabricate defined lotus designs with the effect of a controlled motion of droplets.

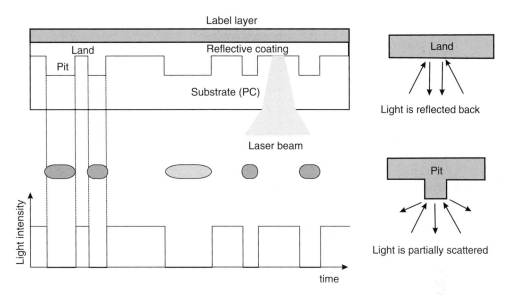

Figure 10.27 Principle of the storage of information on CD. Characteristic are molded positive and negative structures, called pitch and land. A laser beam is focused through the molded polycarbonate substrate onto these structures. The backsides of the structures are coated with a reflective layer. The arrangement of the structures results in a reflection with different intensities that can be used for the digitalizing of information.

Nevertheless, these replication experiences can be transferred under further development to the replication of CDs and DVDs. Compared to the structure size of records, in the range of approximately 20 μm, the structure size decreases to the range of 0.83 μm for CDs and 0.4 μm for DVDs. The track width is 1.6 μm for CD and 0.8 μm for DVD. The size of the structures (pits and lands, Fig. 10.27) corresponds to the laser wavelength and the focusing unit, with the result that the aspect ratio of these structures is in the range of one. This moderate aspect ratio allows easy demolding of the structures, especially under the boundary conditions that for the storage of digital information, accurate vertical sidewalls are not required. The typical components of a CD are shown in Fig. 10.28. The disc and its structures are typically molded in polycarbonate. The structured surface is, in a second step, coated with a reflective layer and finally coated with a protective layer by spin coating. The arrangement of pits and lands is responsible for the difference in reflective behavior, which results finally in digital information in the form of different intensities (light and dark). Depending on the size of the structures, the information is stored by time-dependent high- and low-level signals.

A commercial hot embossing machine described above is theoretically well suited for the replication of these 1.2 mm thick microstructured discs with 120 mm diameter, but it is not cost effective enough for mass production of the required number of discs. Therefore, the embossing cycle is combined with an

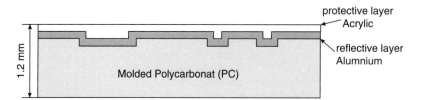

Figure 10.28 Schematic view of the different layers of a compact disc. The carrier layer of PC carbonate with a typical thickness of 1.2 mm with the structures called pits and lands. The structured side is coated by a reflective aluminum layer and finally protected with an acrylic layer by spin coating.

injection cycle, which results in injection compression molding steps. Today, cycle times of approximately three seconds for the whole fabrication line can be achieved. This fabrication line includes the molding of the disc in polycarbonate, the application of an aluminum reflective layer by vapor deposition or sputtering, the spin coating of a final protective lacquer, the printing of a label, and finally the packaging of the CD. In the early days of CD fabrication, all these steps were done separately by several machines. By 1988 the developments of fully automated fabrication lines started with a linking of all these processes by handling systems and disc transfers between each module. The maximal numbers of discs achievable by these systems is determined by the slowest process of the whole chain. Nevertheless, today cycle times for the whole process of a disc are in the range of approximately 3 seconds. Present replication lines are able to fabricate about 30,000 DVDs per day.

Due to this large number of replications, the mold inserts, typically nickel shims, are fabricated by several copy steps. From a master, a first level of different copies (fathers) is fabricated (Fig. 10.29). From these copies, further copies (sons) are fabricated that finally can be used in the fabrication lines for replication. This principle guarantees a large number of replication steps and a continuous fabrication. After the lifespan of the mold insert is finished, the use of another copy from the second level guarantees a continuous fabrication of the discs. Continuous fabrication is underlined by an efficient fixation of the nickel shims by a vacuum chuck, allowing a quick change of the mold insert.

10.5.2 Film Fabrication

To forgo time-consuming cyclic heating and cooling, some applications can be fabricated by roll-to-roll processes that allow a continuous structuring of structures. Regarding the structuring method, the aspect ratios of the structures achievable by this technology depend on the diameter of the roll and the fabrication technology of the related microstructured mold inserts. Typically structures with low aspect ratios are therefore replicated. Nevertheless, a large bandwidth

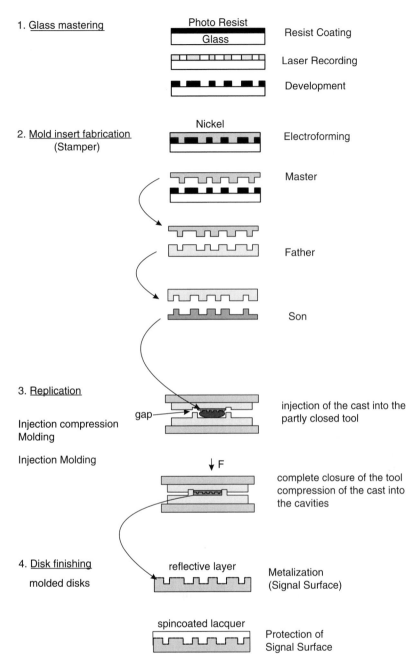

Figure 10.29 Fabrication steps of compact discs. The electroplated master is copied to a first-level mold (father) and finally to a second-level mold (son). These second-level mold inserts are used for replication by injection molding or injection compression molding.

of applications are fabricated by this technology, for example, holographic structures or anti-reflective surfaces [1].

10.5.3 Fresnel Lenses

Remembering record fabrication in the past, where, in the case of long-play records, microstructured discs with a diameter of approximately 300 mm were replicated by hot embossing, another application shows some similarities. A thin layer as carrier layer of microstructures on a large area is a characteristic of both records and Fresnel lenses. Nevertheless, the structures are different. These lenses are especially required if lenses with large diameters are needed, where the use of glass lenses would be very expensive, or if flat lenses are required. These kinds of lenses can be fabricated with a large bandwidth of diameter up to 19 inch [2]. A commercial application of theses lenses is, for example, lenses in overhead projectors or lenses to concentrate sunlight for solar cells. The mold inserts are typically fabricated by mechanical machining and require high-precision machining to achieve an optically smooth surface. These molds are finally replicated in typically PMMA by hot embossing. The embossing machines are usually optimized for this kind of replication and use polymer granules instead of polymer foils (Fig. 10.30).

(a) Hot embossing machine for the fabrication of Fresnel-lenses

(b) Granulate raw material

Figure 10.30 Hot embossing of Fresnel lenses. Fresnel lenses are fabricated by large-area hot embossing. Instead of polymer foils, granules are used. With courtesy of Fresnel-Optics.

10.6 The Future

The examples named above are only a few from a pool of applications, but they underline the relevance of the hot embossing or thermal nanoimprint process as established replication technologies. Further developments in hot embossing and nanoimprinting will be supported by new applications, especially if large series are required. Here the kind of automation and maybe standardization of, for example, molding formats will help to establish these technologies also in industry. Especially the cycle times have to be minimized in the future, which corresponds to an efficient heating and cooling system. Optimized molding tools are therefore a key issue for cost-efficient molding. In combination with already well-established handling systems in the macroscopic arena, process times can be minimized and hot embossing can be established as an industrial replication technology. Cost-effective replication also requires large molding areas. These molding areas correspond to the fabrication methods of the mold inserts that are individually limited. To bypass these limits, molding tools with multiple mold inserts can be used. These aspects are mainly in the foreground of interest for a cost-effective use of hot embossing. Independent of the potential to optimize cost effectiveness, hot embossing will still be a well-suited process for first replications of prototypes. In this case, the future may shift the replication to structures with smaller structure sizes in the nano range, in combination with high aspect ratios, for example, larger than 5. Further, those structures will be replicated over areas as large as 8 inches or more. Nevertheless, all replication processes are inspired by the requirements of future applications and these applications will eventually determine the structure sizes, molding areas, and kind of automation.

Today the knowledge of hot embossing fuses mainly on experience. The diversity of different designs, fabrication methods of mold inserts, polymers, molding areas, and machine properties especially requires an individual and often experimental determination of the process parameters. Simulation tools help us to understand basic relationships, but they are not yet suited to simulating a complete molding cycle with a real microstructured mold insert. In the near future with typical personal computers, modeling and process simulation will be challenging. Here, further requirements in CPU power and developments in simulation tools are necessary to obtain a user-friendly simulation tool. The simulation results are linked to material models valid in the microscopic or nanoscopic range that are still in development and not implemented in simulation tools. Independent of the development of material models and CPU power, the diversity of hot embossing processes with individual boundary conditions will further increase the requirements on a simulation tool. Here, an increase in commercial use of hot embossing may increase the interest in commercial software development.

Nevertheless, development in hot embossing is linked to future applications, and with this theoretical background the reader may be part of the further development of this replication technology.

References

1. 3M. http//www.solutions.3m.com, 2008.
2. Fresnel Optics. http//www.fresnel-optics.de, 2008.
3. P. Abgrall, L.-N. Low, and N.-T. Nguyen. Fabrication of planar nanofluidic channels in a thermoplastic by hot-embossing and thermal bonding. *Lab on a Chip*, 7:520–522, 2007.
4. M. Baer. μ FEMOS *Mikro-Fertigungstechniken für hybride mikrooptische Sensoren*, volume 10 of ISBN 3-937300-95-3. Universitätsverlag Karlsruhe, 2006.
5. P. Binkowska, B. Cord, and P. Wohlfart. Mass production of DVDs: faster, more complex but cheaper and simpler. *Microsystem Technologies*, 13:139–144, 2007.
6. N. Bogdanski, H. Schulz, M. Wissen, H.-C. Scheer, J. Zajadacz, and K. Zimmer. 3D-hot embossing of undercut structures—an approach to microzippers. *Microelectronic Engineering*, 73–74:190–195, 2004.
7. M. Bruendel. *Herstellung photonischer Komponenten durch Heissprägenund UV-induzierte Brechzahlmodifikation von PMMA*. PhD thesis, University of Karlsruhe (TH), Institute for Microstructure Technology, 2008.
8. A. B. D. Cassie and S. Baxter. Wettability of porous surfaces. *Transactions of the Faraday Society*, 40:546–551, 1944.
9. C.-Y. Chang, S.-Y. Yang, L.-S. Huang, and J.-H. Chang. Fabrication of plastic microlens array using gas-assisted micro-hot-embossing with a silicon mold. *Infrared Physiscs & Technology*, 48:163–173, 2006.
10. C. Y. Chang, S. Y. Yang, and J. L. Sheh. A roller embossing process for rapid fabrication of microlens arrays on glass substrates. *Microsystem Technologies*, 12:754–759, 2006.
11. C. L. Chen and F. Jen. Fabrication of polymer splitter by micro hot embossing technique. *Tamkang Journal of Science and Engineering*, 7(1):5–9, 2004.
12. C. H. Chien and Z. P. Chen. Fabrication of light guiding plate of double-sided hot-embossing. *Materials and Science Forum*, 505–507:211–216, 2006.
13. Ch.-G. Choi. Fabrication of optical waveguides in thermosetting polymers using hot embossing. *Journal of Micromechanics and Microengineering*, 14:945–949, 2004.
14. A. Gerlach, G. Knebel, A. E. Guber, M. Heckele, D. Herrmann, A. Muslija, and Th. Schaller. Microfabrication of single-use plastic microfluidic devices for high-throughput screening and DNA analysis. *Microsystem Technologies*, 7:265–268, 2002.
15. A. E. Guber, M. Heckele, D. Herrmann, A. Muslija, V. Saile, L. Eichhorn, T. Gietzelt, W. Hoffmann, P. C. Hauser, J. Tanyanyiwa, A. Gerlach, N. Gottschlich, and G. Knebel. Microfluidic lab-on-a-chip systems based on polymers—fabrication and application. *Chemical Engineering Journal*, 101:447–453, 2004.
16. M. Han, D.-H. Hyun, H.-H. Park, S. S. Lee, C.-H. Kim, and C. G. Kim. A novel fabrication process for out-of-plane microneedle sheets of biocompatible polymer. *Journal of Micromechanics and Microengineering*, 17:1184–1191, 2007.
17. M. Heckele and F. Anna. Hot embossing of microstructures with integrated conduction paths for the production of lab-on-chip systems. In *Design, Test, Integration and Packaging of MEMS/MOEMS, Proceedings of SPIE*, Volume 4755, 2002.
18. M. Heckele, A. E. Guber, and R. Truckenmueller. Replication and bonding techniques for integrated microfluidic systems. *Microsystem Technologies*, 12:1031–1035, 2006.
19. M. Heckele, M. Worgull, P. Fugier, J. Gavillet, G. Tosello, H. H. Hansen, T. Metz, and P. Koltay. Micro- and nanostructured surfaces for the passive liquid and gas management in microfluidic structures. In *3rd Internat. Conf. on Multi-Material Micro Manufacture (4M2007)*, October 2007.
20. Kees A. Schouhamer Immink. The CD story. *Journal of the AES*, 1998.

21. S. C. Kaiser. *Entwicklung eines magnetisch-induktiven Mikroventils nach dem AMANDA Verfahren.* PhD thesis, University of Karlsruhe (TH), Institute for Microstructure Technology (IMT), 2000.

22. J. T. Kim, K. B. Yoon, and C.-G. Choi. Passive alignment method of polymer PLC devices by using a hot embossing technique. *IEEE Photonics Technology Letters,* 16(7):1664–1666, 2004.

23. S. H. Kim, K.-D. Lee, J.-Y. Kim, M.-K. Kwon, and S.-J. Park. Fabrication of photonic crystal structures on light emitting diodes by nanoimprint lithography. *Nanotechnology,* 18:1–5, 2007.

24. M. T. Koesdjojo, Y. H. Tennico, and V. T. Remcho. Fabrication of a microfluidic system for capillary electrophoresis using a two-stage embossing technique and solvent welding on poly (methymethacrylate) with water as a sacrificial layer. *Analytical Chemistry,* 80(7): 2311–2318, 2008.

25. J. Kouba, M. Kubenz, A. Mai, G. Ropers, W. Eberhardt, and B. Loechel. Fabrication of nanoimprint stamps for photonic crystals. *Journal of Physics: Conference Series,* 34: 897–903, 2006.

26. D. Krabe and W. Scheel. Optical interconnects by hot embossing for module and PCB technology—the EOCB approach. In *Proc. of 49th Electronic Components and Technology Conference,* pages 1164–1166, San Diego, CA, USA, 1999.

27. L. J. Kricka, P. Fortina, N. J. Panaro, P. Wilding, G. Alonso-Amigo, and H. Becker. Fabrication of plastic microchips by hot embossing. *Lab on a Chip,* 2:1–4, 2002.

28. A. Last and J. Mohr. *Fehllicht in LIGA Spektrometern.* PhD thesis, University of Karlsruhe (TH), Institute for Microstructure Technology, FZKA Report 6885, 2002.

29. P. W. Leech, R. A. Lee, and T. J. Davis. Printing via hot embossing of optically variable images in thermoplastic acrylic lacquer. *Microelectronic Engineering,* 83:1961–1965, 2006.

30. S. Lehmacher and A. Neyer. Integration of polymer optical waveguides into printed circuit boards. *Electronics Letters,* 36(12):1052–1053, 2000.

31. K. F. Lei, R. H. W. Lam, J. H. M. Lam, and W. J. Li. Polymer based vortex micropump fabricated by micro molding replication technique. In *Proceedings of 2004 IEEE/RJS International Conference on Intelligent Robots and Systems,* pages 1740–1745, September 2004.

32. D. Maas, B. Büstgens, J. Fahrenberg, W. Keller, and D. Seidel. In *Proceedings ACTUATOR94,* pages 75–78, Bremen, Germany, 1994.

33. W. Menz and J. Mohr. *Mikrosystemtechnik für Ingenieure.* ISBN 3-527-29405-8. VCH Verlagsgesellschaft, 2nd edition, 1997.

34. H. Mizuno, O. Sugihara, T. Kaino, N. Okamoto, and M. Hosino. Low-loss polymeric optical waveguides with large cores fabricated by hot embossing. *Optics Letters,* 28(23):2378–2380, 2003.

35. S.-D. Moon, N. Lee, and S. Kang. Fabrication of a microlens array using micro-compression molding with an electroformed mold insert. *Journal of Micromechanics and Microengineering,* 13:98–103, 2003.

36. S. J. Moon, S. S. Lee, H. S. Lee, and T. H. Kwon. Fabrication of microneedle array using LIGA and hot embossing process. *Microsystem Technologies,* 11:311–318, 2005.

37. C.-J. Moran-Iglesias, J. Mohr, and A. Last. *Grossflächige quasi freistrahloptische Mikrospektrometer.* PhD thesis, University of Karlsruhe (TH), Institute for Microstructure Technology, FZKA Report 7211, 2002.

38. S. H. Ng and Z. F. Wang. Hot roller embossing for the creation of microfluidic devices. In *Proceedings of DTIP Conference 2008,* Nice, France, EDA Publishing, 2008.

39. F. Sammoura, Y.-C. Su, Y. Cai, C.-Y. Chi, B. Elamaran, L. Lin, and J.-C. Chiao. Plastic 95-GHz rectangular waveguides by micro molding technologies. *Sensors and Actuators A,* 127:270–275, 2006.

40. H. Schift, S. Park, B. Jung, C.-G. Choi, C.-S. Kee, S.-P. Han, K.-B. Yoon, and J. Gobrecht. Fabrication of polymer photonic crystals using nanoimprint lithography. *Nanotechnology*, 16:261–265, 2005.

41. J. Seekamp, S. Zankovych, A. H. Helfer, P. Maury, C. M. Sotomayor Torres, G. Böttger, C. Liguda, M. Eich, B. Heidari, L. Montelius, and J. Ahopelto. Nanoimprinted passive optical devices. *Nanotechnology*, 13:581–586, 2002.

42. P. Shao, Z. Rummler, and W. K. Schomburg. Polymer micro piezo valve with a small dead volume. *Journal of Micromechanics and Microengineering*, 14:305–309, 2004.

43. G. Sharpless. CD and DVD disc manufacturing. *Deluxe Global Media Services*, www.distronics.com, 2003.

44. Ch. Solf. *Entwicklung von miniaturisierten Forurier-Transformations - Spektrometern und ihre Herstellung mit dem LIGA-Verfahren*. PhD thesis, University of Karlsruhe (TH), Institute for Microstructure Technology, FZKA Report 6964, 2004.

45. L. H. Thamdrup, A. Klukowska, and A. Kristensen. Stretching DNA in polymer nanochannels fabricated by thermal imprint in PMMA. *Nanotechnology*, 19: 125301, 2008.

46. C.-J. Ting, M.-C. Huang, H.-Y. Tsai, C.-P. Chou, and C.-C. Fu. Low cost fabrication of large-area anti-reflection films from polymer by nanoimprint/hot-embossing technology. *Nanotechnology*, 19:1–5, 2008.

47. R. Ulrich, H. P. Weber, E. A. Chandross, W. J. Tomlinson, and E. A. Franke. Embossed optical waveguides. *Applied Physical Letters*, 20(6):213–215, 1972.

48. A. J. Waddie, M. R. Taghizadeh, J. Mohr, V. Piotter, Ch. Mehne, A. Stuck, E. Stijns, and H. Thienpont. Design, fabrication and replication of microoptical components for educational purposes within the network of excellence in micro optics (nemo). In *SPIE Proceedings Series No. 6185*, 2004.

49. M. Worgull, M. Heckele, T. Mappes, B. Matthis, G. Tosello, T. Metz, J. Gavillet, P. Koltay, and H. N. Hansen. Sub-μ structured lotus surfaces manufacturing. In *Conference of Design, Integration and Packaging (DTIP)*, Nice, France, April 2008.

50. Y. Zhao and T. Cui. Fabrication of high-aspect-ratio polymer-based electrostatic comb drives using the hot embossing technique. *Journal of Micromechanics and Microengineering*, 13:430–435, 2003.

51. K. B. Yoon, C.-G. Choi, and S.-P. Han. Fabrication of multimode polymeric waveguides by hot embossing lithography. *Japanese Journal of Applied Physics*, 43(6A):3450–3451, 2004.

52. P. Zhang, G. Londe, J. Sung, E. Johnson, M. Lee, and H. J. Cho. Microlens fabrication using an etched glass master. *Microsystem Technologies*, 13:339–342, 2007.

Index

Acceleration sensor, 329

Additional structures, 303

Adhesion, 146, 264

Aggregate states of polymers, 99

Air pressure demolding, 265

Alignment, 150

Alignment system, 229, 257

Aluminum embossing, 61

Amorphous polymers, 84, 99

Anti adhesive coating, 300

Applications, 307

Arrhenius, 73

Assembled mold inserts, 297

Atomic force microscope (AFM), 117

Battenfeld Microsystem 50, 25

Bearing elements, 258

Bearings, 231

Bending of Machine, 134

Berliner, Emil, 3

Borosilicate glass, 58

Breakaway torque, 96, 147, 196, 281

Bulk plastics, 103

Calorimetric data, 108

Capillary electrophoresis system, 319

Capillary flow shear thinning fluid, 74

Capillary rheometer, 79

Cavity filling, 207

Ceramic injection molding (CIM), 62

Ceramic powder, 62

Ceramics, 62

Characteristic properties of
    processes, 52

Clamping of mold inserts, 262, 297

Comb drive, 329

Commercial hot embossing
    machines, 240

Compact disk (CD), 331

Comparison of processes, 45

Complex elastic modulus, 93

Components for hot embossing, 139

Compressibility, 86

Conducting path, 173, 323

Conduction, 250

Conductivity, 250

Contact stress, 217, 303

Control system, 238

Convection, 249

Cooling, 215

Cooling fluid, 256

Cost effectiveness, 50

Costs, 51
    hot embossing, 34
    injection molding, 28

Creep modulus, 89

Creeping, 89

Cross viscosity model, 70

Cross-WLF model, 210

Cutting tools, 285

Deep reactive ion etching (DRIE), 289

Demolding, 141, 146, 217, 280, 302

Demolding by air pressure, 265

Demolding stress, 219

Demolding systems, 263

Density, 86, 105

Design rules for molds, 301

Differential scanning calorimetry
    (DSC), 109

Differential thermal analysis
    (DTA), 108

Double-sided molding, 150, 276, 328

Drive unit, 231

Dwell pressure, 25, 125

Dynamic friction, 95

Dynamic mechanical analysis
    (DMA), 92

E-beam lithography, 295
Ejector pins, 264
Elastic mold insert, 172
Elastomeres, 66
Electric discharge machining
    (EDM), 288
Electroforming of mold inserts, 292
Embossing cycle, 234
Embossing of Al, 61
Embossing of metallic glass, 61
Embossing of metals, 60
Embossing of Pb, 60
Energy elastic state, 100
Etching processes, 289
Evacuation, 144
EVG, 242
EVG520HE, 242
EVG750HE, 242

Failures, 120
Father–Son copy, 334
Feedstocks, 62
Filling pressure, 211
Fixation of mold inserts, 261
Fixed costs, 51
Flow in capillaries, 73
Flow transition range, 100
Force controlled molding, 146
Fourier, 197
Fresnel lenses, 336
Fresnel-Optics, 336
Friction, 94, 195
Friction measurement, 96

Gas-assisted embossing, 167
Glass, 58
Glass transition range, 100
Glass transition temperature, 86,
    99, 109
Gluing of mold inserts, 262
Guiding columns, 230

Heat capacity, 106
Heat conduction, 106, 197

Heat transfer, 250
Heat transfer coefficient, 255
Heat transition, 198
Heating, 197
Heating concepts, 249
Heating element, 250
Heating fluid, 249
Helical cooling system, 255
HEX01–HEX04, 240
High frequency heating, 251
High performance plastics, 103
History of hot embossing, 3
Holding time, 140
Hot embossing costs, 34
Hot embossing cycle, 140, 143
Hot embossing principles, 137
Hot embossing process, 137
Hot embossing technique, 227
Hot embossing tool
    basic, 267
    double sided, 271
    simple, 266
    universal, 270
Hot punching, 158
Hover cushion, 258
Hydraulic diameter, 256
Hydraulic drive, 233
Hydrophobic surfaces, 331

Influencing factors, 143, 148
Injection compression molding, 30
Injection cycle, 24
Injection molding, 19
Injection molding costs, 28
Injection molding materials, 26

Jenoptik, 240

Laminar flow, 255
Lands, 333
Laser structuring, 286
Lenses, 308
LIGA, 9, 15, 261, 293, 295, 315, 330
Lithographic structuring methods, 291

Loss factor, 94
Lost modulus, 93
Lotus structures, 330

Machine components, 227
Macro language, 238
Macromolecules, 64
Maskaligner, 43
Material parameters, 147
Materials micro injection molding, 26
Maxwell model, 77, 88
Measurement process parameter, 236
Measurement systems, 117
Mechanical machining, 285
Mechanical models, 77
Mechanical stiffness, 230
Melt index, 79
Melting state, 100
Memory programmable controller, 238
Meshing, 186
Metal embossing, 60
Metal injection molding (MIM), 62
Metal substrates, 156
Micro contact printing, 41
Micro injection molding, 19
Micro needles, 327
Micro powder injection molding, 62
Micro pumps, 323
Micro thermoforming, 36
Micro valves, 324
Micro zippers, 329
Microfluidic channel, 320
Microfluidic devices, 319
Microlenses, 164, 308
Micro-optical devices, 308
Microspectrometer, 315
Microtiterplate, 320
Modeling geometry, 189
Modeling material, 192
Modeling process steps, 189
Modular mold inserts, 298
Mold coating, 298
Mold design, 301
Mold insert fixation, 260

Mold inserts, 283
Molded interconnected devices (MID), 173
Molded part quality, 120
Molded parts, 113
Molding analysis, 199
Molding forces, 228
Molding material ceramic, 62
Molding material glass, 58
Molding materials, 57
Molding velocities, 228
Molding velocity, 146
Molding window, 102
Molecular architecture of polymers, 64
Molecular chains, 64
Molecular orientation, 81
Molecular weight, 65
Multi layer molding, 157

Nanoimprint lithography (NIL), 40
Nanoimprint of thermoplastic polymers, 43
Nanoimprint processes, 39
Nanoimprint UV curing, 42
Newtonian flow, 67
Nusselt number, 255

Optical bench, 312
Optical gratings, 310, 316
Orientation of molecules, 81
Overdrawn edges, 122
Overstretched structures, 122

p-v-T-diagram, 85, 127
Packing pressure, 140
Packing time, 140
PDMS, 40, 42, 72, 170, 291, 300, 329
Phase angle, 93
Photonic structures, 315
PIM, 62
Pits, 333
Plastifying unit, 19
Polymer melts, 66

Polymer mold inserts, 291
Polymers for hot embossing, 103
Position controlled molding, 150
Powder molding, 62
Prandtl number, 255
Pressure distribution, 181, 203, 301
Pressure drop, 207, 210
Pressure profile injection molding, 24
Prestructured polymer films, 37
Process parameter, 143
Process simulation, 179
Process steps, 140
Process variations, 149
Programming language, 238
Properties of polymers, 63
PTFE, 300
Pumps, 323
Pyrex, 59

Quality control, 132

Radiation, 249
Reaction injection molding (RIM), 15
Record fabrication, 3
Record grooves, 7
Record replication, 4
Refractive index, 317
Relaxation modulus, 90
Relaxation time, 78, 88
Relaxation, 82, 89
Release agent, 300
Replication processes, 13
Representative point, 74
Requirements micro injection
    molding, 22
Requirements mold inserts, 283
Requirements technique, 227
Requirements tools, 247
Residual layer, 146, 206, 302
Residual layer thickness, 206
Retardation, 82
Reynolds number, 256
Rheology, 66
Role to role embossing, 163

Roller embossing, 163, 310, 320
Rotation rheometer, 80
Rubber elastic state, 100

Self-assembled monolayers
    (SAMs), 41
Semicrystalline polymers, 84, 101
Shear deformation, 67
Shear rheologic behavior, 67
Shear thinning, 69
Shear velocity, 210
Shear viscosity, 69
Shellac, 4
Shim mold insert, 262, 293, 314
Shrinkage, 125, 215, 301
Simulation programs, 187
Sink erosion, 289
Sink marks, 122
Soft embossing, 170
Solid polymers, 88
Solidification, 84
Spectrometer, 315
Spindle drive, 231
Squeeze flow, 179, 200
Stack of foils, 155
Static friction, 95, 146
Stefan equation, 182
Storage modulus, 93
Strain rheology, 80
Stress
    residual, 130
Stress at offset yield point, 91
Stress-strain behavior, 91
Structure viscosity, 69
Structuring methods, 284
Substrate plate, 140
Substrate plates, 264
Suess, 244
Surface quality, 123
Surface tension, 208

Tactile measurement systems, 117
Technical plastics, 103
Temperature distribution, 252, 257

Tensile strength, 91
Tensile testing machines, 10
Tensile tests, 90
Thermal aggregate states, 99
Thermal behavior of polymers, 99
Thermal bonding, 323
Thermal coefficient, 86
Thermal diffusivity, 108
Thermal images, 252
Thermal nanoimprint, 43
Thermal properties of polymers, 105
Thermoforming, 36, 159, 167, 253
Thermoplastic polymers, 66
Thermosets, 65
3D embossing, 172
3D structures, 329
Through holes, 152
Time temperature shift, 72
Tools, 247
Total costs, 51
Touch force, 145
Transition ranges, 100
Turbulent flow, 255

Ultrasonic embossing, 165
Undercuts, 172, 221, 329
UV embossing, 170
UV lithography, 294
UV nanoimprint, 42

Valves, 324
Variable costs, 51
Variotherm process, 23
Velocity controlled molding, 146
Viscoelastic behavior, 67, 75, 88
Viscoelasticity, 75, 88
Viscosity, 208
Viscosity measurement, 79
Viscosity model, 210
Viscosity models, 69
Visualization of flow, 201
Volumetric shrinkage, 86

Wall slip, 208
Warpage, 123, 129, 302
Waveguides, 312, 317
Weldlines, 121
Wet etching, 289
Wickert Press, 242
Wickert WMP1000, 234
Williams, Landel, and Ferry (WLF),
    71, 73
Wire cutting, 289
WMP1000, 242

X-ray lithography, 294

Yield strain, 90
Yield stress, 90